AF739315

Carbon Nanotubes Mediated Bioactive Delivery

Carbon nanotubes (CNTs) have emerged as novel carbon-based nanomaterials with unique physicochemical properties for pharmaceutical and biomedical applications. This book provides the fundamental concept, basic physicochemical properties, functionalization, method of synthesis, applications, and regulatory considerations of CNTs. The primary objective of the editors and authors is to exhaustively cover the fundamentals of carbon nanomaterials, carbon nanotubes, functionalization and chemistries involved, drug delivery aspects, and applications along with their regulatory aspects.

Carbon Nanotubes Mediated Bioactive Delivery

Edited by
Neelesh Kumar Mehra and N. K. Jain

CRC Press
Taylor & Francis Group
Boca Raton London New York

CRC Press is an imprint of the
Taylor & Francis Group, an **informa** business

Designed cover image: Getty Images

First edition published 2026
by CRC Press
2385 NW Executive Center Drive, Suite 320, Boca Raton FL 33431

and by CRC Press
4 Park Square, Milton Park, Abingdon, Oxon, OX14 4RN

CRC Press is an imprint of Taylor & Francis Group, LLC

© 2026 selection and editorial matter, Neelesh Kumar Mehra and N. K. Jain; individual chapters, the contributors

Reasonable efforts have been made to publish reliable data and information, but the author and publisher cannot assume responsibility for the validity of all materials or the consequences of their use. The authors and publishers have attempted to trace the copyright holders of all material reproduced in this publication and apologize to copyright holders if permission to publish in this form has not been obtained. If any copyright material has not been acknowledged please write and let us know so we may rectify in any future reprint.

Except as permitted under U.S. Copyright Law, no part of this book may be reprinted, reproduced, transmitted, or utilized in any form by any electronic, mechanical, or other means, now known or hereafter invented, including photocopying, microfilming, and recording, or in any information storage or retrieval system, without written permission from the publishers.

For permission to photocopy or use material electronically from this work, access www.copyright.com or contact the Copyright Clearance Center, Inc. (CCC), 222 Rosewood Drive, Danvers, MA 01923, 978-750-8400. For works that are not available on CCC please contact mpkbookspermissions@tandf.co.uk

Trademark notice: Product or corporate names may be trademarks or registered trademarks and are used only for identification and explanation without intent to infringe.

ISBN: 978-1-032-47916-3 (hbk)
ISBN: 978-1-032-47919-4 (pbk)
ISBN: 978-1-003-38653-7 (ebk)

DOI: 10.1201/9781003386537

Typeset in Times
by codeMantra

Contents

Preface

Carbon nanotubes have emerged as novel carbon-based nanomaterials with unique physicochemical properties for pharmaceutical and biomedical applications. The proposed book on carbon nanotubes presents the concept, theory, fundamental aspects, and applications based on the basic aspects for diverse applications, including drug delivery in pharmaceutical and biomedical sciences. In our opinion, there is a need for the scientific understanding of the different aspects of the pristine and functionalized carbon nanotubes for application purposes. The primary objective of the editors and authors would be to exhaustively cover the fundamentals of carbon nanomaterials, carbon nanotubes, functionalization and chemistry involved, drug delivery aspects and applications along with their regulatory aspects. This book provides the comprehensive overview of the latest research efforts in designing, optimizing, development, scale up of carbon nanotubes-mediated delivery systems. It analyzes the key challenge, synthesis, design, molecular modeling, fundamental concepts, drug delivery aspects, analytical tools, and biological fate and regulatory consideration to the practical use of CNTs application.

Editors

Neelesh Kumar Mehra, PhD
N. K. Jain, PhD

About the Editors

Dr. Neelesh Kumar Mehra is working in the Department of Pharmaceutics, National Institute of Pharmaceutical Education & Research (NIPER), Hyderabad, India. He earned his PhD from Dr. H.S. Gour University, Sagar, and his PostDoc from Irma Lerma Rangel College of Pharmacy, Texas A & M Health Science Centre, Kingsville, TX, USA. He served as a Manager in Product Development, Sentiss Research Centre, Sentiss Pharma Pvt Ltd, Gurgaon, for the development, scale-up, and technology transfer of complex ophthalmic, inhalation, and optic pharmaceutical products. He received a team award for the successful commercialization of an ophthalmic suspension product. He has authored more than 120 peer-reviewed publications in high-repute international journals and more than 30 book chapter contributions. He has 4 granted patents on manufacturing processes and composition to improve the therapeutic efficacy of topical delivery. He has guided 3 PhD students and 30 MS students and is currently guiding 12 PhD and 5 MS students for their dissertations/research projects. He has received numerous outstanding awards, including Young Scientist and team awards for their research output. He is a peer reviewer of various international journals and publishers. He recently published three edited books on multifunctional nanocarriers, biopharmaceuticals and nanotechnology. He has rich research and teaching experience in the formulation and development of complex, innovative ophthalmic and injectable biopharmaceutical products, including micro- and nanotechnologies for regulated markets. According to an announcement by Stanford University scientists based on the database and basis of standardized citation indicators, Dr. Mehra has been included in the top 2% of scientists in the world in the area of pharmacology and pharmacy, consecutively for 3 years.

Prof. N. K. Jain is an internationally renowned academic and researcher and a senior professor in India who retired in June 2014 after more than 40 years of dedicated and distinguished service as a teacher and 25 years as a full professor. Prof. Jain is the author of two dozen celebrated books on pharmaceutical sciences in India and has contributed several chapters in national and international books. He has been a reviewer for several international and national research journals. His current research interests include various aspects of controlled, novel, and targeted drug delivery; nanotechnology; and nanomedicine. He is globally known for his excellent research in the field of hydrotropic solubilization, resealed erythrocytes-based drug delivery, and dendrimer drug delivery. Currently, he is exploring the pharmaceutical potential of carbon nanotubes and quantum dots. He is a popular speaker at conferences, seminars, faculty development programs, and quality improvement programs, and he has traveled extensively abroad.

Contributors

Sweta Acharya
Department of Pharmacy
Birla Institute of Science and Technology
Pilani, Rajasthan, India

Sumedh Bahadure
Pharmaceutical Nanotechnology Research Laboratory
Department of Pharmaceutics
National Institute of Pharmaceutical Education & Research (NIPER)
Hyderabad, Telangana, India

Meenakshi Bajpai
Institute of Pharmaceutical Research
GLA University, NH-2
Mathura, Uttar Pradesh, India

Sonali Batra
Department of Pharmaceutical Sciences
Indira Gandhi University
Meerpur, Haryana, India

Atreyee Bhattacharyya
Department of Pharmaceutics
National Institute of Pharmaceutical Education & Research (NIPER)
Hyderabad, Telangana, India

Saptarshee Bhattacharjee
Department of Pharmaceutics
National Institute of Pharmaceutical Education & Research (NIPER)
Hyderabad, Telangana, India

Rohit Bhawale
Pharmaceutical Nanotechnology Research Laboratory
Department of Pharmaceutics
National Institute of Pharmaceutical Education & Research (NIPER)
Hyderabad, Telangana, India

Ramesh Chandra
Dr B.R. Ambedkar Centre for Biomedical Research
Department of Chemistry
Institute of Nano Medical Sciences (INMS)
University of Delhi
Delhi, India
and
Maharaja Surajmal Brij University
Bharatpur, Rajasthan, India

Sri Pada Datta Chigurupati
Pharmaceutical Nanotechnology Research Laboratory
Department of Pharmaceutics
National Institute of Pharmaceutical Education & Research (NIPER)
Hyderabad, Telangana, India

Abhijeet Chinchane
Department of Pharmaceutics
National Institute of Pharmaceutical Education & Research (NIPER)
Hyderabad, Telangana, India

Pawan Devangan
Department of Pharmaceutics
National Institute of Pharmaceutical Education & Research (NIPER)
Hyderabad, Telangana, India

Devaraj
Department of Pharmaceutics
National Institute of Pharmaceutical Education and Research (NIPER)
Ahmedabad, Gujarat, India

Shubham Gajdhane
Department of Pharmaceutics
National Institute of Pharmaceutical Education & Research (NIPER)
Hyderabad, Telangana, India

Spandana Gollapalli
Department of Pharmaceutics
National Institute of Pharmaceutical Education & Research (NIPER)
Hyderabad, Telangana, India

Gulsheen
Chitkara College of Pharmacy
Chitkara University
Rajpura, Punjab, India

Aakanchha Jain
Department of Pharmaceutics
National Institute of Pharmaceutical Education and Research (NIPER)
Ahmedabad, Gujarat, India

Ankit Jain
Department of Pharmacy
Birla Institute of Science and Technology-Pilani
Pilani, Rajasthan, India

Keerti Jain
Drug Delivery and Nanomedicine Research Laboratory
Department of Pharmaceutics
National Institute of Pharmaceutical Education and Research (NIPER)
Lucknow, Uttar Pradesh, India

N. K. Jain
Department of Pharmaceutical Sciences
Dr. HS Gour University
Sagar, India

Roshani Jaiswal
Department of Pharmaceutics
National Institute of Pharmaceutical Education & Research (NIPER)
Hyderabad, Telangana, India

Sapna Joshi
School of Health and Allied Sciences
Department of Pharmacy
G D Goenka University
Sohna, Haryana, India

Abhinav Kanwal
Department of Pharmacology
All India Institute of Medical Sciences
Bathinda, Punjab, India

Navjot Kanwar
Department of Pharmaceutical Sciences and Technology
Maharaja Ranjit Singh Punjab Technical University
Bathinda, Punjab, India

Anju Katyal
Dr B.R. Ambedkar Centre for Biomedical Research
University of Delhi
Delhi, India

Omar Khan
Pharmaceutical Nanotechnology Research Laboratory
Department of Pharmaceutics
National Institute of Pharmaceutical Education & Research (NIPER)
Hyderabad, Telangana, India

Renat R. Khaydrov
Institute of Nuclear Physics
Uzbekistan Academy of Sciences
Tashkent, Uzbekistan

Praveen T. Krishnamurthy
Department of Pharmacology
JSS College of Pharmacy
JSS Academy of Higher Education & Research
Ooty, Tamil Nadu, India

Sagar Kulkarni
Department of Pharmaceutics
National Institute of Pharmaceutical Education & Research (NIPER)
Hyderabad, Telangana, India

Jitendra Kumar
Department of Pharmaceutics
National Institute of Pharmaceutical Education & Research (NIPER)
Hyderabad, Telangana, India

Lalit Kumar
Department of Pharmaceutics
GNA School of Pharmacy
GNA University
Phagwara, India

Pankaj Kumar
Department of Pharmaceutics
National Institute of Pharmaceutical Education & Research (NIPER)
Hyderabad, Telangana, India

Garikapati Kusuma Kumari
Department of Pharmacology
JSS College of Pharmacy
JSS Academy of Higher Education & Research
Ooty, Tamil Nadu, India

Nalla Usha Kumari
Pharmaceutical Nanotechnology Research Laboratory
Department of Pharmaceutics
National Institute of Pharmaceutical Education & Research (NIPER)
Hyderabad, Telangana, India

Rajan Kumar Kurmi
Metro College of Health Sciences and Research
Knowledge Park III
Greater Noida, Uttar Pradesh, India

Kshitiz Luthra
Pharmaceutical Nanotechnology Research Laboratory
Department of Pharmaceutics
National Institute of Pharmaceutical Education & Research (NIPER)
Hyderabad, Telangana, India

Jitender Madan
Department of Pharmaceutics
National Institute of Pharmaceutical Education & Research (NIPER)
Hyderabad, Telangana, India

Sandeep Kr Maharana
Department of Pharmaceutics
National Institute of Pharmaceutical Education and Research (NIPER)
Ahmedabad, Gujarat, India

Pooja Mehra
Department of Pharmacy
Sushant University
Haryana, Delhi-NCR, India

K. M. Gijith Mohan
Drug Delivery and Nanomedicine Research Laboratory
Department of Pharmaceutics
National Institute of Pharmaceutical Education and Research (NIPER)
Lucknow, Uttar Pradesh, India

Shiv Kumar Prajapati
Institute of Pharmaceutical Research
GLA University
Mathura, Uttar Pradesh, India

Naveen Rajana
Pharmaceutical Nanotechnology Research Laboratory
Department of Pharmaceutics
National Institute of Pharmaceutical Education & Research (NIPER)
Hyderabad, Telangana, India

Ammu V.V.V. Ravi Kiran
Department of Pharmacology
JSS College of Pharmacy
JSS Academy of Higher Education & Research
Ooty, Tamil Nadu, India

Biswajit Rout
Department of Pharmaceutics
National Institute of Pharmaceutical Education and Research (NIPER)
Ahmedabad, Gujarat, India

Sachin
School of Pharmaceutical Sciences
Delhi Pharmaceutical Sciences and Research University
New Delhi, India

Soyal Sayyed
Department of Pharmaceutics
National Institute of Pharmaceutical Education & Research (NIPER)
Hyderabad, Telangana, India

Sumit Sharma
School of Pharmaceutical Sciences
Delhi Pharmaceutical Sciences and Research University
New Delhi, India

Naveen Singh
Department of Pharmaceutics
National Institute of Pharmaceutical Education and Research (NIPER)
Ahmedabad, Gujarat, India

Snigdha Singh
Department of Chemistry
University of Delhi
Delhi, India
and
Institute of Nano Medical Sciences (INMS)
University of Delhi
Delhi, India

Dimpal Suthar
Department of Pharmaceutics
National Institute of Pharmaceutical Education and Research (NIPER)
Ahmedabad, Gujarat, India

Ravindra Vasave
Pharmaceutical Nanotechnology Research Laboratory
Department of Pharmaceutics
National Institute of Pharmaceutical Education & Research (NIPER)
Hyderabad, Telangana, India

Abbreviations

Abbreviation	Full form
^{18}FDG	^{18}F-fluorodeoxyglucose
AFM	atomic force microscope
CNTs	carbon nanotubes
CS	chitosan
DLS	dynamic light scattering
DOPC	1,2-dioleoyl-sn-glycero-3-phosphocholine
DOTAP	2-dioleoyl-3-trimethylammonium propane
DOX	doxorubicin
DSC	differential scanning calorimetry
DSPE-mPEG	1,2-distearoyl-sn-glycero-3-phosphoethanolamine N-[amino (poly-ethylene glycol)-5000]
DWCNTs	double-walled carbon nanotubes
ECM	extracellular matrix
EDA	ethylene diamine
EDC	1-ethyl-3-(3-dimethylaminopropyl)carbodiimide
EDX	energy dispersive X-ray analysis
FA	folic acid
FTIR	Fourier transform infrared spectroscopy
GLUT	glucose transporters
IC_{50}	inhibitory concentration 50
kV	kilovolts
Lys	lysine
MDR	multidrug resistance
MWCNTs	multi-walled carbon nanotubes
NHS	N-hydroxysuccinimide
NIR irradiation	near-infrared irradiation
NPs	nanoparticles
PDI	polydispersity index
PEG	polyethylene glycol
PLA	polylactic acid
PLA-co-mPEG	poly(lactide)eco-methoxy polyethylene glycol
PMSA	prostate-specific membrane antigen
PNIPAAm	poly(N-isopropyl acrylamide)
PXRD	powder XRD
SCXRD	single crystal XRD
SEM	scanning electron microscope
siRNA	small interfering RNA
SWCNTs	single-walled carbon nanotubes
TEM	transmission electron microscope
TGA	thermogravimetric analysis
XRD	X-ray diffraction

1 Carbon Nanomaterials
Introduction

Rohit Bhawale, Ravindra Vasave, Kshitiz Luthra, N. K. Jain, and Neelesh Kumar Mehra

1.1 INTRODUCTION

The advancement of new technology has greatly aided in providing quick, suitable, acceptable, and competitively cost medical care. Advances in nanoscience have led to the emergence of a new class of nanostructures. Nanomaterials exhibit remarkable physicochemical features due to their diminutive size. These include improved absorption and reactivity, a larger surface area, molar extinction coefficients, adjustable properties, quantum effects, and optical and magnetic characteristics. Scientists are intrigued by carbon-based nanomaterials because of their distinct chemical and physical characteristics, which include thermodynamic, biomechanical, electrical, optical, and structural features. A wide variety of carbon nanomaterials (CNMs), such as fullerenes, graphene, carbon nanotubes (CNTs), and carbon nanofibers (CNFs), have been the subject of extensive research into potential medicinal uses because of their unique characteristics [1]. CNMs are highly adaptable materials that may serve as frameworks for building nanocarrier systems with anticancer properties [2]. An efficient anticancer system may be designed using these nanoparticles, which have inherent therapeutic characteristics, are biocompatible, and are easy to chemically functionalize [3].

CNMs are a group of materials with tiny dimensions that have attracted significant attention over the last three decades. This interest was sparked by the remarkable discovery of fullerene, the first member of this group, which was reported by Smalley, Kroto, Curl, and others in 1985 at Rice University and the University of Sussex [4]. Two more allotropes that have caused quite a stir in science are graphene and CNTs, which were discovered in 2004 and 1991, respectively [5,6]. These three CNMs possess a unique composition, with a network of conjugated π-electrons formed by sp^2 carbon atoms. However, there are also carbon dots and nanodiamonds that contain a mixture of sp^2 and sp^3 carbon atoms, along with defects and heteroatoms. These alternative compositions have also garnered significant interest (Figure 1.1) [7,8]. More recently, it was revealed that a novel form of one-dimensional (1D) nanostructures made entirely of crystalline sp^3 carbon, known as carbon nanothread, may be synthesized using a high-pressure solid-state polymerization process with benzene [9]. Because of the quantum confinement effect, these low-dimensional CNMs have intriguing properties not found in bulk carbon materials like diamond and graphite. These properties encompass a variety of unique optical, electrical, magnetic, and chemical characteristics and have practical uses in biomedicines [10–13]. Utilizing

DOI: 10.1201/9781003386537-1

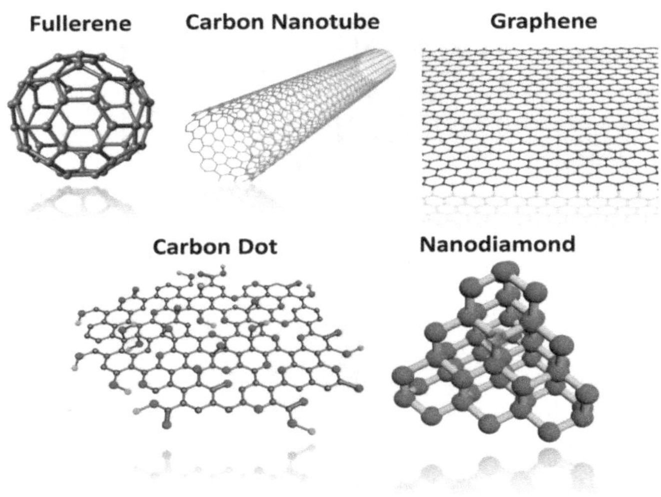

FIGURE 1.1 Various carbon nanomaterials incl. Fullerene CNTs, graphene, carbon dot, and nanodiamond [3].

CNMs, which are beneficial largely because of their tiny size, distinctive optical features, and huge surface area biomedical research, has exploded in the last 15–20 years. The symbiotic relationship between nanoparticles and numerous essential proteins that are essential to sustaining life's core activities is the source of the exquisite marriage of nanoscience and biology. The typical dimensions of CNMs are between 1 nm and 1 μm. This is similar to the widths of DNA and proteins (both around 1–100 nm and 2–3 nm, respectively) and ion channels (a few nanometers) found in biological environments, as well as the glomerular filtration barrier (approximately 5–10 nm) [14]. The advantageous dimensions of carbon nanostructures have rendered them ideal nanocapsules and nanocarriers for the loading and transportation of medicines and genes to precise destinations in living organisms. In addition, the distinctive optical characteristics shown by some carbon nanostructures have generated significant attention for various biological uses. In the classic near-infrared window (NIR-I window, 750–1,000 nm), graphitic CNMs like graphene and CNTs often exhibit substantial optical absorption, resulting in an *in situ* photothermal effect for photoacoustic imaging and photothermal treatment [15,16]. The utilization of fluorescent contrast agents in deep-tissue fluorescence imaging has been made possible by the distinctive intrinsic fluorescence emission exhibited by single-walled carbon nanotubes (SWCNTs) within the second near-infrared window (NIR-II window, 1,000–1,700 nm), which is characterized by diminished photon scattering [17–19]. Moreover, the expansive surface area of CNMs, along with the presence of π-electron conjugated surfaces, allows for the supramolecular binding of hydrophobic molecules to the surfaces of SWCNTs and graphene through π–π interactions. This interaction serves as the foundation for the effective transport of water-insoluble drug molecules in an aqueous biological setting [20,21]. In addition to the previously described characteristics of CNMs, SWCNTs with unaltered carbon networks and elongated structures possess exceptional mechanical qualities that improve the durability of tissue scaffolds [21–24].

In this book, we will embark on a comprehensive exploration of CNTs and their groundbreaking potential in cancer treatment. The upcoming chapters will delve into various facets of this innovative technology, starting with a fundamental understanding of CNMs and their unique properties. We will then examine the synthesis methods and critical characterization techniques essential for developing high-quality CNTs. A significant focus will be placed on the functionalization approaches that enhance the biocompatibility and efficacy of CNTs for medical applications. We will explore hybrid CNTs-nanomaterials, an emerging frontier in biopharmaceuticals, highlighting their synergistic benefits. The role of CNTs in cancer therapy will be scrutinized, detailing their mechanisms and therapeutic outcomes. Additionally, we will investigate their applications in diagnostic imaging, offering insights into their dual role in treatment and detection.

The discussion will extend to the use of CNTs in combating infectious diseases, showcasing their versatility. A deep dive into the cellular trafficking mechanisms will elucidate how CNTs navigate biological environments to deliver drugs effectively. We will also address pharmacokinetic and toxicological considerations, ensuring a balanced view of their safety and efficacy. Lastly, the book will cover current regulatory considerations for nanomedicines, providing a holistic overview of the challenges and advancements in bringing CNT-based therapies from the lab to the clinic. Each chapter is designed to provide a detailed, informative, and up-to-date perspective on the multifaceted role of CNTs in modern medicine.

1.2 CLASSIFICATION OF CARBON NANOMATERIALS

Carbon ranks fourth in the cosmos in terms of mass, behind only oxygen, helium, and hydrogen, and fifteenth in terms of abundance on Earth's surface. This element is essential for all known forms of life because of the wide variety of organic compounds it can make and the unusual temperatures at which it can do it, rendering it capable of polymerization. Approximately 18.5% of the mass of the human body is comprised of this element, making it the second most plentiful element after oxygen [25,26].

Carbon may exist in several dimensionalities, ranging from zero-dimensional (0D) to three-dimensional (3D) forms. 2D carbon may be transformed into 0D fullerenes, 1D nanotubes, or 3D graphite by wrapping, rolling, or stacking processes, respectively [27]. This different carbon allotropes are shown in diagram below Figure 1.2. Nano-carbons can be categorized into two groups based on how they are bonded: sp^2 and sp^3 CNMs. A few examples of CNMs according to their dimensionalities are given below:

1. Zero-dimensional (0D) CNM: fullerene
2. One-dimensional (1D) CNM: CNT
3. Two-dimensional (2D) CNM: graphene
4. Three-dimensional (3D) CNMs

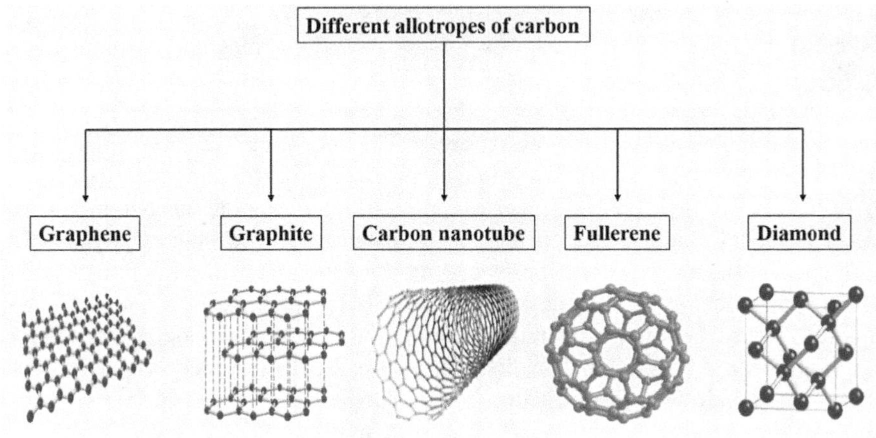

FIGURE 1.2 Diagrammatic representation of different allotropes of carbon..

1.2.1 Zero-Dimensional CNM

The first carbon-based nanostructure that drew attention from the public is fullerene. Due to its exceptional physical, chemical, and biological characteristics, fullerene and several fullerene derivatives via rich covalent and noncovalent chemistry have been extensively investigated for energy, biosensing, and medical applications since the discovery of C60 fullerene almost 30 years ago [28–31]. Fullerene has been the subject of significant attention on the scientific stage since the Nobel Prize for Chemistry 1996 was granted to Smalley, Kroto, and Curl. Even though the scientific pursuit of fullerene has experienced a slight decline in popularity in recent years due to the more rapid ascent of CNTs and graphene, the ultrasmall size, uniform dispersity, and rich chemistry of fullerene continue to captivate attention in a diverse range of biomedical research fields, including imaging, diagnostics, and clinical applications [32].

The C60 molecule is a very balanced sphere with a van der Waals width of 1.1 nm. It is made up of only carbon atoms. It can be made by evaporating graphite with a laser or heat or directly from smaller aromatic molecules through bottom-up chemical synthesis [33]. Even though C60 molecules don't dissolve in water, they can be modified with hydrophilic surface groups to make them more water-soluble and biocompatible. This is done by covalently reacting with the carbon atoms in the sphere [34]. Alternatively, C60 molecules may be enclosed inside cyclodextrins (CDs) or micelles via noncovalent interactions between the hydrophobic C60 molecule and the amphiphilic host molecules [35]. The easy modification of C60 molecules on their surface has enabled their conjugation with antibodies, facilitating their specific targeting toward certain biomarkers [36]. Apart from the abovementioned characteristics of C60 molecules that enable different imaging and therapeutic uses, fullerenes have other general uses depending on their inherent photoluminescence for fluorescence imaging, protein activity inhibition for antiviral therapy, and anticancer action for cancer treatment [37–40].

1.2.2 One-Dimensional CNM: CNTs

Ijima developed CNTs in 1991. They are a new class of synthetic nanomaterials with unique hollow and cylindrical shapes [41]. The fundamental component of CNTs at the nanoscale is a cylinder of graphene with a diameter of a few nanometers and a length ranging from micrometers to millimeters [42]. CNTs may be seen as either individual molecules due to their tiny size, or as quasi-1D crystals having periodicity along the axis of the tube [43]. Graphene sheets may be arranged in several configurations, such as SWCNTs, double-walled carbon nanotubes (DWCNTs), triple-walled carbon nanotubes (TWCNTs), or multi-walled nanotubes (MWCNTs) (Figure 1.3) [41,44].

Figure 1.4 displays the three most prevalent arrangements. The sheet may be folded along a symmetry axis to form either an armchair tube or a zigzag tube. Furthermore, it is possible to twist the sheet in a way that diverges from a central axis, leading to the formation of a chiral nanotube. The atoms in this structure are organized in a helical pattern, with each unit cell having its matching atoms positioned accordingly. Furthermore, the circumference of the cylinder may also be modified, in addition to the chiral angle. Nanotubes may be classified into three categories depending on their diameters: zigzag, armchair, and chiral tubes [2].

1.2.3 Two-Dimensional CNM: Graphene

Graphene is a significantly younger member of the carbon material family than fullerene and CNTs [6]. Graphene is an atomically thin sheet made of hexagonally organized carbon atoms with sp^2 hybridization in two dimensions. It may be well

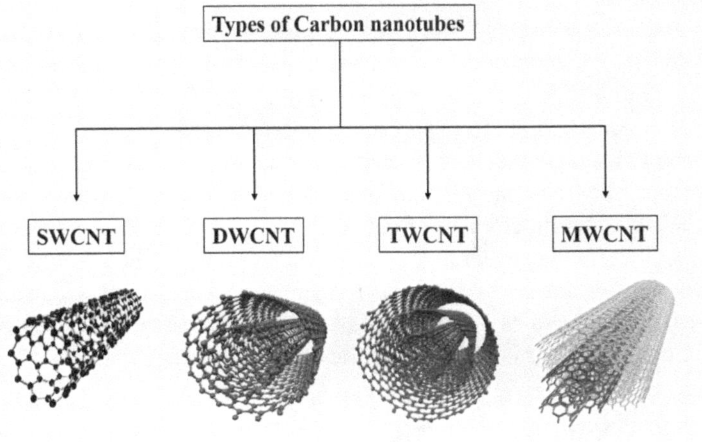

FIGURE 1.3 Different types of CNTs based on a number of walls involved.

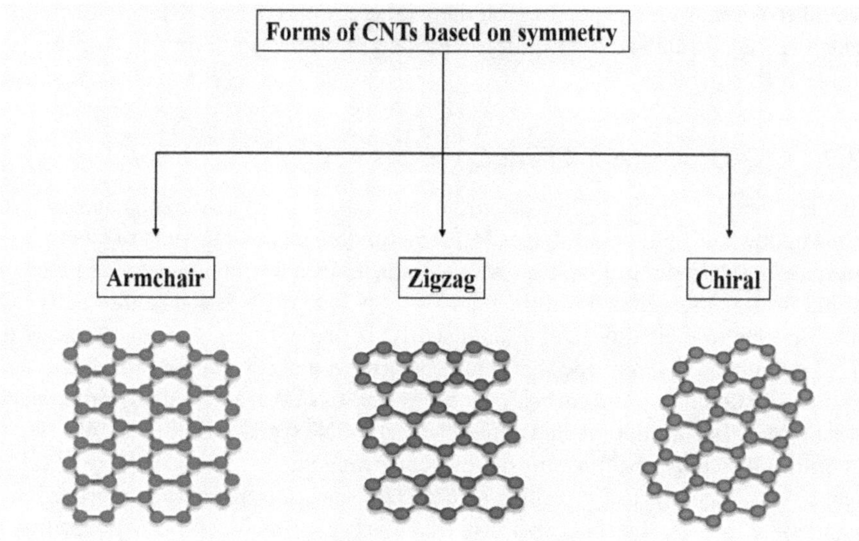

FIGURE 1.4 Different forms of carbon nanotubes are based on symmetry.

separated from its surroundings and hence stand alone. This 2D crystal of carbon may be synthesized in many ways, including early efforts at mechanical and chemical exfoliation, unzipping CNTs, bottom-up epitaxial growth, and chemical synthesis [45–51]. The band structure of a single graphene sheet can be determined by using a tight binding method on the honeycomb hexagonal lattice of the 2D carbon crystal, within the Hückel approximation. This calculation reveals that the valence band and conduction band of graphene meet at the six corners of the first Brillouin zone, leading to the formation of a zero-gap semiconductor [52]. Consequently, electrons that move within the extremely thin layer of graphene exhibit characteristics of massless, ultrarelativistic charge carriers known as Dirac fermions, rather than the conventional Schrödinger fermions. This gives rise to a wide range of extraordinary physical properties, such as high carrier mobility, exceptional thermal conductivity, the ability to be influenced by both positive and negative electric fields, and the occurrence of the quantum Hall effect at room temperature [53–57].

Graphene has unique mechanical characteristics, such as remarkable elasticity and significant flexibility, enabling the graphene sheet to adapt to any substrate with a diverse shape while providing structural reinforcement [58,59]. As a result, graphene might serve as an excellent scaffolding material for successfully reinforcing the structure and shape of synthetic tissues. To that goal, several experiments have been conducted using graphene and its chemical compounds as biocompatible substrates for neuronal development, stem cell differentiation, and osteoblast cell proliferation and osteogenesis [60–62].

1.2.4 Three-Dimensional CNMs

The 3D porous structure is composed of CNTs and graphene. Furthermore, the evenly dispersed CNTs between the graphene layers effectively inhibit the aggregation or stacking of the graphene sheets [63]. A spark plasma process with industrial scalability is used to fabricate CNTs that are interconnected in a 3D structure. Under conditions of sufficient stress and a strong electric field, the nanotubes undergo fusion, resulting in the formation of a compact block. The spectroscopic and microscopic examinations provide a thorough analysis that confirms the creation of linked networks of CNTs. The mechanical characteristics of the 3D CNT block exhibit a significant yield strength and stiffness. An extensive atomistic molecular dynamics simulation elucidates the process of CNT bonding [64]. The 3D designs possess characteristics such as large specific surface areas, hierarchical pores, and interconnected conductive networks. These features enhance mass and electron transport and provide numerous accessible active sites. Consequently, these designs are highly beneficial for a wide range of applications, including sorption, capacitors, and catalysis. The monolithic shape of 3D CNT assemblies may provide new application options for materials such as free-standing electrodes, sensors, and recyclable sorbents [65].

Chemically modified SWCNTs have shown potential in specifically accumulating in tumors in mice, while also being biocompatible, easily excreted, and causing little harm [66]. The primary methods for synthesizing SWCNTs include electric arc discharge, laser ablation, catalytic breakdown of gaseous hydrocarbons, and chemical vapor deposition (CVD) [67–70]. A brief discussion about synthesis and characterization is discussed in the upcoming chapter in detail [71]. Please refer to Chapters 2 and 3.

1.3 SYNTHESIS OF CARBON NANOTUBES

Since their discovery in 1991, CNTs have been regarded as potentially transformational materials because of their exceptional mechanical and electrical capabilities. Because of their remarkable stiffness and tensile strength combined with their extremely light weight, CNTs hold great promise for future uses as biological sensors, transparent conductive films, and fiber sheets. The atomic structure of CNTs determines their properties, as it does for all carbon-based materials. It is especially important to consider the CNTs' length, diameter, and sidewall structure. Thus, a major barrier in the sector continues to be the lack of synthetic methods for creating structurally homogeneous CNT structures [72]. Electric arc discharge is the oldest known method of producing CNTs. R. Bacon had already employed this method in the early 1960s to synthesize whiskers, which are carbon fibers. Krätschmer and Huffman modified the same process in 1990 to manufacture high-quality fullerenes. This approach was then refined and used for manufacturing SWCNTs and MWCNTs. Additional techniques like CVD and laser evaporation/ablation have also been successfully investigated for the synthesis of CNTs. Technically speaking, the arc discharge technique and the laser evaporation procedure are comparable. The goods obtained by these two procedures differ in terms of quality and purity. In large-scale manufacturing of CNTs and similar materials, however, the most promising and widely used methods are arc discharge and various forms of CVD [73]. In contrast, CVD techniques comprise a variety of synthesis procedures, including water-assisted CVD (WA-CVD), microwave plasma-enhanced CVD (MPECVD), radio frequency plasma-enhanced CVD (RF-PECVD), hot filament CVD (HFCVD), and plasma-enhanced CVD (PECVD). Whatever the synthesis method, it is important to keep in mind that there are significant batch-to-batch differences in the length, width, and number of walls between various CNT preparations because it is challenging to control every parameter [74]. Since the majority of these techniques call for vacuum and supporting gases, growth under atmospheric pressure has already been documented. However, because gas-phase procedures are volumetric, they are appropriate for uses like composite materials, which need vast amounts of nanotubes and industrial-scale synthesis techniques to be commercially viable. However, low catalyst yields, where only a tiny portion of catalysts form nanotubes, limited catalyst lifetimes, and low catalyst number density are the drawbacks of gas-phase synthesis methods [75].

1.4 PURIFICATION OF CARBON NANOMATERIALS

Metallic catalyst particles and carbonaceous impurities are invariably present in as-synthesized CNTs made using the aforementioned techniques, and the quantity of impurities often rises as CNT diameter decreases. In general, fullerenes, carbon nanoparticles (CNPs), and amorphous carbon are considered carbonaceous impurities. Graphite rod vaporization provides the carbon source for arc discharge and laser ablation; however, unvaporized graphitic particles that have fallen off the rods frequently persist as impurities in the finished product. Furthermore, graphitic polyhedrons with confined metal particles coexist with CNTs synthesised by arc discharge, laser ablation, and high temperature (>1,000°C) CVD. Because fullerenes are soluble in some organic solvents, they are easily removed. Due to its high defect density, which permits mild oxidation, amorphous carbon is also comparatively easy to remove. The most challenging issue is getting rid of graphitic particles and polyhedral carbons, particularly SWCNTs, which oxidize at a rate comparable to that of CNTs. Usually, transition metal catalyst residues are what we refer to as metal impurities [76]. Therefore, it is necessary to tackle this impurity contamination dispute with different purification techniques. Larger fragments like soot are frequently removed from CNP (carbon nanoparticle) solutions using centrifugation and filtration. The challenge of CNP solution purification from smaller structures, such as molecules and untreated chemicals, can be addressed via dialysis. These techniques, however, have not been able to isolate CNPs with consistent characteristics (or even uniform size) or CNT fractions with the desired characteristics [77]. The purification methods of CNTs are classified into different two types: chemical purification and physical purification, illustrated in Figure 1.5. The chemical methods yield high-purity CNMs but subsequently result in product loss and delicate structural defects in CNTs. Whereas controlled reactions can eliminate the metal catalyst particles to achieve high-purity CNTs. Alternatively, physical purification methods are more popular because of the adsorption of different functional groups, implemented when a higher weight fraction of CNTs is required, leaving the functional group behind and using a comparable (π)-graphene structure. The yield of purified products in physical methods depends on the size of CNTs. The physical method results in minimum damage to the CNT structure but is less effective as compared to chemical purification techniques [77–79].

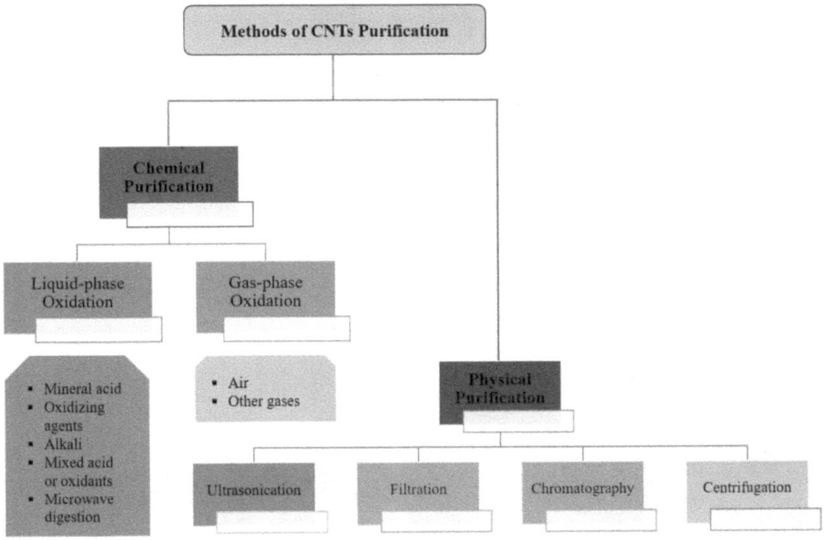

FIGURE 1.5 Diagrammatic representation of chemical and physical purification method for carbon nanomaterials [76,79].

1.5 FUNCTIONALIZATION OF CNMs

Although graphene-based materials possess distinct properties, they seldom exhibit the necessary surface qualities for particular applications. Therefore, surface functionalization is crucial, and scientists have developed many covalent and noncovalent chemical methods [80]. The insoluble nature of CNTs and their challenging handling in many solvents have severely restricted their applications. In both aqueous solutions and all organic solvents, CNTs are manufactured to be insoluble. They can be sonicated into various solvents to be distributed, but if this process is stopped, they precipitate immediately. Nonetheless, it has been shown that CNTs may interact with many chemical types [81,82]. The development of supramolecular complexes allows for improved CNT processing, resulting in the manufacture of unique nanoscale devices. Furthermore, CNTs may undergo chemical processes that increase their solubility, allowing them to be integrated into inorganic, organic, or biological systems [82]. In order to facilitate the fabrication of potential uses for these nanostructures, the carbon nanotubes must first be functionalised, meaning that suitable chemical functions must be attached to their conjugated sp2 carbon scaffold. The derivatised tubes have better characteristics in terms of solubility, dispersion, manipulation, and processibility, and may be regarded a real subdivision of organic molecules [83].

The primary methods for modifying CNTs include (i) chemically bonding chemical groups onto the p-conjugated structure of CNTs, (ii) adsorbing or wrapping other functional molecules onto CNTs without forming chemical bonds, and (iii) the endohedral filling of their empty inner cavity [82]. Figure 1.6 illustrates a common

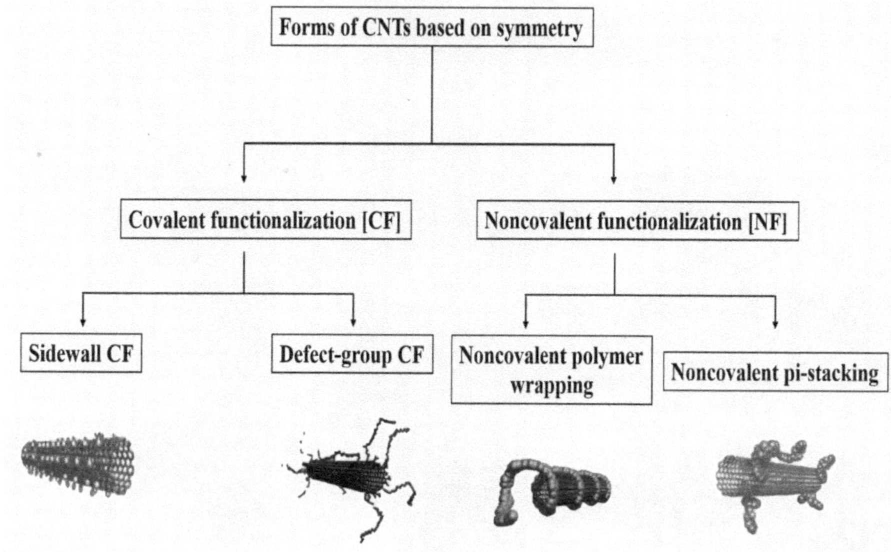

FIGURE 1.6 Surface modification of carbon nanotubes with polymers.

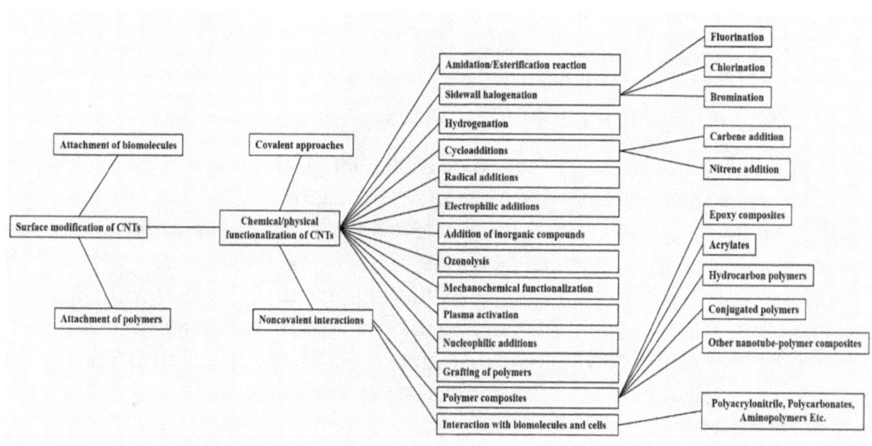

FIGURE 1.7 Various approaches for chemical functionalization of CNTs.

depiction of these surface alterations. Figure 1.7 illustrates many distinct methods of chemical modification.

CDs constitute a fascinating variation of CNMs. CDs have been used as chelating agents to augment the solubility of pharmaceuticals that have low solubility in water, as well as to improve their bioavailability and stability [84]. CDs have a high affinity for carbonaceous supports, leading to easy adsorption. The adsorption of CDs leads to a greater dispersion of support in the aqueous phase [85]. As it is improbable that CNTs or nanotube ropes would form the rotaxane structures with the CDs, the CDs are thought to be adsorbed at the surface of nanotube ropes via the van der Waals force [86,87]. Chen et al. discovered that CDs had much greater CNT-dispersing capabilities. SWCNTs may be aqueously dispersed using γ-CD in deionized water [86].

1.6 NANOSTRUCTURE PROPERTIES

Nanostructures serve as often-used colloidal nanocarriers for targeted transportation of medications or other compounds to precise locations. These nanostructures are often considered a secure technique because of their inactivity, which also contributes to their remarkable biocompatibility. By circumventing the endosome–lysosome systems, the medications will have a prolonged circulation duration and controlled release [88,89]. Furthermore, alterations to the surface morphology, shape, and component composition augment the effectiveness and productivity of nanostructures as vehicles for delivering drugs [90–92]. There are two distinct types of nanostructures: biological and inorganic nanostructures. Organic nanostructures exhibit biocompatibility and possess superior medication-loading capacity. Organic nanostructures can physically or chemically integrate active medications into their structure. Polymeric micelles are not only more stable than surfactant micelle systems, but they also have superior drug release control, are less prone to dissociate, have longer retention periods, and lower critical micelle concentrations. Furthermore, polymeric micelles outperform liposomes in terms of regulated drug release and simplicity of development [93]. Dendrimers, on the other hand, may improve the drug's solubility and bioavailability. In addition, they have shown an extended period of medication exposure and reduced negative reactions by reducing the necessary dosage [94]. Inorganic nanostructures exhibit nanoscale physical characteristics and are characterized by their tiny size. They are made of several elements, including selenium, gold, cadmium, and silica. This composition is chosen because metal nanocarriers, such as silver- and gold-coated nanoparticles, may be easily included. They have been hired as couriers for transporting drugs. Carbon-based nanomaterials have gained significant traction in the fields of medicinal chemistry and biomedical engineering due to their distinct mechanical and structural properties (Figure 1.8) [95,96]. Carbon-based nanomaterials have been used for biological and medical applications, namely in advanced cancer therapy [97]. Porous silica-based nanoparticles have been extensively studied because of their unique characteristics, including high pore volume, excellent compatibility, and bigger surface area compared to other nanomaterials. Nevertheless, when compared to other materials with pores, mesoporous silica nanoparticles (MSNs) have pore sizes ranging from around 2–50 nm, making them very attractive for drug delivery and biological applications [98]. Aside from medication delivery, they may also be used for diverse applications such as diagnostics, radiation, and magnetic resonance imaging (MRI) [99]. In recent years, there has been a growing interest in inorganic–organic hybrid nanostructures due to their desirable features and wide range of applications. By merging two separate analogs, the stability of the nanocarrier may be improved while enhancing or integrating the spatial controllability and pliability of the organic fragment [100].

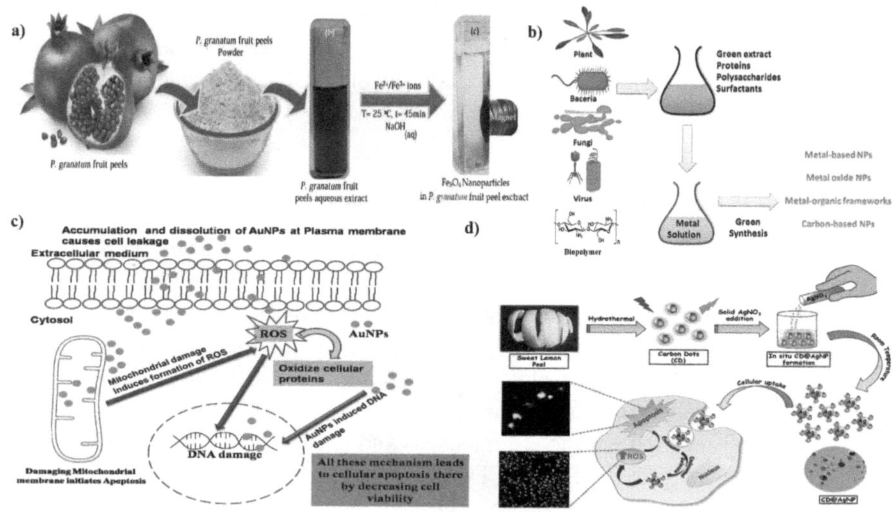

FIGURE 1.8 (a) Fe_3O_4 NPs preparation using fruit peel extract. (Reproduced with permission from Ref. [96].) (b) Schematic representation of synthesis of NMs using natural resources. (Reproduced with permission from Ref. [97].) (c) Process that green synthetic NPs perform and their actions on cancer cells. (Reproduced with permission from Ref. [95].) (d) Preparation process of NPs using lemon peel and their activity on breast cancer cells. (Reproduced with permission from Ref. [98].)

1.7 TOXICITY

The toxicity of CNMs may be reduced both *in vitro* and *in vivo* by using several approaches. Initially, modifications to the surface may be used to enhance the compatibility with living organisms and reduce the harmful effects on cells. By incorporating biocompatible polymers or biomolecules into carbon nanostructures, the negative effects may be reduced, and the occurrence of nonspecific interactions can be prevented. Furthermore, the potential harm caused by CNMs may be significantly mitigated via meticulous control of their dosage, dimensions, and morphology. The potential adverse outcomes may be mitigated by enhancing these configurations while preserving the positive attributes. Finally, it is important to conduct thorough toxicity assessments and studies to fully understand the causes of toxicity and identify any potential dangers. This will facilitate the establishment of appropriate protective norms and regulations. Biocompatible polymers can enclose the active pharmaceutical ingredient in a drug delivery system, therefore preventing it from interacting with unintended targets in the body and reducing unwanted side effects. Furthermore, by carefully controlling the size and shape of CNMs used in biomedical applications like drug delivery or imaging, it is possible to minimize their potential toxicity while still making use of their distinctive properties for efficient and precise treatment. Comprehensive toxicity studies are required to fully understand the risks associated with the use of carbon nanoparticles in biological applications. This research evaluates numerous factors, such as the concentration, size, and surface chemistry of nanomaterials, in relation to different cell lines and animal models. Thorough toxicity assessments enable researchers to identify possible harmful effects and develop strategies to minimize them. Moreover, the development of biomedical devices and therapies that are both safer and more effective requires a deep understanding of the basic principles underlying nanomaterial toxicity.

1.8 SUMMARY AND CONCLUSION

Carbon-based nanomaterials as one of the most employed kinds of nanomaterials have been the subject of extensive research over the last two decades. Carbon-based nanomaterials have been widely used in a variety of applications due to their inherent mechanical, optical, electrochemical, and electrical capabilities. Furthermore, throughout the last decade, CBNs have received a lot of attention in biomedical engineering because of their variable surface characteristics, size, and form. Interestingly, carbon-based nanomaterials are emerging as interesting materials due to the presence of both inorganic semiconducting and organic stacking capabilities. As a result, it could successfully interface with biomolecules while also responding to light. By combining such properties in a single organism, carbon-based nanomaterials could be exploited to produce biomedical applications in the future. Regarding their toxic effect on biological systems, numerous chemical modifying procedures have been devised and effectively implemented in biomedical applications such as drug delivery, tissue engineering, biomolecule identification, and cancer treatment. This chapter discusses some of the advancements made in the use of carbon-based nanomaterials for biomedical applications.

ACKNOWLEDGMENT

The Department of Pharmaceutics of the National Institute of Pharmaceutical Education and Research (NIPER), Hyderabad (Ministry of Chemicals and Fertilisers, India) is acknowledged by the writers for providing facilities for the authoring of this publication.

REFERENCES

1. Abu Owida H, Turab NM, Al-Nabulsi J. Carbon nanomaterials advancements for biomedical applications. *Bull Electr Eng Inform.* 2023;12:891–901.
2. Chary PS, Bhawale R, Vasave R, Rajana N, Singh PK, Madan J, et al. A review on emerging role of multifunctional carbon nanotubes as an armament in cancer therapy, imaging and biosensing. *J Drug Deliv Sci Technol.* 2023;85:104588.
3. Hong G, Diao S, Antaris AL, Dai H. Carbon nanomaterials for biological imaging and nanomedicinal therapy. *Chem Rev.* 2015;115:10816–906.
4. Kroto HW, Heath JR, O'Brien SC, Curl RF, Smalley RE. C60: buckminsterfullerene. *Nature.* 1985;318:162–3.
5. Iijima S. Helical microtubules of graphitic carbon. *Nature.* 1991;354:56–8.
6. Novoselov KS, Geim AK, Morozov SV., Jiang D, Zhang Y, Dubonos SV., et al. Electric field effect in atomically thin carbon films. *Science (1979).* 2004;306:666–9.
7. Baker SN, Baker GA. Luminescent carbon nanodots: emergent nanolights. *Angew Chem Int Ed.* 2010;49:6726–44.

8. Mochalin VN, Shenderova O, Ho D, Gogotsi Y. The properties and applications of nanodiamonds. *Nat Nanotechnol.* 2012;7:11–23.
9. Fitzgibbons TC, Guthrie M, Xu E, Crespi VH, Davidowski SK, Cody GD, et al. Benzene-derived carbon nanothreads. *Nat Mater.* 2015;14:43–7.
10. Liu Z, Tabakman S, Welsher K, Dai H. Carbon nanotubes in biology and medicine: in vitro and in vivo detection, imaging and drug delivery. *Nano Res.* 2009;2:85–120.
11. Liu Z, Robinson JT, Tabakman SM, Yang K, Dai H. Carbon materials for drug delivery & cancer therapy. *Mater Today.* 2011;14:316–23.
12. Liang Y, Li Y, Wang H, Dai H. Strongly coupled inorganic/nanocarbon hybrid materials for advanced electrocatalysis. *J Am Chem Soc.* 2013;135:2013–36.
13. Wang H, Dai H. Strongly coupled inorganic–nano-carbon hybrid materials for energy storage. *Chem Soc Rev.* 2013;42:3088.
14. Ruggiero A, Villa CH, Bander E, Rey DA, Bergkvist M, Batt CA, et al. Paradoxical glomerular filtration of carbon nanotubes. *Proc Natl Acad Sci.* 2010;107:12369–74.
15. De La Zerda A, Zavaleta C, Keren S, Vaithilingam S, Bodapati S, Liu Z, et al. Carbon nanotubes as photoacoustic molecular imaging agents in living mice. *Nat Nanotechnol.* 2008;3:557–62.
16. Robinson JT, Welsher K, Tabakman SM, Sherlock SP, Wang H, Luong R, et al. High performance in vivo near-IR (>1 µm) imaging and photothermal cancer therapy with carbon nanotubes. *Nano Res.* 2010;3:779–93.
17. Welsher K, Sherlock SP, Dai H. Deep-tissue anatomical imaging of mice using carbon nanotube fluorophores in the second near-infrared window. *Proc Natl Acad Sci.* 2011;108:8943–8.
18. Hong G, Lee JC, Robinson JT, Raaz U, Xie L, Huang NF, et al. Multifunctional in vivo vascular imaging using near-infrared II fluorescence. *Nat Med.* 2012;18:1841–6.
19. Hong G, Diao S, Chang J, Antaris AL, Chen C, Zhang B, et al. Through-skull fluorescence imaging of the brain in a new near-infrared window. *Nat Photonics.* 2014;8:723–30.
20. Liu Z, Fan AC, Rakhra K, Sherlock S, Goodwin A, Chen X, et al. Supramolecular stacking of doxorubicin on carbon nanotubes for in vivo cancer therapy. *Angew Chem Int Ed.* 2009;48:7668–72.
21. Sherlock SP, Tabakman SM, Xie L, Dai H. Photothermally enhanced drug delivery by ultrasmall multifunctional FeCo/graphitic shell nanocrystals. *ACS Nano.* 2011;5:1505–12.
22. Nayak TR, Andersen H, Makam VS, Khaw C, Bae S, Xu X, et al. Graphene for controlled and accelerated osteogenic differentiation of human mesenchymal stem cells. *ACS Nano.* 2011;5:4670–8.
23. Li N, Zhang Q, Gao S, Song Q, Huang R, Wang L, et al. Three-dimensional graphene foam as a biocompatible and conductive scaffold for neural stem cells. *Sci Rep.* 2013;3:1604.
24. Shin SR, Bae H, Cha JM, Mun JY, Chen Y-C, Tekin H, et al. Carbon nanotube reinforced hybrid microgels as scaffold materials for cell encapsulation. *ACS Nano.* 2012;6:362–72.
25. Topping TM, Williams GA. Xenon endophotocoagulation under air using sodium hyaluronate. *Am J Ophthalmol.* 1984;97:785–6.
26. Hosnedlova B, Kepinska M, Fernandez C, Peng Q, Ruttkay-Nedecky B, Milnerowicz H, et al. Carbon nanomaterials for targeted cancer therapy drugs: a critical review. *Chem Rec.* 2019;19:502–22.
27. Dasgupta A, Rajukumar LP, Rotella C, Lei Y, Terrones M. Covalent three-dimensional networks of graphene and carbon nanotubes: synthesis and environmental applications. *Nano Today.* 2017;12:116–35.

28. Kordatos K, Da Ros T, Bosi S, Vázquez E, Bergamin M, Cusan C, et al. Novel versatile fullerene synthons. *J Org Chem*. 2001;66:4915–20.
29. Thompson BC, Fréchet JMJ. Polymer–fullerene composite solar cells. *Angew Chem Int Ed*. 2008;47:58–77.
30. Ramuz MP, Vosgueritchian M, Wei P, Wang C, Gao Y, Wu Y, et al. Evaluation of solution-processable carbon-based electrodes for all-carbon solar cells. *ACS Nano*. 2012;6:10384–95.
31. Lanzellotto C, Favero G, Antonelli ML, Tortolini C, Cannistraro S, Coppari E, et al. Nanostructured enzymatic biosensor based on fullerene and gold nanoparticles: preparation, characterization and analytical applications. *Biosens Bioelectr.* 2014;55:430–7.
32. Cha C, Shin SR, Annabi N, Dokmeci MR, Khademhosseini A. Carbon-based nanomaterials: multifunctional materials for biomedical engineering. *ACS Nano*. 2013;7:2891–7.
33. Scott LT, Boorum MM, McMahon BJ, Hagen S, Mack J, Blank J, et al. A rational chemical synthesis of C_{60}. *Science (1979)*. 2002;295:1500–3.
34. Brettreich M, Hirsch A. A highly water-soluble dendro[60]fullerene. *Tetrahedron Lett.* 1998;39:2731–4.
35. Torres VM, Posa M, Srdjenovic B, Simplício AL. Solubilization of fullerene C60 in micellar solutions of different solubilizers. *Colloids Surf B Biointerfaces*. 2011;82:46–53.
36. Ashcroft JM, Tsyboulski DA, Hartman KB, Zakharian TY, Marks JW, Weisman RB, et al. Fullerene (C60) immunoconjugates: interaction of water-soluble C60 derivatives with the murine anti-gp240 melanoma antibody. *Chem Commun*. 2006;3004.
37. Jeong J, Jung J, Choi M, Kim JW, Chung SJ, Lim S, et al. Color-tunable photoluminescent fullerene nanoparticles. *Adv Mater.* 2012;24:1999–2003.
38. Jeong J, Cho M, Lim YT, Song NW, Chung BH. Synthesis and characterization of a photoluminescent nanoparticle based on fullerene–silica hybridization. *Angew Chem Int Ed*. 2009;48:5296–9.
39. Ma H, Liang X-J. Fullerenes as unique nanopharmaceuticals for disease treatment. *Sci China Chem*. 2010;53:2233–40.
40. Yang D, Zhao Y, Guo H, Li Y, Tewary P, Xing G, et al. $[Gd@C_{82}(OH)_{22}]_n$ nanoparticles induce dendritic cell maturation and activate Th1 immune responses. *ACS Nano*. 2010;4:1178–86.
41. Kostarelos K, Lacerda L, Pastorin G, Wu W, Wieckowski S, Luangsivilay J, et al. Cellular uptake of functionalized carbon nanotubes is independent of functional group and cell type. *Nat Nanotechnol*. 2007;2:108–13.
42. Donaldson K, Aitken R, Tran L, Stone V, Duffin R, Forrest G, et al. Carbon nanotubes: a review of their properties in relation to pulmonary toxicology and workplace safety. *Toxicol Sci*. 2006;92:5–22.
43. Thomsen C, Reich S, Maultzsch J. *Carbon Nanotubes: Basic Concepts and Physical Properties*. Wiley, Hoboken, NJ, 2004.
44. Huang JY, Chen S, Ren ZF, Wang Z, Kempa K, Naughton MJ, et al. Enhanced ductile behavior of tensile-elongated individual double-walled and triple-walled carbon nanotubes at high temperatures. *Phys Rev Lett*. 2007;98:185501.
45. Zhang D, Fu L, Liao L, Liu N, Dai B, Zhang C. Preparation, characterization, and application of electrochemically functional graphene nanocomposites by one-step liquid-phase exfoliation of natural flake graphite with methylene blue. *Nano Res*. 2012;5:875–87.
46. Del Rio-Castillo AE, Merino C, Díez-Barra E, Vázquez E. Selective suspension of single layer graphene mechanochemically exfoliated from carbon nanofibres. *Nano Res*. 2014;7:963–72.
47. Khalil I, Julkapli N, Yehye W, Basirun W, Bhargava S. Graphene–gold nanoparticles hybrid—synthesis, functionalization, and application in a electrochemical and surface-enhanced raman scattering biosensor. *Materials*. 2016;9:406.

48. Jiao L, Xie L, Dai H. Densely aligned graphene nanoribbons at ~35 nm pitch. *Nano Res.* 2012;5:292–6.
49. Jiao L, Wang X, Diankov G, Wang H, Dai H. Facile synthesis of high-quality graphene nanoribbons. *Nat Nanotechnol.* 2010;5:321–5.
50. Ma D, Zhang Y, Liu M, Ji Q, Gao T, Zhang Y, et al. Clean transfer of graphene on Pt foils mediated by a carbon monoxide intercalation process. *Nano Res.* 2013;6:671–8.
51. Li Z, Xu Y, Cao B, Qi L, Zhao E, Yang S, et al. The interface of epitaxial nanographene on GaN by PECVD. *AIP Adv.* 2019;9.
52. Nakada K, Fujita M, Dresselhaus G, Dresselhaus MS. Edge state in graphene ribbons: nanometer size effect and edge shape dependence. *Phys Rev B.* 1996;54:17954–61.
53. Geim AK. Graphene: status and prospects. *Science (1979).* 2009;324:1530–4.
54. Bolotin KI, Sikes KJ, Jiang Z, Klima M, Fudenberg G, Hone J, et al. Ultrahigh electron mobility in suspended graphene. *Solid State Commun.* 2008;146:351–5.
55. Du X, Skachko I, Barker A, Andrei EY. Approaching ballistic transport in suspended graphene. *Nat Nanotechnol.* 2008;3:491–5.
56. Pop E, Varshney V, Roy AK. Thermal properties of graphene: fundamentals and applications. *MRS Bull.* 2012;37:1273–81.
57. Westervelt RM. Graphene nanoelectronics. *Science (1979).* 2008;320:324–5.
58. Stöberl U, Wurstbauer U, Wegscheider W, Weiss D, Eroms J. Morphology and flexibility of graphene and few-layer graphene on various substrates. *Appl Phys Lett.* 2008;93.
59. Lee C, Wei X, Kysar JW, Hone J. Measurement of the elastic properties and intrinsic strength of monolayer graphene. *Science (1979).* 2008;321:385–8.
60. Serrano MC, Patiño J, García-Rama C, Ferrer ML, Fierro JLG, Tamayo A, et al. 3D free-standing porous scaffolds made of graphene oxide as substrates for neural cell growth. *J Mater Chem B.* 2014;2:5698.
61. Depan D, Girase B, Shah JS, Misra RDK. Structure–process–property relationship of the polar graphene oxide-mediated cellular response and stimulated growth of osteoblasts on hybrid chitosan network structure nanocomposite scaffolds. *Acta Biomater.* 2011;7:3432–45.
62. Kalbacova M, Broz A, Kong J, Kalbac M. Graphene substrates promote adherence of human osteoblasts and mesenchymal stromal cells. *Carbon N Y.* 2010;48:4323–9.
63. Zhu Z, Jiang H, Guo S, Cheng Q, Hu Y, Li C. Dual tuning of biomass-derived hierarchical carbon nanostructures for supercapacitors: the role of balanced meso/microporosity and graphene. *Sci. Rep.* 2015;5(1):15936.
64. Ozden S, Tiwary CS, Yao J, Brunetto G, Bhowmick S, Asif S, Vajtai R, Ajayan PM. Highly ordered carbon-based nanospheres with high stiffness. *Carbon* 2016;105:144–150.
65. Du Y, Guo S. Chemically doped fluorescent carbon and graphene quantum dots for bioimaging, sensor, catalytic and photoelectronic applications. *Nanoscale* 2016;8(5):2532–2543.
66. Liu Z, Chen K, Davis C, Sherlock S, Cao Q, Chen X, et al. Drug delivery with carbon nanotubes for *in vivo* cancer treatment. *Cancer Res.* 2008;68:6652–60.
67. Mehra NK, Jain AK, Nahar M. Carbon nanomaterials in oncology: an expanding horizon. *Drug Discov Today.* 2018;23:1016–25.
68. Mehra NK, Palakurthi S. Interactions between carbon nanotubes and bioactives: a drug delivery perspective. *Drug Discov Today.* 2016;21:585–97.
69. Mehra NK, Jain K, Jain NK. Pharmaceutical and biomedical applications of surface engineered carbon nanotubes. *Drug Discov Today.* 2015;20:750–9.
70. Mehra NK, Mishra V, Jain NK. A review of ligand tethered surface engineered carbon nanotubes. *Biomaterials.* 2014;35:1267–83.
71. Krasley AT, Li E, Galeana JM, Bulumulla C, Beyene AG, Demirer GS. Carbon nanomaterial fluorescent probes and their biological applications. *Chem Rev.* 2024;124:3085–185.
72. Szabó A, Perri C, Csató A, Giordano G, Vuono D, Nagy JB. Synthesis methods of carbon nanotubes and related materials. *Materials.* 2010;3:3092.

73. Stepek IA, Nagase M, Yagi A, Itami K. New paradigms in molecular nanocarbon science. *Tetrahedron*. 2022;123:132907.
74. Facciolà A, Visalli G, La Maestra S, Ceccarelli M, D'Aleo F, Nunnari G, et al. Carbon nanotubes and central nervous system: environmental risks, toxicological aspects and future perspectives. *Environ Toxicol Pharmacol*. 2019;65:23–30.
75. Prasek J, Drbohlavova J, Chomoucka J, Hubalek J, Jasek O, Adam V, et al. Methods for carbon nanotubes synthesis-review. *J Mater Chem*. 2011;21:15872–84.
76. Hou PX, Liu C, Cheng HM. Purification of carbon nanotubes. *Carbon N Y*. 2008;46:2003–25.
77. Kokorina AA, Sapelkin A V., Sukhorukov GB, Goryacheva IY. Luminescent carbon nanoparticles separation and purification. *Adv Colloid Interface Sci*. 2019;274:102043.
78. Carbon Nanotubes: Redefining the World of Electronics - Google Books [Internet]. [cited 2024 Jun 27]. Available from: https://books.google.co.in/books?hl=en&lr=&id=jbZaEAAAQBAJ&oi=fnd&pg=PA29&dq=purification+of+carbon+nanotubes&ots=uViD9nJoVu&sig=CKW3idiY_EHBnnc2lTKJ_Zq1SHk#v=onepage&q=purificationofcarbonnanotubes&f=false
79. He S, Wu L, Li X, Sun H, Xiong T, Liu J, et al. Metal-organic frameworks for advanced drug delivery. *Acta Pharm Sin B*. 2021;11:2362–95.
80. Dai L. Functionalization of graphene for efficient energy conversion and storage. *Acc Chem Res*. 2013;46:31–42.
81. Son KH, Hong JH, Lee JW. Carbon nanotubes as cancer therapeutic carriers and mediators. *Int J Nanomedicine*. 2016;11:5163–85.
82. Tasis D, Tagmatarchis N, Bianco A, Prato M. Chemistry of carbon nanotubes. *Chem Rev*. 2006;106:1105–36.
83. Syrgiannis Z, Melchionna M, Prato M. *Covalent Carbon Nanotube Functionalization. Encyclopedia of Polymeric Nanomaterials*. Springer, Berlin, Heidelberg, 2015, pp. 480–7.
84. Loftsson T, Jarho P, Másson M, Järvinen T. Cyclodextrins in drug delivery. *Expert Opin Drug Deliv*. 2005;2:335–51.
85. Cassez A, Ponchel A, Hapiot F, Monflier E. Unexpected multifunctional effects of methylated cyclodextrins in a palladium charcoal-catalyzed suzuki–miyaura reaction. *Org Lett*. 2006;8:4823–6.
86. Szejtli J. Introduction and general overview of cyclodextrin chemistry. *Chem Rev*. 1998;98:1743–54.
87. Ikeda A, Nobusawa K, Hamano T, Kikuchi J. Single-walled carbon nanotubes template the one-dimensional ordering of a polythiophene derivative. *Org Lett*. 2006;8:5489–92.
88. Malode SJ, Pandiaraj S, Alodhayb A, Shetti NP. Carbon nanomaterials for biomedical applications: progress and outlook. *ACS Appl Bio Mater*. 2024;7:752–77.
89. Chamundeeswari M, Jeslin J, Verma ML. Nanocarriers for drug delivery applications. *Environ Chem Lett*. 2019;17:849–65.
90. Sun T, Zhang YS, Pang B, Hyun DC, Yang M, Xia Y. Engineered nanoparticles for drug delivery in cancer therapy. *Angew Chem Int Ed*. 2014;53:12320–64.
91. Avramović N, Mandić B, Savić-Radojević A, Simić T. Polymeric nanocarriers of drug delivery systems in cancer therapy. *Pharmaceutics*. 2020;12:298.
92. Escudero A, Carregal-Romero S, Miguel-Coello AB, Ruíz-Cabello J. Engineered polymeric nanovehicles for drug delivery. *Front Nanosci*. 2020;16:201–32.
93. Amin MCIM, Butt AM, Amjad MW, Kesharwani P. Polymeric micelles for drug targeting and delivery. In: Mifshra V, Kesharwani P, Amin MCIM, Iyer A (eds.) *Nanotechnology-Based Approaches for Targeting and Delivery of Drugs and Genes*. Elsevier, Amsterdam, Netherlands, 2017, 167–202.
94. Santos A, Veiga F, Figueiras A. Dendrimers as pharmaceutical excipients: synthesis, properties, toxicity and biomedical applications. *Materials*. 2019;13:65.

95. Singh AK, Tiwari R, Singh VK, Singh P, Khadim SR, Singh U, et al. Green synthesis of gold nanoparticles from *Dunaliella salina*, its characterization and in vitro anticancer activity on breast cancer cell line. *J Drug Deliv Sci Technol.* 2019;51:164–76.
96. Yusefi M, Shameli K, Ali RR, Pang S-W, Teow S-Y. Evaluating anticancer activity of plant-mediated synthesized iron oxide nanoparticles using punica granatum fruit peel extract. *J Mol Struct.* 2020;1204:127539.
97. Rabiee N, Ahmadi S, Iravani S, Varma RS. Natural resources for sustainable synthesis of nanomaterials with anticancer applications: a move toward green nanomedicine. *Environ Res.* 2023;216:114803.
98. Ghosal K, Ghosh S, Ghosh D, Sarkar K. Natural polysaccharide derived carbon dot based in situ facile green synthesis of silver nanoparticles: synergistic effect on breast cancer. *Int J Biol Macromol.* 2020;162:1605–15.
99. Amiri M, Salavati-Niasari M, Akbari A. Magnetic nanocarriers: evolution of spinel ferrites for medical applications. *Adv Colloid Interface Sci.* 2019;265:29–44.
100. Hu T, Li Z, Lu L, Dai K, Zhang J, Li R, et al. Inorganic-organic CdSe-diethylenetriamine nanobelts for enhanced visible photocatalytic hydrogen evolution. *J Colloid Interface Sci.* 2019;555:166–73.

2 Carbon Nanotubes (CNTs)

Fundamental and Physicochemical Properties

Soyal Sayyed, Saptarshee Bhattacharjee, Jitendra Kumar, Abhijeet Chinchane, Pankaj Kumar, Sagar Kulkarni, Shubham Gajdhane, Roshani Jaiswal, Atreyee Bhattacharyya, Spandana Gollapalli, Pawan Devangan, Anju Katyal, Snigdha Singh, Ramesh Chandra, Neelesh Kumar Mehra, and Jitender Madan

ABBREVIATIONS

CNTs	carbon nanotubes
DWCNTs	double-walled carbon nanotubes
EDA	electron-donating acceptors
GA	Gum arabic
MWCNTs	multiwalled carbon nanotubes
PEG	Polyethylene glycol
SWCNTs	single-walled carbon nanotubes
TEM	transmission electron microscopy
XRD	X-ray diffraction
DSPE	1,2-Disteroyl-sn-Glycero-3-Phosphoethanolamine

2.1 INTRODUCTION

Carbon is a remarkable element that exists in various allotropic forms [1]. Carbon allotropes have appeared in different forms: zero dimension (0D), such as quantum dots and nanoclusters; one dimension (1D), such as nanofiber and nanotubes; two dimensions (2D), such as nanocoating; and three dimensions (3D), such as bulk material and powders. Newly developed nanostructures with remarkable versatility and exceptional properties are known as carbon nanomaterials [2]. In the past, attention has increased due to the impact of the physical and chemical properties of nanomaterials,

DOI: 10.1201/9781003386537-2

including the stiffness and shape of their interactions with cells. Carbon nanomaterials serve as highly adaptable substrates for theranostic and biomedical applications [3,4]. Nanomaterials have an extensive number of structures and geometries, including nanoshells, nanowires, nanoparticles, and nanotubes [5]. It has been demonstrated that the key properties of nanomaterials, like shape, surface chemistry, size, and stiffness, have an impact on drug release [6], targeted efficiency [7], and cytotoxicity [8].

2.1.1 Overview of Carbon Nanotubes

Carbon nanotubes (CNTs) are 1D carbon allotropes with sp^2 hybridization. CNTs are inert and chemically stable nanomaterials [9]. CNTs are nanostructure forms with a high aspect ratio, and among their many unique properties such as nano range size, distinctive fiber structure, high surface area, and mechanical stiffness [10]. These nanotubes exhibit miraculous optical, physical, chemical, and electrical properties that make them promising candidates for bioactive delivery [11]. CNTs are hollow, tubular structures, more drug-loading, biocompatible, and do not require fluorescent labeling, further enhancing their potential for biomedical applications [12].

CNTs are produced by heating graphite and black carbon in a regulated flame. However, the primary problem with this method is that it produced CNT with inconsistencies in size, shape, quality, and purity [13]. Therefore, CNTs were synthesized using alternative techniques like vapor-phase growth, electric arc discharge, chemical vapor deposition (CVD), laser ablation, and flame synthesis. The electric arc discharge was a widely used and old approach. This technique has been recognized as conventional and requires high temperatures to evaporate carbon atoms. The laser ablation method was the same as arc discharge, in which carbon is first vaporized, followed by deposition onto a substrate. In the CVD method, solid materials are deposited on a heated surface as a result of a chemical reaction occurring in the vapor phase [14]. Since various routes are employed, resulting CNTs vary in terms of length, diameter, and quality [15].

CNTs have distinct benefits over other drug delivery systems. Their extended inner volume can be occupied by specific chemicals and biochemical species [16]. The unique inner and outer surfaces of CNTs may be altered for modification that enhances their functionalization property. This provided the ability to load compounds inside the CNTs while the outer surface was chemically characterized to improve biocompatibility. The ends of CNTs are open, making it especially simple to incorporate molecules within the nanotubes [17]. CNTs are recognized as effective nanocarrier systems due to their high blood circulation time in the bloodstream and ability to easily cross cell membranes [18]. Drug delivery occurs by absorption of drugs within CNTs mesh, covalent or noncovalent bonding of drugs, nucleic acid, and peptides to the CNTs outer walls, and also utilization as catheters are potential pathways for CNT-drug interaction that have been suggested [19].

The characterization of CNTs is performed using several techniques, including scanning electron microscopy (SEM), transmission electron microscopy (TEM), Raman spectroscopy, atomic force microscopy (AFM), and absorption spectroscopy [20]. Raman spectroscopy is an easy, fast, and sensitive method for characterizing nanotubes and analyzing the changes in the properties of nanotubes produced under various processes [21]. TEM gives detailed information about the structure, shape, size, and imaging cellular uptake of the drug [22]. SEM is employed for the morphology of nanotubes, while proton NMR examines the different functional groups attached to CNTs [23]. AFM is a microscopic technique that investigates the CNT length and diameter of the bunch [24].

CNTs offers a wide range of potential advantages for targeted drug delivery, including encapsulation of both hydrophilic and hydrophobic compounds, improving medication stability, and site-specific administration [5]. CNTs are used in artificial implants [25], sensors [26], semiconductor instruments, conductivity, and radiation sources [27]. Several studies have reported that CNTs can enhance both the uptake and penetration of protein into cells, making them effective tools for protein delivery [28]. Additionally, CNTs show applicability in DNA delivery, regenerative medicine, and cancer therapy.

2.1.2 History of Carbon Nanotubes

The discovery of CNTs has been a subject of debate. However, most scientific communities consider that Iijima first discovered the multiwalled CNTs in 1991 [29]. The true identity of CNTs remains contested, with some research suggesting that the discovery of CNTs dates back a long way. It had been hypothesized that the carbon filament may have formed as early as 1889 during Thomas Alva Edison's times when a light bulb filament was being looked for in order to create incandescent lamps [30]. In 1952, the structural nature of carbon filaments was published and characterized by TEM and showed the tubular structure of carbon filaments [31]. The carbon monoxide was catalytically disproportionated at 480°C–700°C, resulting in carbon fibers, which were described by Boehm in 1973 [32]. A vapor-growth method was used in 1976 to prepare carbon fibers with diameters on the nanometer scale [33]. Hollow carbon was formed via the arc discharge method on carbon anodes in 1979 [34]. The patent was granted to H.G. Tennent for synthesizing cylindrical carbon fibrils in 1987 [35]. The patent described these fibrils as having features of a length of 102 times larger than the diameter and a constant diameter ranging from approximately 3.5 to about 70 nm, with an outside region consisting of several essentially continuous layers [29]. However, the majority of the research conducted before 1991 was not well recognized in the scientific world.

Scientists have become increasingly interested in nanomaterials since the discovery of fullerene via the arc discharge method [36]. In 1991, the arc discharge method was used to prepare C60, which observed this method for the formation of CNTs, and this study was published by Iijima. CNTs were first noticed by scientists throughout the world in 1991. After Iijima discovered the MWCNTs and different metal catalysts, such as Fe or Co, they were added to arc discharge methods to synthesize single-walled carbon nanotubes (SWCNTs) [37,38]. Arc discharged methods produced low-quality CNTs because of a mixture of CNTs, graphite, and amorphous carbon, which required more purification methods [39,40]. In 1995, Guo et al. developed the laser vaporization method as a substitute for arc discharge, producing small-walled CNTs, and observed a higher yield of SWCNTs [41]. However, other structures of CNTs were discovered, such as CNT nanobuds, carbon microtubes, CNTs Y junctions, and CNTs nanotorus (Figure 2.1) [29].

2.1.3 Types of Carbon Nanotubes

CNTs may be classified based on the presence of layers: (i) single-walled carbon nanotubes (SWCNTS), (ii) double-walled carbon nanotubes (DWCNTs), and (iii) multiwalled carbon nanotubes (MWCNTs) (Figure 2.2).

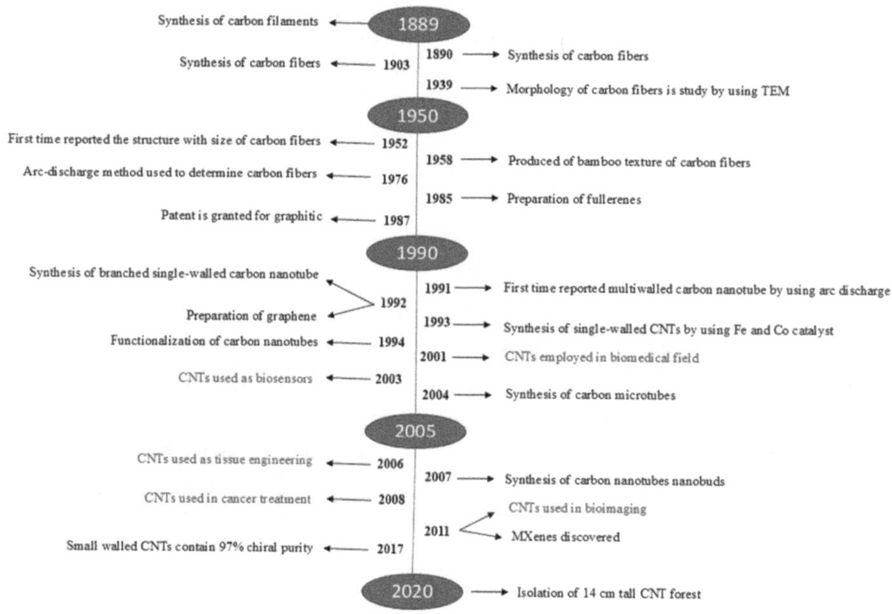

FIGURE 2.1 Schematic representation of the history of carbon nanotubes.

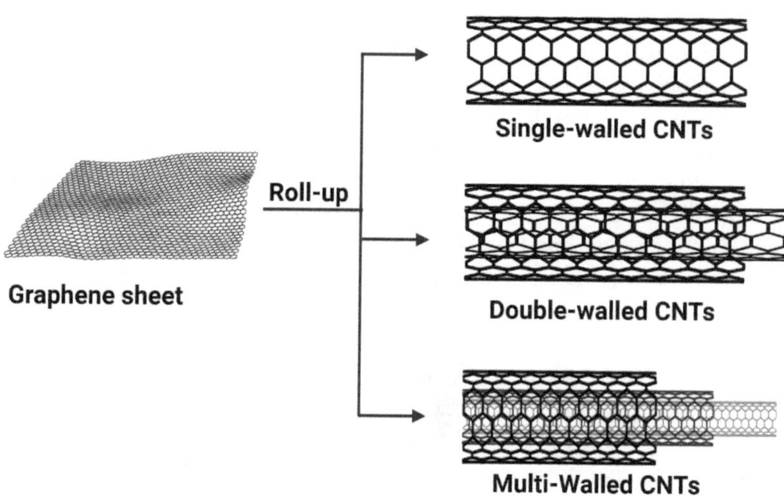

FIGURE 2.2 Different types of carbon nanotubes.

2.1.3.1 Single-Walled CNTs

Iijima et al. first reported the SWCNTs in 1993 [37]. SWCNTs are composed of a single atomic sheet of carbon, which consists of a hollow structure and incredibly high aspect ratios. SWCNTs exhibit marvelous properties such as electrical, thermal, and mechanical. Therefore, they are recognized as among the most desirable nanomaterials for fundamental research [42]. The catalyst is necessary for the synthesis of CNTs to be well controlled over improvement, and their diameter is 1–2 nm. Depending on the preparation techniques, SWCNT length may change, and its purity may decrease [14]. CNT ends can be joined by two caps, which are made of hemi fullerene. A number of rings are present at the tips of nanotubes, such as hexagonal and pentagonal rings [43].

2.1.3.2 Double-Walled CNTs

DWCNTs were identified in 1991. However, the first synthesis was published in 1998. DWCNTs are the latest class of CNTs [44]. DWCNTs are composed of two CNTs that are separated from one another by the outer tube enclosing the inner tube. The diameter of both the inner and outer tubes are 1–3 nm and 2–4 nm, respectively [14].

2.1.3.3 Multiwalled Carbon Nanotubes

MWCNTs are made of numerous layers of graphene rolled over one another, which can range in diameter from 2 to 50 nm [45]. The tubes feature an inner layer radius of 0.34 nm and an outside layer radius ranging from 3 nm to 20–30 nm [46]. The structure of MWCNTs shows two types of the model depending on the arrangement of the graphene sheet: (i) composed of a graphene sheet that has been folded up around it, seen as resembling parchment model, (ii) graphene sheets are produced within a concentric structure known as Russian doll model [47,48]. MWCNTs are differentiated from other CNTs because their synthesis doesn't require a catalyst or complicated structure. It is obtained in a very pure form and not easily twisted [23]. SWCNTs have a reputation for being more effective than MWCNTs in drug delivery. The SWCNTs are a 1D shape with a high surface area, which allows for effective drug-loading capacity [49].

2.2 CHEMICAL PROPERTIES OF CARBON NANOTUBES

CNTs possess distinct physical and chemical properties (Figure 2.3). CNTs are a type of carbon material derived from graphene, and chemical modification is required during their synthesis. The chemical properties of CNTs are influenced by the difference in pi-orbitals, which results from their larger curvature, and this response can be easily observed in CNTs. Consequently, there is a notable difference in reactivity between the end cap and the sidewall of CNTs. The reactivity tends to increase with a decreasing diameter of CNT, and reaction occurs on the sidewall and end top of CNT [14]. These reactions enable the formation of various derivatizations through chemical functionalization [50].

2.2.1 Functionalization of CNTs

CNTs have shown some limitations, such as poor solubility in solvents, formation of agglomeration, and challenges with dispersibility [51]. To overcome these limitations,

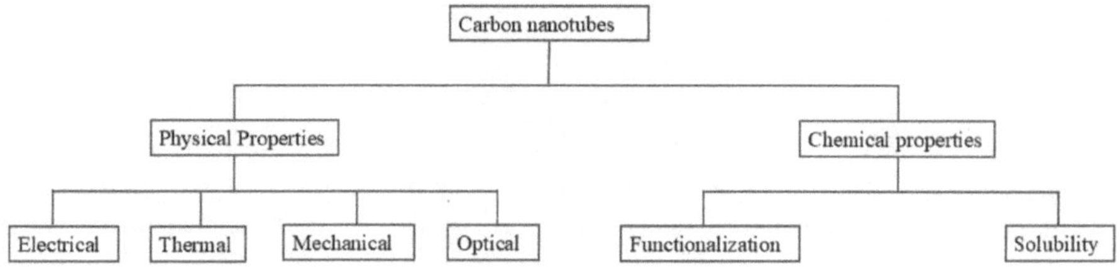

FIGURE 2.3 Different properties of carbon nanotubes.

introducing an aliphatic functional group on the outside of CNTs can significantly improve their solubility and controllability [52].

2.2.1.1 Covalent Functionalization

Covalent interactions are stronger than noncovalent interactions, while functional groups and the carbon structure of CNTs are the basis required for covalent functionalization. Covalent functionalization has been utilized to improve the wetting and dispersion in the various mediums. This process consists of two methods: direct modification (on the sidewall) and indirect modification (at the defect site). However, covalent interaction has significant drawbacks, such as disrupting the surface of π conjugation and decreasing electrical conductivity [51]. Kordzadeh et al. studied the covalent functionalization with doxorubicin (DOX) using molecular dynamics [53].

2.2.1.2 Noncovalent Functionalization

An additional technique for altering the interfacial characteristics of nanotubes is known as noncovalent functionalization. This interaction depends on several forces, including π-stacking, van der Waals, electrostatic, and hydrogen bonds. For CNTs, it is crucial to retain their pristine structure and properties; noncovalent surface functionalization usually results in fewer defects in the graphitic structure and does not damage the conjugated system of the CNT sidewalls. The disadvantage of this interaction is the different forces between CNTs, which can lead to the formation of weak nanotubes [51]. Alam et al. successfully formed noncovalently stable shellac-coated CNTs using shellac bio-resin, which was used at 5% wt concentration. In the investigation, shellac imparts COOH and OH functional groups to the surface of MWCNT and prepares shellac-coated CNTs. The noncovalent functionalization lowered the cytotoxicity, minimized impurities, and enhanced the thermal stability of CNTs. Due to the better thermal stability, the resulting nanocomposite is an attractive option for usage in the biomedical industry and at high temperatures [54].

CNTs typically functionalize with polymer molecules to enhance their dispersion and create CNT-based composites to explore novel properties. Covalent and noncovalent attachment are the two primary methods for modifying CNTs with polymers [55]. There is a lot of promising use of CNT functionalization in biological molecules, such as nanotechnology and bioengineering. They are primarily attentive to interactions between CNTs and biological molecules, such as proteins, peptides, and nucleic acids [51]. Madni et al. created a new technique to prepare highly dispersed MWCNTs using a combination of the anionic surfactant (sodium octanoate) and the cationic surfactant (dodecyl trimethylammonium bromide) in a 1:1 ratio. The surfactants were fastened onto the surface of MWCNTs via ultrasonic techniques, resulting in the physical attachment of CNT and surfactants. After the surfactant coating was applied to the MWCNTs, the nanotubes were detached and separated. In the mixed surfactant solution, the MWCNTs demonstrated good dispersion, comparable to systems containing larger concentrations of anionic and cationic surfactants. The study proved that the mixed surfactant system synergistic activity in a mixture of cationic and anionic surfactants made it beneficial for stable dispersions [56].

2.2.2 Grafting Approaches

An exceptional method for enhancing compatibility with several polymeric matrices is chemical functionalization. CNTs can be functionalized through covalently and noncovalently, reacting with organic molecules, such as synthetic polymers and biomolecules like proteins, enzymes, and polysaccharides. These approaches increase their hydrophilicity and solve the dispersion issue. For this reason, the grafting of molecules onto CNTs has created a new avenue for applying biomolecules and polymers to their surface. The chemical approaches employed for polymer functionalization in CNTs via grafting approaches include grafting to, grafting from, and grafting through. Prefabricated polymers are attached to functionalized or nonfunctionalized surfaces of CNTs, known as the "grafting to" technique. This method interacts with coupling methods like click, nitrene, and radical addition, which do not yield thickness and high-density grafting. Using this approach, copolymers such as poly (propionyl ethylenimine-co-ethylenimine) and polystyrene, biomolecules such as proteins, DNA, and polysaccharides can be successfully introduced using the "grafting to" technique. As a result, "grafting to" permits the insertion of complicated structural natural polymers, which may be grafted into the surface of CNTs after prior activation. CNTs are modified by polymer via an *in situ* polymerization process involving "grafting from" and "grafting through" approaches. The "grafting from" approach is particularly effective in forming high-density grafting. "Grafing through" is a method that has not been reported much. Still, it involves the functionalization of CNT with unsaturated bonds during propagation reaction to make the polymer chain attached to the nanotube surface. Furthermore, these synthesized polymers are used to functionalize nanotubes [57,58].

2.3 SOLUBILITY/AQUEOUS DISPERSION OF CNTs

CNTs have some drawbacks related to their low solubility and dispersibility in various solvents such as acetone, alcohol, and aqueous solutions. This limitation can restrict their use in drug delivery systems. To address this issue, certain modifications are required to enhance the solubility and dispersibility of CNTs. Aggregation of CNT is prevented through surface modifications and to increase the dispersibility by their interaction with the surrounding polymeric matrices. Surface modification methods for CNTs, such as surfactants, solvent interaction, and polymers, have been employed to improve solubility and dispersibility, making them more suitable for drug delivery applications [59,60].

2.3.1 Solvent Interaction

Different organic solvents, such as tetrahydrofuran (THF), dimethylformamide (DMF), chloroform, and toluene, are utilized to augment the dispersion quality of CNTs. The organic solvents are rapidly dissolved polymers and improve the dispersion of CNTs (Figure 2.4) [61]. Surfactants and other chemical molecules can change the charge distribution on the surface of CNTs, thereby enhancing their stability and interaction with solvents [56]. The surfactants covered the surface of the nanotubes and improved compatibility with the surrounding solvent [62]. The potentiometry method analyzes the interactions between surfactants, solvents, and CNTs (SWCNTs and MWCNTs) in the dispersion phase. However, this technique helps to explain how surfactants attach to the surface of CNTs [63]. Alkali metals are utilized to disperse the CNTs in polar solvents, such as lithium and sodium [64]. The separation of CNTs in solvent was done using acids. Strong acids such as H_2SO_4 and HNO_3 are used in the oxidation process, which leads to surface modifications in CNTs [65]. Meanwhile, organic solvents can disperse to a specific limit. Giordani et al. have studied the SWCNTs dispersed in the organic solvent of N-methyl pyrrolidinone (NMP). Nanotubes were characterized by AFM and spectroscopy. AFM analysis reported that the concentration decreased as it reduced the diameter of the bundle. Spectroscopy shows the presence of nanotubes in the organic solvent [66].

2.3.2 Use of Surfactant, Polymers, Biopolymers

The alterations in the CNT tubular surface can be achieved using biopolymers, surfactants, and polymers, which can form the π–π bonding and van der Waals forces [67]. Polymer coating on the surface of CNTs provides thermodynamic stability and polymer

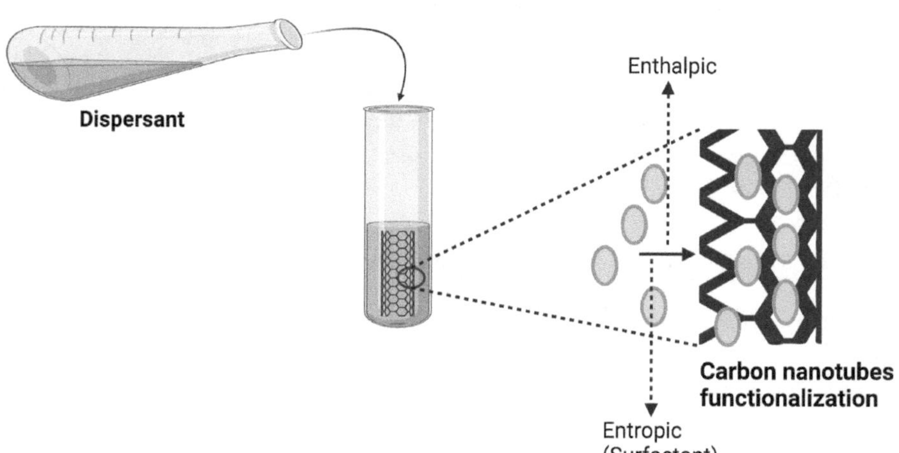

FIGURE 2.4 Carbon nanotubes interacted with the dispersant.

content that remains at the surface of nanotubes. Various methods are used to eliminate unbound polymer from the surface of nanotubes, such as dialysis, precipitation, decantation cycle, filtration, ultracentrifugation, and chromatographic separation [68]. Polymers can maintain the stability of CNT dispersion by forming complexes through polymer chain wrapping on the surface of nanotubes [63]. The interaction between polymers and CNTs contains a donor–receptor interaction mechanism where electron-rich CNTs interact with electron-deficient polymers. Another one is noncovalent interactions containing the functional groups in polymers and the electrons of CNTs, which are responsible for coating the polymer onto the CNT surfaces [69]. Several interactions, such as CH–π, π–π, and cation–π, are used for the polymer wrapping of CNTs [68]. The modifications in CNTs with polymers can be performed using chemical and physical methods. The chemical method involves polymers binding to the surface of CNTs, which form a stable composite of CNT and polymer. The chemical approaches employed for polymer functionalization in CNTs via grafting include grafting to, grafting from, and grafting through [57]. The physical methods are nondestructive, simple, and suitable for functionalizing nanotubes with different polymers. These methods rely on interactions including hydrogen bonding, π–π interaction, and electrostatic interaction [70].

Polymers are flexible in nature, coat the surface of CNT, and improve the dispersion ability (Table 2.1) [71]. Organometallic reagents such as butyllithium and sodium hydride combined with a polymer-like polyvinyl carbazole, and these mixtures are grafted to nanotube interaction to enhance dispersibility. Polymers such as polypropylene, polystyrene, and polyethylene have been dispersed in CNTs [72]. When a polymer binds to the surface of the CNT, conformational changes occur due to solvent–solute interaction [73]. Polymers, CNTs, and solvents formed a ternary system that can enhance desirable polymer conformations on the CNT surface after evaporation (Figure 2.5) [74].

Biopolymers are widely noted as binders for nanotubes in the membrane, and this ability serves as linkers between the CNTs [75]. Gum arabic (GA) is a biopolymer that is employed to stabilize the SWCNTs dispersions. GA is particularly advantageous as it can break SWCNT bundles into single ones [76]. Polysaccharide molecules play a role in stabilizing the CNT surface via hydrophobic impacts on CNT external parts. These polymers lower surface tension by projecting their hydrophilic tails toward the surface of CNTs, thereby limiting the aggregation of CNTs with the help of electrostatic repulsion [77]. Chitosan is a polysaccharide polymer with outstanding biocompatibility and has been considered a dispersant for nanotubes. Researchers have studied the MWCNTs that interacted with chitosan and reported that it improved the dispersion of nanotubes in aqueous media [78]. Gelatin is a biopolymer composed of 18 types of amino acids and is a biodegradable, biocompatible, and nonimmunogenic polymer. This polymer can functionalize the external layers of the CNTs. Zheng et al. reported that the CNTs are functionalized with gelatin, showing better solubility in an aqueous phase and enhanced biocompatibility with hemoglobin. It results in effectivly separation of aggregated CNTs, which greatly facilitates the use of CNTs in bioelectrochemical processes. Electrochemical, UV-vis spectroscopy and FTIR technques were used to investigate the properties of haemoglobin/gelatin-CNT [79]. Xanthan gum is a water-soluble biopolymer with a large molecular mass obtained from *Xanthomonas campestris*. Xanthan gum complexes have been used to improve the dispersibility of

TABLE 2.1
The Various Polymers and Their Role in CNTs

S.No	Polymer	CNTs Type	Role	References
1.	Poly (9,9-dialkylfluorene)	SWCNTs	Conjugated polymers improve the dispersion of CNTs.	[86]
2.	Polycarbazole	SWCNTs	A stable carbazole–SWCNT hybrid complex is formed, which leads to noncovalent functionalization of the CNT surface	[87]
3.	Poly(N-isopropylacrylamide)	MWCNTs	This polymer can disperse MWCNTs at certain conditions, such as at room temperature and pH 7.	[88]
4.	Polyacrylonitrile (PAN)	SWCNTs	Under the right circumstances, the admicellar polymerization technique successfully created the PAN ultrathin layer on the CNT surface. This layer increases the interaction of the CNT surface with solvent to improve the solubility of CNTs.	[89]
5.	Polyethylene glycol	SWCNTs	CNTs can be made more hydrophilic and readily dissolved in aqueous solvents.	[90]
7.	Poly [p-{2,5-bis(3-propoxysulfonic acid sodium salt)}phenylene] ethynylene (PPES)	SWCNTs	PPES is a hydrophilic polymer that consists of negatively charged groups and can augment the miscibility of CNTs in aqueous medium.	[71]
8.	Polystyrene and polyurethane	SWCNTs	These polymers affect the dispersion of SWCNTs in polymer matrices by using organic solvents.	[91]
9.	Poly(phenylacetylene)	MWCNTs	The noncovalent functionalization of CNTs by pyrene-containing PPAs enhances the solubility of MWCNTs in solvents.	[92]

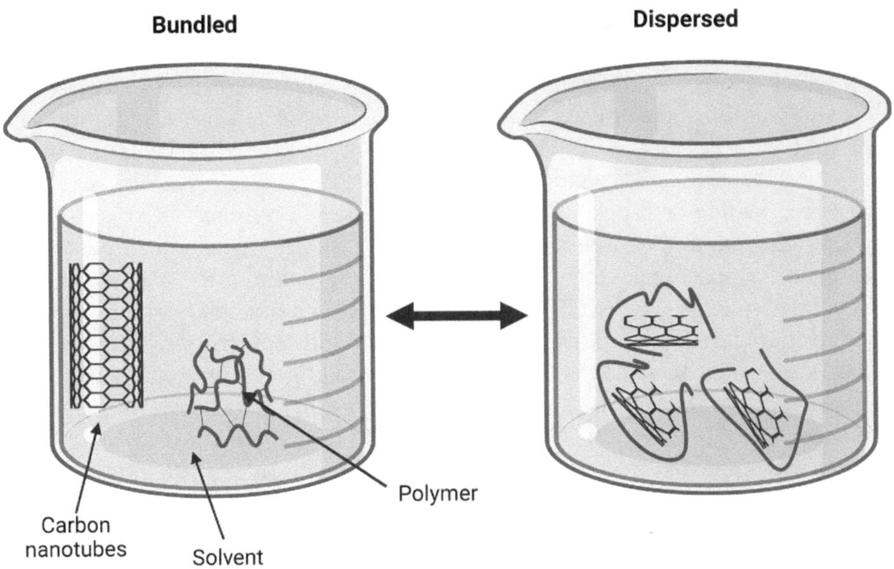

FIGURE 2.5 Carbon nanotubes disperse with polymer.

CNTs in aqueous suspensions. This polymer avoids DWCNT clumping in aqueous media and provides stability through hydrophobic interactions [80]. Surfactants have been crucial for forming a stable aqueous dispersion of CNTs [81]. Surfactants act by reducing the surface tension on the surface of nanotubes, which improves the wetting ability of the polymer and prevents the bundles of CNTs [82]. According to reported results, the hydrophobic surface of nanotubes tends to draw in the lipophilic tail part of surfactant, which stabilizes the dispersion [83]. Linear and ionic surfactants were added to the MWCNTs dispersion in water to prevent nanotubes from aggregation, such as cationic surfactant cetyltrimethylammonium bromide (CTAB) and the anionic surfactant sodium dodecyl sulfate (SDS) [84]. However, ionic surfactants are ineffective when dispersing CNTs in organic solvents [85]. Foroutan et al. explained the application of Triton X-100 for dispersing clustered CNTs in both an aqueous and nonaqueous solvent. The interaction of the water molecules with the hydrophilic part of surfactants was studied with the help of molecular dynamic simulation. The study determined the impact of water molecules on the adherence of surfactants to CNT surfaces. Surfactants create a void between two CNTs, and it causes more water molecules to enter this void space. Thus, it creates space among two CNTs due to it being readily dispersed in the presence of water molecules. Results from this study showed that the intermolecular interaction between CNTs and Triton X-100 was not affected by temperature [73].

2.4 CNTs-BASED ADSORBENT

CNTs have excellent mechanical, chemical, and thermal properties, therefore showing considerable adsorption capability for pharmaceutically active compounds. CNTs have been utilized as drug adsorbents in various fields because of their inherent and dynamic properties [93–95]. CNTs are showing adsorbent properties for biomolecules [96]. They have shown better sorption capacity due to higher chemical stability, thermal stability, and surface area. The various characteristics of CNTs make them an exceptional adsorbent, such as high available surface area, high porosity, large ratio of accessible micropores, hydrophobicity, high resistance to corrosion, layered structure, ease of chemical activation, tunable surface properties, excellent support for coating or deposition of other materials, and chemical inert surface for physical adsorption [97]. The adsorption capacity of CNTs was enhanced by different modifications, including graphitization, hydroxylation, carboxylation, etching with potassium hydroxide (KOH), and other appropriate functional groups. Ji et al. reported the alteration of the surface of SWCNTs and MWCNTs by etching with KOH. Investigated their ability to adsorb drugs such as tylosin, tetracycline, and sulfamethoxazole. In comparison to SWCNTs and MWCNTs, the adsorption performance of KOH-etched SWCNTs and MWCNTs for sulfamethoxazole and tetracyclines was improved by about 56% and 84%, respectively [98]. In another study by Goscinska et al., MWCNTs were modified with silver nanoparticles to investigate their adsorbent capacity of amino acids. Researchers chose L-phenylalanine as the model adsorbate amino acid; it contains both hydrophobic and hydrophilic groups. These samples were characterized by nitrogen adsorption at low temperatures, X-ray diffraction (XRD), and TEM. According to TEM and XRD results noted that spherical silver nanoparticles (~20 nm) were presence on the surface of CNTs. The modified MWCNTs sorption capacity was increased with amino acids at increased concentrations. [99].

Adsorption sites on CNTs can be categorized into four groups, as shown in Figure 2.6. Internal sites are found in the channels between each individual nanotube and inside the hollow, cylindrical structure of the CNTs. Interstitial pathway sites are present in the channels between the individual CNTs. Outer surface sites are

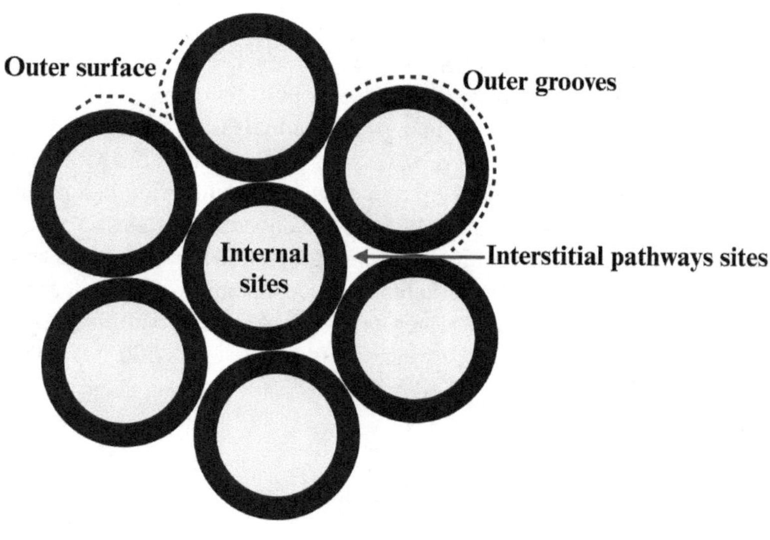

FIGURE 2.6 The various types of adsorbent sites for carbon nanotubes.

present on the outer surface of the CNTs. Outer groove sites connect the outermost nanotubes [100]. However, Pan and Xing have investigated the external surface, and groove areas are generally available for adsorption. Inner pores are not available for adsorption because of amorphous carbon and metals on both ends [101].

2.4.1 Adsorption Mechanism of CNTs

The mechanism in which adsorbate was adsorbed into CNTs may involve electron-donating acceptors interaction, hydrophobic interaction, hydrogen bonding, pi-hydrogen bonding, and the Lewis acid–base interaction.

2.4.1.1 Electron-Donating Acceptors Interaction

Electron-donating acceptors (EDA) interaction serves as the primary adsorption mechanism for the adsorption process and involves both n–π and π–π interaction. The π-electron-depleted region on the graphene surface of CNTs and the π-electron-rich aromatic ring of the adsorbate are the primary sites for π–π interaction. CNTs could act as both π-acceptors and π-donors [102]. Ali-Boucetta et al. investigated the adsorption behavior of DOX onto MWCNTs and observed the π–π interaction between DOX and MWCNT, which caused the drug to adsorb on the surface of nanotubes [103]. Yan et al. reported that the atrazine adsorbed onto surface CNTs was driven by π–π interaction [104]. Another possible mechanism for adsorption was n–π EDA interaction, wherein an unshared pair of electrons from nitrogen and oxygen, such as the amino and hydroxyl groups, was used as n-electron donors. These electron pairs directly interacted with the π-electron-deficient site on CNTs, which act as π-acceptors. This n–π EDA interaction may enhance the ionization of amino and hydroxyl groups at high pH. At such conditions, these groups become even more stronger electron-donor [102].

2.4.1.2 Hydrophobic Interaction

Hydrophobic interaction helps to increase the adsorption affinity of CNTs. Many targeted pharmaceutical compounds exhibit stronger nonhydrophobic interactions due to the attendance of functional groups, such as hydroxy, carboxyl, and amino groups [104,105]. The study of the hydrophobic effect of the *n*-hexadecane and *n*-octanol in water partition. Their results showed that *n*-hexadecane had more hydrophobic effects than *n*-octanol. Although this interaction was not the direct cause of the adsorption of CNTs, its influence cannot be disregarded [106]. Hu Chengguo et al. investigated the adsorption behavior of various surfactants, including Congo red and CTAB, on the surface of CNTs. They observed that Congo red, a nonionic surfactant, adsorbed onto the surface of CNTs through hydrophobic interaction between Cango red and CNTs [107].

2.4.1.3 Hydrogen Bonding and pi-Hydrogen Bonding

Hydrogen bonding was prevalent during the adsorption process of polar organic compounds onto CNTs. The acceptor and donor of hydrogen bonds were present on the surface of CNTs. The target compounds possess both donor groups and hydrogen-bonding receptors. Likewise, target molecules and CNTs contained pi donors. Consequently, the pi-hydrogen bond and hydrogen bond interaction may occur simultaneously during the adsorption process. The pH impacted both on pi-hydrogen interaction and hydrogen bonding [108].

2.4.1.4 The Lewis Acid–Base Interaction

Acid–base interactions are considered another significant mechanism regulating the adsorption of therapeutic moiety onto CNTs [102]. Research suggests that the Lewis acid–base interaction plays an additional role in the adsorption of molecules containing an amino group. Wang et al. investigated the adsorption behavior of different heterocyclic aromatic compounds on CNTs. They concluded that $-NH_2$ substituted compounds (2-aminopyrimidine, 4,6-diaminopyrimidine, and aniline) adsorb onto CNT surfaces by Lewis acid–base interactions. In this process, the NH_2-containing compounds serve as the Lewis base, and the oxygen-containing groups on the CNTs serve as the Lewis acid [106].

When CNTs are treated with acids or oxidizing agents, they form negatively charged functional groups on their surface, including hydroxyl, carboxylic acid, aldehyde, and ketone. In addition to hydroxylated or hydrogen bonding, these modified CNTs can be employed for the adsorption of positively charged metal ions. When CNTs are treated with acid, their impurities are eliminated, and their closed ends are helped to open. According to reports, the surface oxygen concentration of MWCNTs increased from 4.8% to 13.9% after being treated with nitric acid [109,110]. In addition, other mechanisms such as surface complexation, ion exchange, physical adsorption, electrostatic contact, and precipitation play a role in the adsorption of the adsorbate onto the CNTs [111].

2.5 STRUCTURAL PROPERTIES OF CNTs

2.5.1 Surface Area

CNTs have improved surface performance and greater surface area. CNTs have been widely used in several fields and industries, including drug delivery, cosmetics, biomedical sciences, nanotherapeutics, tissue engineering, spaceflight, and maritime transportation [23]. CNTs are present in various forms depending on the orientations of their graphene sheet, each form characterized by a more surface area and aspect ratio. The surface area of CNT is calculated through different factors such as the diameter, number of nanotubes in a bundle, and length [112]. The surface area of MWCNTs and SWCNTs is shown in the range of between 100 and 1,300 m^2/g, respectively. SWCNTs are present in bundles, and their surface area is drastically reduced to less or about 300 m^2/g, which is still a high surface area [113,114]. This large surface area of CNT is considered an efficient vehicle for gene delivery [115,116]. After several years of study on CNTs, researchers concluded that surface area, size, and shape are responsible for the toxic nature of nanotubes [117]. CNTs interacting with biological media alter the surface area and aggregation due to size reduction, which also causes an exponential rise in surface area [118].

Active therapeutic moiety with low permeability in cells, moiety enters into cells by interacting with CNTs and showing a nanoneedle shape of CNTs. They act as nanoneedles for cellular internalization because of changes in surface chemistry with significant internal and external surface areas. Therefore, these properties make CNTs more effective carriers in anticancer drug transporters [119]. DOX is an anticancer drug that faces challenges such as toxicity, side effects, and low permeability. Hence, CNTs are preferred for the delivery of DOX due to the decrease in the side effects, ease of transport of the drug, stability, and more surface area [120]. Graphene was introduced with CNTs to produce a 3D structure with a large surface area [121]. Graphene nanotubes are indicated to have exceptionally high density due to exposed graphene edges and maintaining the porosity of the CNT scaffold [119]. Distinct and truncated conical graphene sheets are stacked in a patchy pattern formed by cup-stacked CNTs (CS-CNTs). The CS-CNTs are the structural framework that offers a reactive edge through surface floating bonds. Therefore, greater adsorption capacity for CNTs via hydrogen bonding, pore filling, electrostatic, and π–π interaction is possible by higher surface area and numerous functional groups [122,123].

2.5.2 Tubular Inner Volumes for Endohedral Entrapment

CNTs are widely used as drug delivery vehicles because of their hollow structural nature of nanotubes, which allow them to payload drugs effectively [124]. The entrapment efficiency study was conducted on carboplatin as chemotherapeutic drug loading in CNTs [125,126]. The reaction occurs between the inner side and catalysts or nanoreactors, known as the confinement effect, which causes the endohedral functionalization of CNTs [127]. Factors affecting the confinement effect are as follows: (i) various electrical interactions between metal nanoparticles within and outside CNTs, (ii) molecule's classical arrangements might change due to space constraints. (iii) the enrichment of reactants in the inner walls, and (iv) the length and thickness of CNTs affect endohedral functionalization [128].

2.5.3 Aspect Ratio and Hydrogen Bonding

A high aspect ratio of carbon nanomaterials may provide better drug delivery potential than spherical nanoparticles [3,129]. CNTs represent a class of nanomaterial having a high aspect ratio [130]. CNTs have a high aspect ratio, indicating that they possess the properties of both nanoparticles and fibers. CNTs contain a length-to-diameter ratio of 100 [47]. The aspect ratio is a parameter that can impact cell viability. Some research groups have studied the toxicity of CNTs at different concentrations. It reported that 40 µg/mL showed less toxicity, and nanotubes contain an average diameter and length of 40 nm and 1 µm, respectively. CNT toxicity is based on the nature of compounds, length, diameter, physical form, and concentration [131]. Magrez et al. conducted a toxicity study of CNTs based on their chemistry and aspect ratio. The aspect ratio was characterized by SEM, in which MWCNTs show an aspect ratio of 80–90 with a diameter of 20 nm. Carbon nanofibers (CNFs) and carbon black are indicated to have an aspect ratio of 30–40 and 1, respectively. Their finding suggests the inhibition of cell proliferation and cell death, which have some correlation with poor aspect ratio [132]. Its lengths and diameters are particular characteristics that make CNT such a significant parameter from a research theory and practical standpoint [131]. Hydrogen bonds are important for improving the mechanical and interfacial strength of CNTs. The computer simulation method has been used extensively in recent years to investigate the water structure within CNTs [133–135]. Wide-spectral range infrared spectroscopy is the preferred method for examining the hydrogen bond network [136]. Ren and Yu described the effect of hydrogen bonding on the properties of CNTs and polyvinyl pyrrolidone (PVP) composites. MWCNTs are functionalized to form MWCNTs-polyvinyl alcohol (PVA) and MWCNTs-OH and disperse in a PVA matrix. The study revealed more interaction between MWCNTs-OH and PVA due to the formation of stronger hydrogen bonds between them. These bonds significantly enhanced composites' mechanical and interfacial performance [137].

2.6 PHYSICAL PROPERTIES OF CNTs

CNTs have demonstrated remarkable physical properties, such as mechanical strength, electrical and thermal conductivities. Different types of nanotubes indicate different physical properties due to their structural variability. The variation in two CNT diameters exhibits different levels of interaction between them. DWCNTs exhibit better stability in electrical and thermal conductivities than the SWCNTs and MWCNTs [50]. The excellent physical characteristics of CNTs are shown due to their tubular graphene composition and nanoscale structure. CNTs are ideal for reinforcement in materials because of their high tensile strength, elasticity, and Young's modulus, which allow them to bear significant strains before breaking [138].

2.6.1 Electrical Properties

CNTs possess some electronic properties such as metallic, semiconductor, electronic structure, and transport. These electronic properties of CNTs are changed with hexagon orientation and structure configuration. Depending on the diameter and chirality of CNTs, they can be either metallic or semiconducting type [139]. The electrical properties of CNTs are mainly related to helicity and diameter. Their electrical resistance is incredibly low [50]. There are three orientations of CNTs: (i) armchair orientation: in this arrangement, the hexagons align parallel to the tube axis, resulting in metallic properties; (ii) zigzag orientation: here, hexagons are arranged circularly around the tube, which gives the nanotubes semiconducting characteristics; (iii) chiral orientation: in this arrangement, the hexagons twist without forming a line, yielding in semiconducting behavior. These unique properties arise from the specific atomic arrangement of CNTs, enabling them to function as either conductors or semiconductors. This versatility makes them valuable in a variety of electronic applications [140].

2.6.2 Thermal Properties

CNTs have high thermal conductivity and reach up to 6,600 W/mK [141]. SWCNTs and MWCNTs exhibit thermal conductivities in the range of 3,000–6,000 W/mK and 2,000–3,000 W/mK, respectively. Thermal conductivity depends on their size and structure. The thermal conductivity of nanotubes is affected by internal and external temperatures. Various factors affect the thermal characteristics of CNT, including diameter, length, atomic structure, impurities, and morphology [50]. The thermal conductivity depends on the highly architectural arrangement of CNTs. Another factor is temperature, in which thermal conductivity increases with temperature until a certain point and again declines at that point. This maximizes the heat-conducting benefits of CNTs and enables more effective cooling solutions. Researchers are investigating many strategies to improve their shapes and arrangements in composites [142].

2.6.3 Mechanical Properties

The mechanical properties of CNTs consist of incredibly high tensile strength and elasticity. MWCNTs show tensile strength around 11–63 GPa, while SWCNTs show values of up to 22.2 ± 2.2 GPa. The elastic modulus of SWCNTs and MWCNTs are reported in the range of 1,000–3,000 GPa and 300–1,000 GPa, respectively [50]. CNTs have excellent mechanical properties like more elastic modulus (1 TPa) and tensile strength [143]. Stress concentration at structural flaws, which are essentially inevitable in actual nanotubes and cause defects in CNTs. The interfacial interactions between polymer matrices and nanotubes profoundly impact the mechanical properties of CNT/polymer composites. Effective stress transfer is demonstrated by the interfacial shear strength between CNTs and polymer matrices can reach 160 MPa, as demonstrated by molecular dynamics simulations [143,144].

2.6.4 Optical Properties

CNTs have a distinctive optical property, and this property is related to their 1D structure. According to theoretical research, when the CNT size gets bigger, the optical property disappears [14]. The optical properties of CNTs are determined mainly by their electrical structure. The unique optical absorption properties of CNTs contain sharp peaks representing electronic transitions [145]. Raman spectroscopy is an effective technique for characterizing nanotubes to obtain information about their electrical and structural characteristics. CNTs contain certain peaks in Raman spectra, which correspond to vibrational modes. These types of peaks give information about the diameter and chirality of nanotubes [146].

2.7 DRUG DELIVERY ASPECTS OF CNTs

CNTs are applicable in the various biomedical fields of diagnosis, bioimaging, drug delivery, and phototherapy for treating disorders, such as chemotherapy and gene delivery by vectors in the treatment of cancer [147,148]. CNTs are helpful for various biomedical fields due to excellent properties, including high flexibility, more surface area, high drug loading, biocompatibility, stability, and low toxicity [120]. Additionally, CNTs have remarkable attributes, such as optical, thermal, mechanical strength, and electrical conductivity properties, that make them suitable for developing advanced electrochemical and optical biosensors [148]. CNTs are a good choice for tissue engineering due to the nature of nanotubes, such as biocompatible and rigid, limiting the natural nanofibers and provoking the cells to adhere and divide, which facilitates the formation of 3D tissue cultures [149]. CNTs are also applicable for peptide and nucleic acid delivery systems [115]. CNTs are more useful as biomarkers due to their distinctive properties, which give high photostability, ensure less quenching, and low cytotoxicity [147]. Furthermore, CNTs are utilized in various parts like membranes, channel walls, and sensors for lab-on-chips (LOC). In this system, a small volume of fluid flows through various channels for drug screening in multiple disease models and cell growth devices [150].

One of the promising applications of CNTs is in photothermal therapy, in which CNTs can render near-infrared radiation into heat [151]. In drug delivery aspects, CNTs play a role in photothermal and photodynamic therapy for cancer treatment. Behnam et al. mentioned the major role of the combination of MWCNTs and silver nanorods in optical absorbance and cell penetration during photothermal therapy. The silver nanorods possess excellent optical absorption while MWCNTs enhance the cell permeability and dispersibility for the silver nanorods. Polyethylene glycol (PEG) is used for coating of MWCNTs, and this polymer shows benefit to increased hydrophilicity. This study establishes the impact of photothermal therapy on murine melanoma cells B16/F10. The histopathology study shows that the necrosis of the cells is increased by up to 70% in combination of MWCNTs and the silver nanorods [152]. Researchers are focused on the functionalization of MWCNT for photothermal therapy in glioblastoma cells. Chemically functionalization of nanotubes to form an MWCNT 1,2-Disteroyl-sn-Glycero-3-Phosphoethanolamine- Polyethylene Glycol (DSPE-PEG) conjugates to overcome the stability issues. PEG containing terminal group was attached to ligands for the chemical functionalization of CNTs. These functionalization shows an increase in bioavailability and blood circulation time, as well as a decrease in undesirable adverse effects. TEM analysis determines that the CNT cell permeability is 100 µm. Spheroids were treated with DSPE-PEG functionalized MWCNT, which determines the inhibition of glioblastoma tumors after getting the NIR 3–100 µg/mL [153].

CNTs are considered novel carriers to ensure the better delivery of genes and drugs due to the high surface area, the chemistry of the surface, lack of immunological reactions, and distinctive needle-like shapes with an efficiency to penetrate cell membranes [115,116]. Scientists have determined the PH-dependent release of docetaxel (DTX), in which DTX loaded with arginyl glycyl aspartic acid (RGD)-chitosan (CS) on the functionalized CNTs to form arginyl glycyl aspartic acid-chitosan-single wall carbon nanotubes-loading docetaxel (RGD-CS-SWCNT-DTX) complex. These complexes enhanced the drug release with 68% at PH 5.0 and 49% at PH 7.4. *In vitro* studies indicate that functionalized formulation inhibited the tumor cell (A549). The formulations showed excellent anticancer activity both *in vivo* and *in vitro*, more drug loading, and satisfactory animal safety [154]. Hampel and colleagues worked on the anticancer drug (carboplatin), and the drug was filled in CNTs. Multifunctional CNTs were prepared with a wet chemical approach for bladder cancer. The chemical functionalization was determined with the endohedral or interior filling of the drug, and this formulation effectively decreases the growth of bladder cells. *In vitro* studies EJ28 cells demonstrated the anticancer effect of the drug on cells. CNT conjugate was determined to enhance the delivery system efficiency and reduce the side effects [126].

Limlin Wu et al. prepared the chemical functionalization of MWCNT with PEG 600 to encapsulate oxaliplatin drug for colorectal cancer disease. Cytotoxicity studies were conducted on the HT29 cells, which showed extended cytotoxic activity against the cancer cells and sustained drug release [155].

Amir and colleagues developed the MWCNT and anti-EGFP siRNA through chemical functionalization by the polyethyleneimine and pyridinium groups. The positive charge of MWCNT interacts more effectively with the negative charge of the human cells, increasing the permeability of tumor cells in lung cancer. MWCNT of anti-EGFP siRNA has shown enhanced targeting of tumor cells. *In vitro* studies on H1299 cells reveal enhanced permeability and cytotoxicity against tumor cells [156]. Wang et al. developed the SWCNT of cyclin A2 siRNA chemically functionalized by the amine group for leukemia. Amine groups give a positive charge on the surface of SWCNT and interact more efficiently with the negative charge of the tumor cells. It increases the permeability to the tumor cells for cyclin A2 siRNA. The decrease in the proliferation of tumor cells and induced apoptosis effects on tumor cells have been determined against the K562 cells *in vitro* study [157].

2.8 CHALLENGES AND FUTURE PERSPECTIVES

CNTs are showing main challenges in biocompatibility and toxicity due to their extensive use in drug delivery systems [158]. The different properties of CNT, such as the type of functionalization, aseptic ratio and types of nanotubes, the existence of contaminants, and solubilizing agents, are attributed to the toxicity of nanotubes. CNT toxicity is primarily caused by their structural property, which contains more surface area relative to the size ratio. More surface area suggests increased interaction with the cellular membrane as well as increased toxin absorption and transport. Some processes causing nanotube toxicity are damage to the membrane, genetic toxicity, and oxidative stress [159]. The diameter of nanotubes is a factor that affects the toxicity of MWCNT, and researchers studying the cell counting kit assay reported the diameter of nanotubes increased with their cytotoxicity reduced [160]. The tremendous physical strength of CNTs are indicated the break the membranes and cell organelles, resulting shows the tissue damage and inflammatory response [159].

The types of CNTs, such as SWCNTs and MWCNTs, are one of the factors for their toxicity. SWCNTs have high van der Waal forces and surface area, resulting in a propensity for bundle formation and a decrease in their efficiency. However, MWCNTs have comparatively smaller surface areas and numerous active defects in walls, making them less likely to form bundles [161]. Some research studies indicated that MWCNTs are formed more aggregates than SWCNTs. The SWCNT surface is more hydrophilic due to carboxyl and hydroxyl groups. MWCNT is prepared 300 μm size aggregates, and SWCNT is formed 5–30 μm [162]. The solubilizing agent is required for CNT to disperse in water, but it produces toxicity effects. CNTs are dispersed in various solubilizing agents such as amylose, GA, PVP, and Triton; the results show the enhanced toxicity of CNT with solubilizing agents. The length of the nanotube is another factor that affects cellular internalization, with longer CNTs (825 nm) causing more inflammation than small CNTs (220 nm) [161].

The common challenge for preparing CNT is the cost and preparation method. The functionalization is another factor that changes the properties of CNTs. The functionalization of CNTs modified the fundamental properties, such as mechanical and electrical properties. The larger surface area and dimensions of CNTs can cause toxicity issues to organisms and the environment. Developing a proper synthesis method was needed to improve the quality, purity, yields, and nontoxic effects. Appropriate solubilizing agents, surfactants, and polymers are used to enhance the dispersibility of CNT. All the above-mentioned challenges of CNTs are overcome and significantly utilized in various fields such as drug delivery, gene delivery, and biosensors [163].

It has impressive impacts in various research fields, including analytical instruments, pharmaceutical sciences, medicine, electronics, industrial production processes, and mechanical industries. In nanotechnology, CNTs are one such innovative substance or carrier that has recently gained more interest [164]. In addition to their distinct chemical, physical, and mechanical characteristics and high stability, nanotubes serve as the preferred nanocarrier. CNTs have been strongly suggested for controlled drug release and site-specific medication delivery. Because of their widespread use and high demand, CNTs are typically generated in enormous quantities, which exposes the public to them in both direct and indirect ways. In recent years, safety concerns have been raised about CNT implications on human health and toxicological problems [165].

2.9 CONCLUSION

CNTs possess different physical, structural, mechanical, and functional properties, making them excellent carriers in various fields. CNTs are a type of carbon allotrope that is hollow, biocompatible, and nanostructural. CNTs are synthesized using various methods such as vapor-phase growth, CVD, electrical arc discharge, and flame synthesis. Their characterizations are through different techniques such as SEM, TEM, Raman spectroscopy, and AFM. The chemical properties of CNTs have two properties: grafting approaches and functionalization, which enhance the solubility, biocompatibility, and dispersion of nanotubes. Various surface modification methods, such as solvent interaction, surfactants, polymers, and biopolymers, are used to avoid aggregates of nanotubes. CNTs are indicated as exceptional adsorbent properties due to their greater surface area, high porosity, high resistance, and ease of chemical activation. The different adsorption mechanisms used in CNT are EDA, hydrophobic interaction, Lewis acid–base interaction, and hydrogen bonding. A high aspect ratio and surface area represent the potential for drug delivery and the impact on cell viability. Hydrogen bonding mainly plays a role in enhancing the mechanical strength of CNT. The physical properties of CNTs make them suitable candidates for biomedical applications. Their outstanding physical properties are such as high tensile strength (11–63 GPa), high thermal conductivity (~6,600 W/mK), elastic modulus (300–1,000 GPA), and excellent optical properties. The unique properties of CNTs are used in various applications such as biosensors, drugs, gene approaches, and biomedical technology. It gives information on cancer drugs used in CNTs as carriers. The main challenges of CNTs are toxicity and biocompatibility because of their extensive use. This chapter offers an extensive overview of the physicochemical properties, functionalization strategies, and structural advantages of CNTs.

ACKNOWLEDGMENT

The authors are highly thankful to the Department of Pharmaceuticals, Ministry of Chemicals and Fertilizers, Government of India, New Delhi, India, for providing financial assistance.

REFERENCES

1. Rahman G, Najaf Z, Mehmood A, Bilal S, Shah AHA, Mian SA, et al. An overview of the recent progress in the synthesis and applications of carbon nanotubes. *C*. 2019;5(1):3.
2. Kim M, Jang J, Cha C. Carbon nanomaterials as versatile platforms for theranostic applications. *Drug Discov Today*. 2017;22(9):1430–7.
3. Banerjee A, Qi J, Gogoi R, Wong J, Mitragotri S. Role of nanoparticle size, shape and surface chemistry in oral drug delivery. *J Control Release*. 2016;238:176–85.
4. Meng H, Yang S, Li Z, Xia T, Chen J, Ji Z, et al. Aspect ratio determines the quantity of mesoporous silica nanoparticle uptake by a small GTPase-dependent macropinocytosis mechanism. *ACS Nano*. 2011;5(6):4434–47.
5. Niezabitowska E, Smith J, Prestly MR, Akhtar R, von Aulock FW, Lavallée Y, et al. Facile production of nanocomposites of carbon nanotubes and polycaprolactone with high aspect ratios with potential applications in drug delivery. *RSC Adv*. 2018;8(30):16444–54.
6. Champion JA, Katare YK, Mitragotri S. Particle shape: a new design parameter for micro-and nanoscale drug delivery carriers. *J Control Release*. 2007;121(1–2):3–9.
7. Chen J, Clay NE, Park N, Kong H. Non-spherical particles for targeted drug delivery. *Chem Eng Sci*. 2015;125:20–4.
8. Lewinski N, Colvin V, Drezek R. Cytotoxicity of nanoparticles. *Small*. 2008;4(1):26–49.
9. Anzar N, Hasan R, Tyagi M, Yadav N, Narang J. Carbon nanotube – a review on synthesis, properties and plethora of applications in the field of biomedical science. *Sensors Int*. 2020;1:100003.
10. Cheung W, Pontoriero F, Taratula O, Chen AM, He H. DNA and carbon nanotubes as medicine. *Adv Drug Deliv Rev*. 2010;62(6):633–49.
11. Mauter MS, Elimelech M. Environmental applications of carbon-based nanomaterials. *Environ Sci Technol*. 2008;42(16):5843–59.
12. Mehra NK, Palakurthi S. Interactions between carbon nanotubes and bioactives: a drug delivery perspective. *Drug Discov Today*. 2016;21(4):585–97.
13. Beg S, Rizwan M, Sheikh AM, Hasnain MS, Anwer K, Kohli K. Advancement in carbon nanotubes: basics, biomedical applications and toxicity. *J Pharm Pharmacol*. 2011;63(2):141–63.
14. Rathinavel S, Priyadharshini K, Panda D. A review on carbon nanotube: an overview of synthesis, properties, functionalization, characterization, and the application. *Mater Sci Eng B*. 2021;268:115095.
15. Herrera-Ramirez JM, Perez-Bustamante R, Aguilar-Elguezabal A. An overview of the synthesis, characterization, and applications of carbon nanotubes. In: Yaragalla S, et al. (eds) *Carbon*-Based *Nanofillers and Their Rubber Nanocomposites*. Elsevier, Amsterdam, The Netherlands, 2019;47–75.
16. Mitchell DT, Lee SB, Trofin L, Li N, Nevanen TK, Söderlund H, et al. Smart nanotubes for bioseparations and biocatalysis. *J Am Chem Soc*. 2002;124(40):11864–5.

17. Jain AK, Dubey V, Mehra NK, Lodhi N, Nahar M, Mishra DK, et al. Carbohydrate-conjugated multiwalled carbon nanotubes: development and characterization. *Nanomed Nanotechnol Biol Med.* 2009;5(4):432–42.
18. Liu Z, Davis C, Cai W, He L, Chen X, Dai H. Circulation and long-term fate of functionalized, biocompatible single-walled carbon nanotubes in mice probed by Raman spectroscopy. *Proc Natl Acad Sci.* 2008;105(5):1410–5.
19. Huang H, Yuan Q, Shah JS, Misra RDK. A new family of folate-decorated and carbon nanotube-mediated drug delivery system: synthesis and drug delivery response. *Adv Drug Deliv Rev.* 2011;63(14–15):1332–9.
20. Bati ASR, Yu L, Batmunkh M, Shapter JG. Synthesis, purification, properties and characterization of sorted single-walled carbon nanotubes. *Nanoscale.* 2018;10(47):22087–139.
21. Dresselhaus MS, Dresselhaus G, Jorio A, Souza Filho AG, Saito R. Raman spectroscopy on isolated single wall carbon nanotubes. *Carbon N Y.* 2002;40(12):2043–61.
22. Kim BM, Qian S, Bau HH. Filling carbon nanotubes with particles. *Nano Lett.* 2005;5(5):873–8.
23. Jha R, Singh A, Sharma PK, Fuloria NK. Smart carbon nanotubes for drug delivery system: a comprehensive study. *J Drug Deliv Sci Technol.* 2020;58:101811.
24. Safarova K, Dvorak A, Kubinek R, Vujtek M, Rek A. Usage of AFM, SEM and TEM for the research of carbon nanotubes. *Mod Res Educ Top Microsc.* 2007;2:513–9.
25. Chua M, Chui CK, Chng CB, Lau D. Carbon nanotube-based artificial tracheal prosthesis: carbon nanocomposite implants for patient-specific ENT care. *IEEE Nanotechnol Mag.* 2013;7(4):27–31.
26. Arunachalam S, Gupta AA, Izquierdo R, Nabki F. Suspended carbon nanotubes for humidity sensing. *Sensors.* 2018;18(5):1655.
27. Lee J, Kim T, Jung Y, Jung K, Park J, Lee DM, et al. High-strength carbon nanotube/carbon composite fibers via chemical vapor infiltration. *Nanoscale.* 2016;8(45):18972–9.
28. Karimi M, Solati N, Ghasemi A, Estiar MA, Hashemkhani M, Kiani P, et al. Carbon nanotubes part II: a remarkable carrier for drug and gene delivery. *Expert Opin Drug Deliv.* 2015;12(7):1089–105.
29. Ren Z, Lan Y, Wang Y, Ren Z, Lan Y, Wang Y. *Carbon Nanotubes. Aligned Carbon Nanotubes: Physics, Concepts, Fabrication and Devices.* Springer, Berlin, 2013;7–43.
30. Monthioux M, Kuznetsov VL. Who should be given the credit for the discovery of carbon nanotubes? *Carbon N Y.* 2006;44(9):1621–3.
31. Radushkevich LV, Lukyanovich VM. About the carbon structure, thermal CO decomposition on metal contact synthesized. *J Phys Chem Russ.* 1952;26:88–95.
32. Boehm HP. Carbon from carbon monoxide disproportionation on nickel and iron catalysts: morphological studies and possible growth mechanisms. *Carbon N Y.* 1973;11(6):583–90.
33. Oberlin A, Endo M, Koyama T. High resolution electron microscope observations of graphitized carbon fibers. *Carbon N Y.* 1976;14(2):133–5.
34. Abrahamson J, Wiles PG, Rhoades BL. Structure of carbon fibres found on carbon arc anodes. *Carbon N Y.* 1999;11(37):1873–4.
35. Tennent HG. Carbon fibril, method for producing same and compositions containing same. U.S Patent 4663230, 1987.
36. Kroto HW, Heath JR, O'Brien SC, Curl RF, Smalley RE. C60: Buckminsterfullerene. *Nature.* 1985;318(6042):162–3.
37. Iijima S, Ichihashi T. Single-shell carbon nanotubes of 1-nm diameter. *Nature.* 1993;363(6430):603–5.
38. Bethune DS, Kiang CH, De Vries MS, Gorman G, Savoy R, Vazquez J, et al. Cobalt-catalysed growth of carbon nanotubes with single-atomic-layer walls. *Nature.* 1993;363(6430):605–7.

39. Chaturvedi P, Verma P, Singh A, Chaudhary PK, Basu PK. Carbon nanotube-purification and sorting protocols. *Def Sci J.* 2008;58(5):591–9.
40. Pillai SK, Ray SS, Moodley M. Purification of multi-walled carbon nanotubes. *J Nanosci Nanotechnol.* 2008;8(12):6187–207.
41. Guo T, Nikolaev P, Thess A, Colbert DT, Smalley RE. Catalytic growth of single-walled manotubes by laser vaporization. *Chem Phys Lett.* 1995;243(1–2):49–54.
42. Nanot S, Thompson NA, Kim JH, Wang X, Rice WD, Hároz EH, et al. Single-walled carbon nanotubes. In: Vajtai R (ed) *Springer Handbook of Nanomaterials.* Springer, Berlin, 2013;105–46.
43. Baughman RH, Zakhidov AA, De Heer WA. Carbon nanotubes--the route toward applications. *Science.* 2002;297(5582):787–92.
44. Shen C, Brozena AH, Wang Y. Double-walled carbon nanotubes: challenges and opportunities. *Nanoscale.* 2011;3(2):503–18.
45. Ajayan PM. Bulk metal and ceramics nanocomposites. In: Ajayan PM, Schadler LS, Braun PV (eds) *Nanocomposite Science and Technology.* Wiley, Berlin, 2003;1–75.
46. Prasher RS, Hu XJ, Chalopin Y, Mingo N, Lofgreen K, Volz S, et al. Turning carbon nanotubes from exceptional heat conductors into insulators. *Phys Rev Lett.* 2009;102(10):105901.
47. Madani SY, Naderi N, Dissanayake O, Tan A, Seifalian AM. A new era of cancer treatment: carbon nanotubes as drug delivery tools. *Int J Nanomedicine.* 2011;6:2963–79.
48. Pandey P, Dahiya M. Carbon nanotubes: types, methods of preparation and applications. *Carbon N Y.* 2016;1(4):15–21.
49. Feazell RP, Nakayama-Ratchford N, Dai H, Lippard SJ. Soluble single-walled carbon nanotubes as longboat delivery systems for platinum (IV) anticancer drug design. *J Am Chem Soc.* 2007;129(27):8438–9.
50. Gupta N, Gupta SM, Sharma SK. Carbon nanotubes: synthesis, properties and engineering applications. *Carbon Lett.* 2019;29:419–47.
51. Mallakpour S, Soltanian S. Surface functionalization of carbon nanotubes: fabrication and applications. *RSC Adv.* 2016;6(111):109916–35.
52. Alosime EM. A review on surface functionalization of carbon nanotubes: methods and applications. *Discov Nano.* 2023;18(1):12.
53. Kordzadeh A, Amjad-Iranagh S, Zarif M, Modarress H. Adsorption and encapsulation of the drug doxorubicin on covalent functionalized carbon nanotubes: a scrutinized study by using molecular dynamics simulation and quantum mechanics calculation. *J Mol Graph Model.* 2019;88:11–22.
54. Alam AKMM, Beg MDH, Yunus RM, Islam MR, Shubhra QTH. Tailoring the dispersibility of non-covalent functionalized multi-walled carbon nanotube (MWCNT) nanosuspension using shellac (SL) bio-resin: structure-property relationship and cytotoxicity of shellac coated carbon nanotubes (SLCNTs). *Colloid Interface Sci Commun.* 2021;42:100395.
55. Liu P. Modifications of carbon nanotubes with polymers. *Eur Polym J.* 2005; 41(11):2693–703.
56. Madni I, Hwang CY, Park SD, Choa YH, Kim HT. Mixed surfactant system for stable suspension of multiwalled carbon nanotubes. *Colloids Surf A Physicochem Eng Asp.* 2010;358(1):101–7.
57. Eskandari P, Abousalman-Rezvani Z, Roghani-Mamaqani H, Salami-Kalajahi M. Polymer-functionalization of carbon nanotube by in situ conventional and controlled radical polymerizations. *Adv Colloid Interface Sci.* 2021;294:102471.
58. Martinez-Hernandez AL, Velasco-Santos C, Castaño VM. Carbon nanotubes composites: processing, grafting and mechanical and thermal properties. *Curr Nanosci.* 2010;6(1):12–39.

59. Naqvi STR, Rasheed T, Hussain D, Najam ul Haq M, Majeed S, shafi S, et al. Modification strategies for improving the solubility/dispersion of carbon nanotubes. *J Mol Liq*. 2020;297:111919.
60. Alves Z, Abreu B, Ferreira NM, Marques EF, Nunes C, Ferreira P. Enhancing the dispersibility of multiwalled carbon nanotubes within starch-based films by the use of ionic surfactants. *Carbohydr Polym*. 2021;273:118531.
61. Liu CX, Choi JW. Improved dispersion of carbon nanotubes in polymers at high concentrations. *Nanomaterials*. 2012;2(4):329–47.
62. Yang H, Neal L, Flores EE, Adronov A, Kim NY. Role and impact of surfactants in carbon nanotube dispersions and sorting. *J Surfactants Deterg*. 2023;26(5):607–22.
63. Ostos FJ, Lebrón JA, Moyá ML, Bernal E, Flores A, Lépori C, et al. Potentiometric study of carbon nanotube/surfactant interactions by ion-selective electrodes. Driving forces in the adsorption and dispersion processes. *Int J Mol Sci*. 2021;22(2):826.
64. Gu Z, Liang F, Chen Z, Sadana A, Kittrell C, Billups WE, et al. In situ Raman studies on lithiated single-wall carbon nanotubes in liquid ammonia. *Chem Phys Lett*. 2005;410(4):467–70.
65. López-López M, Bernal E, Moyá ML, Sanchez F, López-Cornejo P. Study of ionic surfactants interactions with carboxylated single-walled carbon nanotubes by using ion-selective electrodes. *Electrochem Commun*. 2016;67:31–4.
66. Giordani S, Bergin S, Nicolosi V, Lebedkin S, Blau WJ, Coleman JN. Fabrication of stable dispersions containing up to 70% individual carbon nanotubes in a common organic solvent. *Phys Status Solidi*. 2006;243(13):3058–62.
67. Ravi Kiran A, Kusuma Kumari G, Krishnamurthy PT. Carbon nanotubes in drug delivery: focus on anticancer therapies. *J Drug Deliv Sci Technol*. 2020;59:101892.
68. Fujigaya T, Nakashima N. Non-covalent polymer wrapping of carbon nanotubes and the role of wrapped polymers as functional dispersants. *Sci Technol Adv Mater*. 2015;16(2):024802.
69. Bilalis P, Katsigiannopoulos D, Avgeropoulos A, Sakellariou G. Non-covalent functionalization of carbon nanotubes with polymers. *RSC Adv*. 2014;4(6):2911–34.
70. Abousalman-Rezvani Z, Eskandari P, Roghani-Mamaqani H, Salami-Kalajahi M. Functionalization of carbon nanotubes by combination of controlled radical polymerization and "grafting to" method. *Adv Colloid Interface Sci*. 2020;278:102126.
71. Kang YK, Lee OS, Deria P, Kim SH, Park TH, Bonnell DA, et al. Helical wrapping of single-walled carbon nanotubes by water soluble poly (p-phenyleneethynylene). *Nano Lett*. 2009;9(4):1414–8.
72. Tasis D, Tagmatarchis N, Bianco A, Prato M. Chemistry of carbon nanotubes. *Chem Rev*. 2006;106(3):1105–36.
73. Fatemi SM, Foroutan M. Study of dispersion of carbon nanotubes by Triton X-100 surfactant using molecular dynamics simulation. *J Iran Chem Soc*. 2015;12:1905–13.
74. Pramanik C, Gissinger JR, Kumar S, Heinz H. Carbon nanotube dispersion in solvents and polymer solutions: mechanisms, assembly, and preferences. *ACS Nano*. 2017;11(12):12805–16.
75. Alshahrani A, Alharbi A, Alnasser S, Almihdar M, Alsuhybani M, AlOtaibi B. Enhanced heavy metals removal by a novel carbon nanotubes buckypaper membrane containing a mixture of two biopolymers: chitosan and i-carrageenan. *Sep Purif Technol*. 2021;276:119300.
76. Li X, Zhang H, Jin Q, Cai Z. Contribution of arabinogalactan protein to the stabilization of single-walled carbon nanotubes in aqueous solution of gum arabic. *Food Hydrocoll*. 2018;78:55–61.
77. Liu Y, Chipot C, Shao X, Cai W. Edge effects control helical wrapping of carbon nanotubes by polysaccharides. *Nanoscale*. 2012;4(8):2584–9.

78. Alshahrani AA, Al-Zoubi H, Nghiem LD, In het Panhuis M. Synthesis and characterisation of MWNT/chitosan and MWNT/chitosan-crosslinked buckypaper membranes for desalination. *Desalination*. 2017;418:60–70.
79. Zheng W, Zheng YF. Gelatin-functionalized carbon nanotubes for the bioelectrochemistry of hemoglobin. *Electrochem Commun*. 2007;9(7):1619–23.
80. Skender A, Hadj-Ziane-Zafour A, Flahaut E. Chemical functionalization of Xanthan gum for the dispersion of double-walled carbon nanotubes in water. *Carbon N Y*. 2013;62:149–56.
81. Abreu B, Rocha J, Fernandes RMF, Regev O, Furó I, Marques EF. Gemini surfactants as efficient dispersants of multiwalled carbon nanotubes: interplay of molecular parameters on nanotube dispersibility and debundling. *J Colloid Interface Sci*. 2019;547:69–77.
82. Vaisman L, Marom G, Wagner HD. Dispersions of surface-modified carbon nanotubes in water-soluble and water-insoluble polymers. *Adv Funct Mater*. 2006;16(3):357–63.
83. Fatemi SM, Foroutan M. Recent developments concerning the dispersion of carbon nanotubes in surfactant/polymer systems by MD simulation. *J Nanostr Chem*. 2016;6:29–40.
84. Mostaani F, Moghbeli MR, Karimian H. Electrical conductivity, aging behavior, and electromagnetic interference (EMI) shielding properties of polyaniline/MWCNT nanocomposites. *J Thermoplast Compos Mater*. 2018;31(10):1393–415.
85. Vaisman L, Wagner HD, Marom G. The role of surfactants in dispersion of carbon nanotubes. *Adv Colloid Interface Sci*. 2006;128–130:37–46.
86. Samanta SK, Fritsch M, Scherf U, Gomulya W, Bisri SZ, Loi MA. Conjugated polymer-assisted dispersion of single-wall carbon nanotubes: the power of polymer wrapping. *ACC Chem Res*. 2014;47(8):2446–56.
87. Mayo ML, Hogle D, Yilmaz B, Köse ME, Kilina S. Morphology and dispersion of polycarbazole wrapped carbon nanotubes. *RSC Adv*. 2013;3(43):20492–502.
88. Pandey S, Siva Prasanna SR V, Karakoti M, Tewari C, SanthiBhushan B, Pandey JK, et al. Dispersion and stability study of carbon nanotubes in pH and temperature responsive polymeric matrix: experiment and dispersion-corrected DFT study. *Mater Today Commun*. 2018;17:187–93.
89. Poochai C, Pongprayoon T. Enhancing dispersion of carbon nanotube in polyacrylonitrile matrix using admicellar polymerization. *Colloids Surf A Physicochem Eng Asp*. 2014;456:67–74.
90. Bhirde AA, Patel S, Sousa AA, Patel V, Molinolo AA, Ji Y, et al. Distribution and clearance of PEG-single-walled carbon nanotube cancer drug delivery vehicles in mice. *Nanomedicine*. 2010;5(10):1535–46.
91. Foster J, Singamaneni S, Kattumenu R, Bliznyuk V. Dispersion and phase separation of carbon nanotubes in ultrathin polymer films. *J Colloid Interface Sci*. 2005;287(1):167–72.
92. Yuan WZ, Sun JZ, Dong Y, Häussler M, Yang F, Xu HP, et al. Wrapping carbon nanotubes in pyrene-containing poly (phenylacetylene) chains: solubility, stability, light emission, and surface photovoltaic properties. *Macromolecules*. 2006;39(23):8011–20.
93. Kamel M, Raissi H, Morsali A, Shahabi M. Assessment of the adsorption mechanism of Flutamide anticancer drug on the functionalized single-walled carbon nanotube surface as a drug delivery vehicle: an alternative theoretical approach based on DFT and MD. *Appl Surf Sci*. 2018;434:492–503.
94. Vashist SK, Zheng D, Pastorin G, Al-Rubeaan K, Luong JHT, Sheu FS. Delivery of drugs and biomolecules using carbon nanotubes. *Carbon N Y*. 2011;49(13):4077–97.
95. Yahyazadeh F, Ghazanfari D, Ahmadi SA, Akhgar MR. Adsorption of Nefazodone on single-wall carbon nanotube as an antidepressant drug delivery: a DFT study. *Carbon Trends*. 2024;16:100394.
96. Feng W, Luo R, Xiao J, Ji P, Zheng Z. Self-assembly of sugar-based amphiphile on carbon nanotubes for protein adsorption. *Chem Eng Sci*. 2011;66(20):4807–13.

97. Sajid M, Asif M, Baig N, Kabeer M, Ihsanullah I, Mohammad AW. Carbon nanotubes-based adsorbents: properties, functionalization, interaction mechanisms, and applications in water purification. *J Water Process Eng.* 2022;47:102815.
98. Ji L, Shao Y, Xu Z, Zheng S, Zhu D. Adsorption of monoaromatic compounds and pharmaceutical antibiotics on carbon nanotubes activated by KOH etching. *Environ Sci Technol.* 2010;44(16):6429–36.
99. Goscianska J, Nowicki W, Pietrzak R. Physicochemical and sorption properties of multi-walled carbon nanotubes decorated with silver nanoparticles. *Chem Eng J.* 2014;250:295–302.
100. Agnihotri S, Mota JPB, Rostam-Abadi M, Rood MJ. Theoretical and experimental investigation of morphology and temperature effects on adsorption of organic vapors in single-walled carbon nanotubes. *J Phys Chem B.* 2006;110(15):7640–7.
101. Pan BO, Xing B. Adsorption mechanisms of organic chemicals on carbon nanotubes. *Environ Sci Technol.* 2008;42(24):9005–13.
102. Chen W, Duan L, Wang L, Zhu D. Adsorption of hydroxyl-and amino-substituted aromatics to carbon nanotubes. *Environ Sci Technol.* 2008;42(18):6862–8.
103. Ali-Boucetta H, Al-Jamal KT, McCarthy D, Prato M, Bianco A, Kostarelos K. Multiwalled carbon nanotube–doxorubicin supramolecular complexes for cancer therapeutics. *Chem Commun.* 2008;(4):459–61.
104. Yan XM, Shi BY, Lu JJ, Feng CH, Wang DS, Tang HX. Adsorption and desorption of atrazine on carbon nanotubes. *J Colloid Interface Sci.* 2008;321(1):30–8.
105. Zhu D, Pignatello JJ. Characterization of aromatic compound sorptive interactions with black carbon (charcoal) assisted by graphite as a model. *Environ Sci Technol.* 2005;39(7):2033–41.
106. Wang L, Zhu D, Duan L, Chen W. Adsorption of single-ringed N- and S-heterocyclic aromatics on carbon nanotubes. *Carbon N Y.* 2010;48(13):3906–15.
107. Hu C, Yang C, Hu S. Hydrophobic adsorption of surfactants on water-soluble carbon nanotubes: a simple approach to improve sensitivity and antifouling capacity of carbon nanotubes-based electrochemical sensors. *Electrochem Commun.* 2007;9(1):128–34.
108. Yang K, Xing B. Adsorption of organic compounds by carbon nanomaterials in aqueous phase: Polanyi theory and its application. *Chem Rev.* 2010;110(10):5989–6008.
109. Rodríguez C, Leiva E. Enhanced heavy metal removal from acid mine drainage wastewater using double-oxidized multiwalled carbon nanotubes. *Molecules.* 2019;25(1):111.
110. Rodríguez C, Briano S, Leiva E. Increased adsorption of heavy metal ions in multi-walled carbon nanotubes with improved dispersion stability. *Molecules.* 2020;25(14):3106.
111. Yang X, Wan Y, Zheng Y, He F, Yu Z, Huang J, et al. Surface functional groups of carbon-based adsorbents and their roles in the removal of heavy metals from aqueous solutions: a critical review. *Chem Eng J.* 2019;366:608–21.
112. Zhang W, Zhang Z, Zhang Y. The application of carbon nanotubes in target drug delivery systems for cancer therapies. *Nanoscale Res Lett.* 2011;6:1–22.
113. Ye Y, Ahn CC, Witham C, Fultz B, Liu J, Rinzler AG, et al. Hydrogen adsorption and cohesive energy of single-walled carbon nanotubes. *Appl Phys Lett.* 1999;74(16):2307–9.
114. Peigney A, Laurent C, Flahaut E, Bacsa RR, Rousset A. Specific surface area of carbon nanotubes and bundles of carbon nanotubes. *Carbon N Y.* 2001;39(4):507–14.
115. Bianco A, Kostarelos K, Prato M. Applications of carbon nanotubes in drug delivery. *Curr Opin Chem Biol.* 2005;9(6):674–9.
116. Karimi M, Sahandi Zangabad P, Ghasemi A, Amiri M, Bahrami M, Malekzad H, et al. Temperature-responsive smart nanocarriers for delivery of therapeutic agents: applications and recent advances. *ACS Appl Mater Interfaces.* 2016;8(33):21107–33.
117. Lanone S, Boczkowski J. Biomedical applications and potential health risks of nanomaterials: molecular mechanisms. *Curr Mol Med.* 2006;6(6):651–63.

118. Soto K, Garza KM, Murr LE. Cytotoxic effects of aggregated nanomaterials. *Acta Biomater.* 2007;3(3):351–8.
119. Chary PS, Bhawale R, Vasave R, Rajana N, Singh PK, Madan J, et al. A review on emerging role of multifunctional carbon nanotubes as an armament in cancer therapy, imaging and biosensing. *J Drug Deliv Sci Technol.* 2023;85:104588.
120. Zare H, Ahmadi S, Ghasemi A, Ghanbari M, Rabiee N, Bagherzadeh M, et al. Carbon nanotubes: smart drug/gene delivery carriers. *Int J Nanomed.* 2021;16:1681–706.
121. Agboola O, Fayomi OSI, Ayodeji A, Ayeni AO, Alagbe EE, Sanni SE, et al. A review on polymer nanocomposites and their effective applications in membranes and adsorbents for water treatment and gas separation. *Membranes (Basel).* 2021;11(2):139.
122. Gong J, Liu J, Jiang Z, Wen X, Mijowska E, Tang T, et al. A facile approach to prepare porous cup-stacked carbon nanotube with high performance in adsorption of methylene blue. *J Colloid Interface Sci.* 2015;445:195–204.
123. Li Y, Liu H, Yang C, Zhu M, Chen T. The activation and hydrogen storage characteristics of the cup-stacked carbon nanotubes. *Diam Relat Mater.* 2019;100:107567.
124. Rieger C, Kunhardt D, Kaufmann A, Schendel D, Huebner D, Erdmann K, et al. Characterization of different carbon nanotubes for the development of a mucoadhesive drug delivery system for intravesical treatment of bladder cancer. *Int J Pharm.* 2015;479(2):357–63.
125. Arlt M, Haase D, Hampel S, Oswald S, Bachmatiuk A, Klingeler R, et al. Delivery of carboplatin by carbon-based nanocontainers mediates increased cancer cell death. *Nanotechnology.* 2010;21(33):335101.
126. Hampel S, Kunze D, Haase D, Krämer K, Rauschenbach M, Ritschel M, et al. Carbon nanotubes filled with a chemotherapeutic agent: a nanocarrier mediates inhibition of tumor cell growth. *Nanomedicine.* 2008;3(2):175–82.
127. Iglesias D, Melchionna M. Enter the tubes: carbon nanotube endohedral catalysis. *Catalysts.* 2019;9(2):128.
128. Pan X, Bao X. The effects of confinement inside carbon nanotubes on catalysis. *ACC Chem Res.* 2011;44(8):553–62.
129. Toy R, Peiris PM, Ghaghada KB, Karathanasis E. Shaping cancer nanomedicine: the effect of particle shape on the in vivo journey of nanoparticles. *Nanomedicine.* 2014;9(1):121–34.
130. Singh B, Lohan S, Sandhu PS, Jain A, Mehta SK. Chapter 15- Functionalized carbon nanotubes and their promising applications in therapeutics and diagnostics. In: Grumezescu AM (ed) *Nanobiomaterials in Medical Imaging.* William Andrew Publishing, Norwich, NY, 2016;455–78.
131. Bottini M, Bruckner S, Nika K, Bottini N, Bellucci S, Magrini A, et al. Multi-walled carbon nanotubes induce T lymphocyte apoptosis. *Toxicol Lett.* 2006;160(2):121–6.
132. Magrez A, Kasas S, Salicio V, Pasquier N, Seo JW, Celio M, et al. Cellular toxicity of carbon-based nanomaterials. *Nano Lett.* 2006;6(6):1121–5.
133. Hummer G, Rasaiah JC, Noworyta JP. Water conduction through the hydrophobic channel of a carbon nanotube. *Nature.* 2001;414(6860):188–90.
134. Alexiadis A, Kassinos S. Molecular simulation of water in carbon nanotubes. *Chem Rev.* 2008;108(12):5014–34.
135. Nakamura Y, Ohno T. Structure of water confined inside carbon nanotubes and water models. *Mater Chem Phys.* 2012;132(2):682–7.
136. Sharma M, Donadio D, Schwegler E, Galli G. Probing properties of water under confinement: infrared spectra. *Nano Lett.* 2008;8(9):2959–62.
137. Ren J, Yu D. Effects of enhanced hydrogen bonding on the mechanical properties of poly (vinyl alcohol)/carbon nanotubes nanocomposites. *Compos Interfaces.* 2018;25(3):205–19.

138. Nurazzi NM, Sabaruddin FA, Harussani MM, Kamarudin SH, Rayung M, Asyraf MRM, et al. Mechanical performance and applications of CNTS reinforced polymer composites—a review. *Nanomaterials.* 2021;11(9):2186.
139. Ando T. The electronic properties of graphene and carbon nanotubes. *NPG Asia Mater.* 2009;1(1):17–21.
140. Chen Y, Lin Y, Liu Y, Doyle J, He N, Zhuang X, et al. Carbon nanotube-based functional materials for optical limiting. *J Nanosci Nanotechnol.* 2007;7(4–5):1268–83.
141. Kumanek B, Janas D. Thermal conductivity of carbon nanotube networks: a review. *J Mater Sci.* 2019;54(10):7397–427.
142. Niu M, Cui C, Tian R, Zhao Y, Miao L, Hao W, et al. Mechanical and thermal properties of carbon nanotubes in carbon nanotube fibers under tension–torsion loading. *RSC Adv.* 2022;12(46):30085–93.
143. Arash B, Wang Q, Varadan VK. Mechanical properties of carbon nanotube/polymer composites. *Sci Rep.* 2014;4(1):6479.
144. Takakura A, Beppu K, Nishihara T, Fukui A, Kozeki T, Namazu T, et al. Strength of carbon nanotubes depends on their chemical structures. *Nat Commun.* 2019;10(1):3040.
145. Saito R, Dresselhaus MS. Chapter 5- Optical properties of carbon nanotubes. In: Tanaka K, Iijima S (eds) *Carbon Nanotubes and Graphene (Second Edition).* Elsevier, Oxford, 2014;7–98.
146. Rao AM, Richter E, Bandow S, Chase B, Eklund PC, Williams KA, et al. Diameter-selective Raman scattering from vibrational modes in carbon nanotubes. *Science.* 1997;275(5297):187–91.
147. Hong G, Diao S, Antaris AL, Dai H. Carbon nanomaterials for biological imaging and nanomedicinal therapy. *Chem Rev.* 2015;115(19):10816–906.
148. Tîlmaciu CM, Morris MC. Carbon nanotube biosensors. *Front Chem.* 2015;3:59.
149. Harrison BS, Atala A. Carbon nanotube applications for tissue engineering. *Biomaterials.* 2007;28(2):344–53.
150. Ghasemi A, Amiri H, Zare H, Masroor M, Hasanzadeh A, Beyzavi A, et al. Carbon nanotubes in microfluidic lab-on-a-chip technology: current trends and future perspectives. *Microfluid Nanofluidics.* 2017;21:1–19.
151. Singh R, Torti SV. Carbon nanotubes in hyperthermia therapy. *Adv Drug Deliv Rev.* 2013;65(15):2045–60.
152. Behnam MA, Emami F, Sobhani Z, Koohi-Hosseinabadi O, Dehghanian AR, Zebarjad SM, et al. Novel combination of silver nanoparticles and carbon nanotubes for plasmonic photo thermal therapy in melanoma cancer model. *Adv Pharm Bull.* 2018;8(1):49.
153. Eldridge BN, Bernish BW, Fahrenholtz CD, Singh R. Photothermal therapy of glioblastoma multiforme using multiwalled carbon nanotubes optimized for diffusion in extracellular space. *ACS Biomater Sci Eng.* 2016;2(6):963–76.
154. Li B, Zhang XX, Huang HY, Chen LQ, Cui JH, Liu Y, et al. Effective deactivation of A549 tumor cells in vitro and in vivo by RGD-decorated chitosan-functionalized single-walled carbon nanotube loading docetaxel. *Int J Pharm.* 2018;543(1):8–20.
155. Wu L, Man C, Wang H, Lu X, Ma Q, Cai Y, et al. PEGylated multi-walled carbon nanotubes for encapsulation and sustained release of oxaliplatin. *Pharm Res.* 2013;30:412–23.
156. Varkouhi AK, Foillard S, Lammers T, Schiffelers RM, Doris E, Hennink WE, et al. SiRNA delivery with functionalized carbon nanotubes. *Int J Pharm.* 2011;416(2):419–25.
157. Wang X, Ren J, Qu X. Targeted RNA interference of cyclin A2 mediated by functionalized single-walled carbon nanotubes induces proliferation arrest and apoptosis in chronic myelogenous leukemia K562 cells. *ChemMedChem Chem Enabling Drug Discov.* 2008;3(6):940–5.
158. Sonowal L, Gautam S. Advancements and challenges in carbon nanotube-based drug delivery systems. *Nano-Str Nano-Objects.* 2024;38:101117.

159. Murjani BO, Kadu PS, Bansod M, Vaidya SS, Yadav MD. Carbon nanotubes in biomedical applications: current status, promises, and challenges. *Carbon Lett.* 2022;32(5):1207–26.
160. Zhao X, Chang S, Long J, Li J, Li X, Cao Y. The toxicity of multi-walled carbon nanotubes (MWCNTs) to human endothelial cells: the influence of diameters of MWCNTs. *Food Chem Toxicol.* 2019;126:169–77.
161. Alshehri R, Ilyas AM, Hasan A, Arnaout A, Ahmed F, Memic A. Carbon nanotubes in biomedical applications: factors, mechanisms, and remedies of toxicity: miniperspective. *J Med Chem.* 2016;59(18):8149–67.
162. Fraczek A, Menaszek E, Paluszkiewicz C, Blazewicz M. Comparative in vivo biocompatibility study of single- and multi-wall carbon nanotubes. *Acta Biomater.* 2008;4(6):1593–602.
163. Tao Z, Zhao Y, Wang Y, Zhang G. Recent advances in carbon nanotube technology: bridging the gap from fundamental science to wide applications. *C.* 2024;10(3):69.
164. Flahaut E, Govindaraj A, Peigney A, Laurent C, Rousset A, Rao CNR. Synthesis of single-walled carbon nanotubes using binary (Fe, Co, Ni) alloy nanoparticles prepared in situ by the reduction of oxide solid solutions. *Chem Phys Lett.* 1999;300(1):236–42.
165. Dai H. Carbon nanotubes: opportunities and challenges. *Surf Sci.* 2002;500(1):218–41.

3 Synthesis and Critical Characterization Aspects of Carbon Nanotubes

Sumedh Bahadure, Omar Khan, and Neelesh Kumar Mehra

ABBREVIATIONS

AFM	**atomic force microscope**
CNTs	carbon nanotubes
CS	**chitosan**
DLS	dynamic light scattering
DOX	doxorubicin
DSC	**differential scanning calorimetry**
DWCNTs	double-walled carbon nanotubes
EDA	ethylene diamine
EDX	energy dispersive X-ray analysis
FA	**folic acid**
FTIR	**Fourier transform infrared spectroscopy**
kV	kilovolts
MWCNTs	multiwalled carbon nanotubes
NPs	nanoparticles
PDI	polydispersity index
PEG	polyethylene glycol
PLA	**polylactic acid**
PXRD	**powder XRD**
SCXRD	**single crystal XRD**
SEM	scanning electron microscope
SWCNTs	small-walled carbon nanotubes
TEM	transmission electron microscope
TGA	**thermogravimetric analysis**
XRD	**X-ray diffraction**

3.1 INTRODUCTION

In the past few decades, nanotechnology has been evolving enormously in terms of discoveries of new versatile materials with unique nanostructures and spatial arrangements. These nanomaterials demonstrate unique properties that are generally absent in microscale materials, developing new possibilities for various chemical

 DOI: 10.1201/9781003386537-3

and biomedical applications [1–3]. Carbon nanotubes (CNTs) have become one of the most extensively studied nanomaterials owing to their remarkable properties, high aspect ratio and tensile strength, and excellent optical and electrical properties. These characteristics make CNTs promising candidates for various potential applications [4,5]. The first CNTs were discovered by the Japanese scientist Sumio Iijima in 1991 using the arc discharge method [6,7]. Graphite comprises stacked layers of sp^2 hybridized carbon atoms arranged in a honeycomb pattern called graphene. These graphenes are extremely strong and rigid. CNTs can be visualized as a graphene sheet that has been seamlessly rolled into a cylinder. Similar to graphene, CNTs exhibit remarkable strength and stiffness and excellent electrical and thermal conductivity due to sp^2 carbon bonds. Depending on the number of concentric walls, CNTs can be categorized into single-walled (SWCNTs), double-walled (DWCNTs), and multiwalled (MWCNTs), with diameters ranging from 0.4 nm (SWCNTs) to 100 nm (MWCNTs) and up to 0.5 m long [7,8].

Since their discovery, CNTs have been widely utilized in engineering and chemical fields for developing thermal interface materials, lightweight electrical wires, sensors, transistors, actuators, electrodes in energy conversion and storage instruments, structural composites and fibers, and water desalination membranes, etc. [9]. In the last two decades, CNTs have begun to gain attention in the pharmaceutical and biomedical fields focusing their use in drug delivery along with developing polymer nanocomposites, electronic and biosensing devices, and tissue engineering. Due to the rapidly increasing applications of CNTs, there is a significant emergence in their improved synthesis methodologies and characterization approaches [8]. This chapter comprehensively reviews the recent methods of synthesis of CNTs along with their critical characterization aspects in the pharmaceutical and biomedical fields.

3.2 SYNTHESIS OF CNTs

CNTs are typically synthesized using the following techniques: arc discharge, laser ablation, chemical vapor deposition, plasma-enhanced chemical vapor deposition, flame-assisted synthesis, and microwave-assisted synthesis. Each synthesis method possesses distinct techniques along with specific advantages and limitations, which are summarized in Table 3.1 [4,10–13].

3.2.1 Arc Discharge

This is the oldest technique of preparation. Two electrodes are assembled in a vacuum compartment filled with inert gas. Direct current is circulated via both electrodes of the assembly. Regulated current and pressure of a gas are requirements of this synthesis. The reaction chamber is pressurized with an inert gas. The positive and negative electrodes are brought into proximity, forming an arc upon activating the electric supply. The production of plasma occurs as a consequence of the electrodes reaching an extremely high temperature due to the presence of this arc. During the formation of plasma, CNTs start to accumulate on the negatively charged electrode. Both SWCNTs and MWCNTs can be synthesized using this method. In contrast to MWCNTs, which may be synthesized by keeping the inert gas pressure constant,

TABLE 3.1
Various Synthesis Methods of CNTs

Sr. No	Method of Preparation	Process	Advantages	Limitations
1	Arc discharge	An electric arc within graphite electrodes evaporates carbon, resulting in the condensation of CNTs	High production rate and Ease of handling	Need of high temperature, formation of metallic impurities over CNT, and poor control on lengths and diameters of CNTs
2	Laser ablation	A powerful laser vaporizes a graphite sample inside a furnace with high temperatures comprising inert gas	Precise regulation on the morphology of CNTs	High-cost machinery and low output
3	Chemical vapor deposition	Carbonaceous gases like methane undergo decomposition over a metallic catalyst at elevated temperatures	Devoid of requirements of high temperature, high yield, and adaptable for many substrates	Impurities in catalysts Demand more regulation of variables
4	Plasma-enhanced chemical vapor deposition	Breakdown of hydrocarbon gases to form CNTs and self-organization around catalyst nanoparticles	Operate at low temperature, chemical vapor deposition, with precise control over growth parameters	Extravagant operational cost
5	Flame-assisted synthesis	A flame rapidly elevates the temperature of hydrocarbon gases and catalyst precursors, favoring the formation of nanoparticles and expediting the growth of carbon nanotubes	Abundant yield	The problem of consistent results
6	Microwave-assisted synthesis	Subjecting carbon precursors and catalysts to microwave radiation to produce heat, facilitating the efficient development of carbon nanotubes by focused heating and plasma interactions	Cost-effective, can avoid usage of solvent	Constricted production range

SWCNTs require a particular catalyst for production. Nanotubes with lengths of several microns and sizes of 2–20 nm can be grown using this method [14]. Without a catalyst, soot and deposits are prepared. Soot mainly comprises fullerenes, and MWCNTs are found in carbon deposits. Numerous factors influence the efficiency and output of this method line, including the quantity of carbon vapor, the ability of gaseous carbon to spread within the chamber, the process heat, and the composition of the catalyst. These variables also affect the morphological parameters of CNTs [6] (Figure 3.1).

3.2.2 Laser Ablation

A high-power laser beam or nonstop pulse is targeted at a graphite rod and a mixture of equal amounts of nickel and cobalt. Argon assists in heating the graphite rod to a temperature of approximately 1,200°C. Laser is used to selectively convert graphite into vapor. A pressure of 500 torr is maintained using gases such as helium and argon. This approach is predominantly chosen for synthesizing SWCNTs and is known for its high cost [16]. The laser ablation technique yields a 70% output. This approach yields over 50% of SWCNTs interwoven with one another. The CNTs generated by this process have more purity than the arc discharge method [6]. Laser ablation also needs catalysts for the formation of CNTs. Using only graphite produces clusters of fullerenes. Catalyst aids in the attachment of extra carbon, thus preventing the closure of the carbon rings. The monoatomic or polyatomic entity holds carbon atoms and converts them into spiral and curlicue structures. This process continues till a stockpile of catalysts is at the posterior edge of CNTs [17] (Figure 3.2).

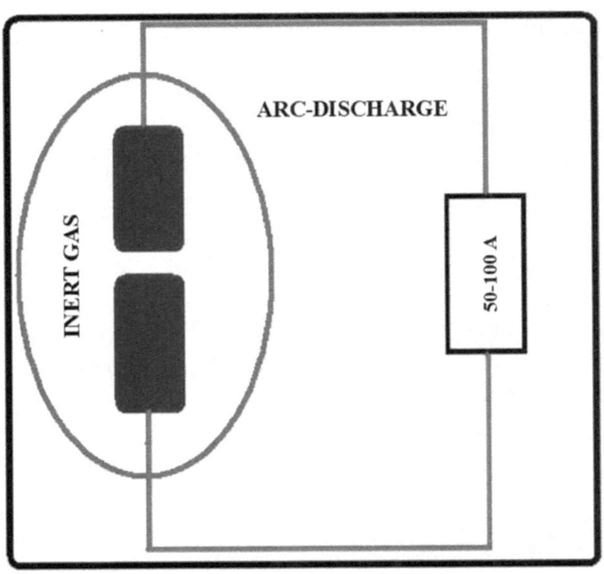

FIGURE 3.1 Schematic representation of arc discharge method of synthesis of CNTs [15].

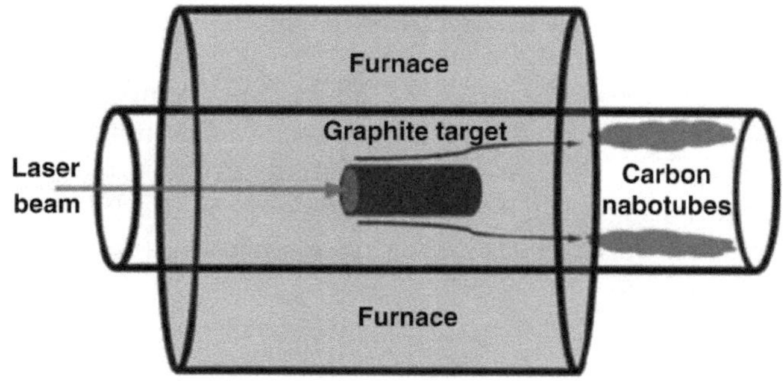

FIGURE 3.2 Schematic representation of laser ablation method for the fabrication of CNTs [18].

3.2.3 Chemical Vapor Deposition

Chemical vapor deposition, unlike the other two methods, does not necessitate elevated temperatures. Arc discharge and laser vaporization methods have significant technological drawbacks in that they produce CNTs that possess inherent instability. Both a conventional and patterned substrate are incapable of facilitating the growth of CNTs [19]. An Radio frequency (RF) heater elevates the furnace temperature to 500°C–900°C following the insertion of the quartz tube. The quartz tube, filled with argon, contains the substrate and catalyst. Heat converts gaseous hydrocarbons into vaporized carbon within a quartz tube. Van der Waal forces bind these carbon vapors to the reacting substance, resulting in the formation of MWCNTs. Iron, cobalt, and nickel serve as catalysts for synthesizing SWCNTs [20] (Figure 3.3).

3.2.4 Plasma-Enhanced Chemical Vapor Deposition

Plasma-enhanced chemical vapor deposition (PECVD) is a contemporary synthesis technique that produces CNTs by generating plasma under low-temperature conditions from gaseous substrates such as carbon dioxide, methane, and acetylene. The concentration of the PECVD precursor increases due to plasma production, leading to an elevated rate and enhanced control over the formation rate. The carbon source becomes ionized, resulting in its plasma state. This plasma subsequently generates reactive carbon entities that interact with the catalyst, facilitating the development of CNTs [22,23] (Figure 3.4).

3.2.5 Flame-Assisted Synthesis

The predominant quantity of carbon black, fumed silica, and other industrial-scale materials is produced using flame-assisted synthesis, a highly established method for material fabrication. Howard et al. were among the pioneers in detecting trace quantities of CNTs in low-pressure premixed hydrocarbon fires utilizing oxygen that reacts with fuels such as C_2H_2, C_2H_4, and C_6H_6. Since that time, other flame configurations

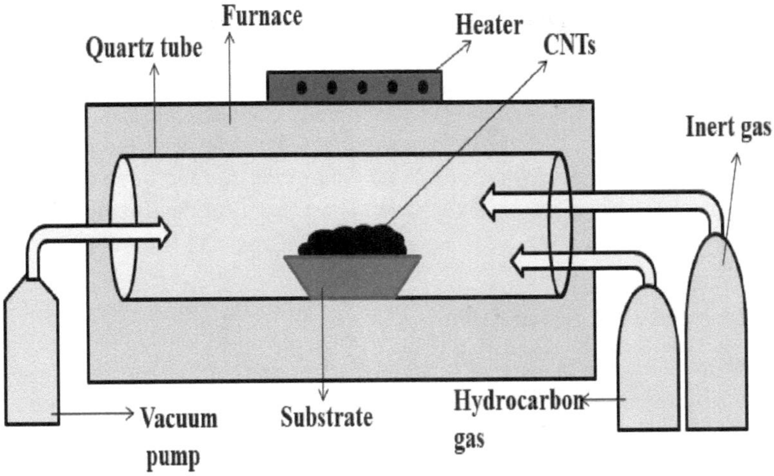

FIGURE 3.3 Schematic representation of the CNTs synthesis using chemical vapor deposition method [21].

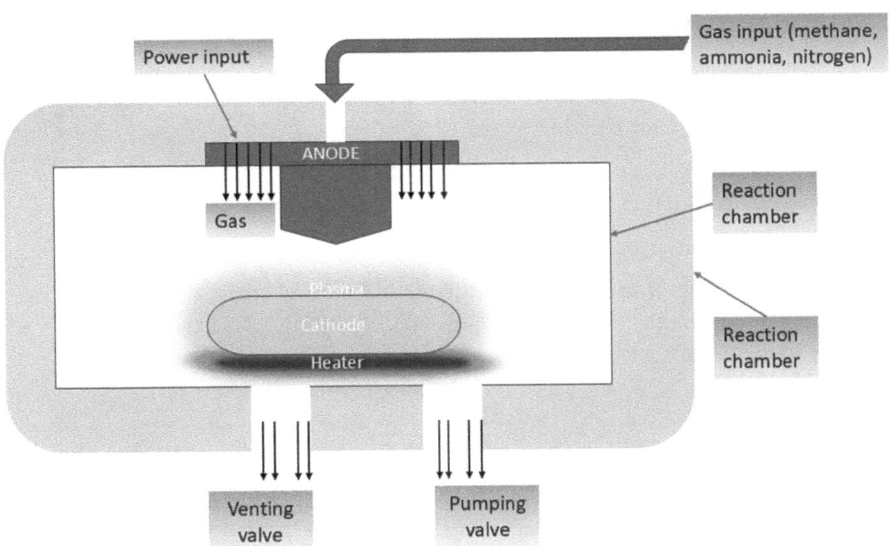

FIGURE 3.4 Plasma-enhanced chemical vapor deposition.

have been developed for CNT synthesis [24,25]. Catalytic metal acts as a support site for fusion and the saturation of CNTs. These catalysts are in the form of coatings or aerosols. The type of CNTs that will deposit on metal particles is influenced by their constituents, surface, and structure. The flame medium is composed of a mixture of carbon fuel and oxidizing gases, which results in the formation of partially oxidized products along with water and carbon dioxide. A flame front is where the temperature is at its peak, and in this area, the value of the local equivalence ratio (ф) is 1. A combination of both components can be operated comfortably by changing their respective feedstock parameters. Heat-related parameters can be adjusted through the manipulation of chimney-related values. Catalysts are often introduced into the flame system as gases, subsequently undergoing nucleation and condensation to generate solid spherical metallic nanoparticles. By adjusting the characteristics of the flame, one can provide optimal flame conditions that facilitate the creation of catalytic particles of appropriate size, which is essential for catalyzing CNT growth and development [26] (Figure 3.5).

3.2.6 Microwave-Assisted Synthesis

Microwave-assisted synthesis is a technique of preparation where CNTs are prepared by using microwave radiation to quicken the process. This technique is rapid and requires less solvent [27]. Microwave heating is the primary stage of preparation, which leads to the disintegration of the source material. Nonreversible carbon sources such as petroleum-based analogs, and carbonaceous gases like methylene and ethylene are heated using iron and nickel catalysts at a temperature range of 700°–1,000° [28,29]. The benefits of microwave heating compared to conventional methods, including uniform heat distribution and increased transfer of volatile substances, have demonstrated an improvement in process efficiency, resulting in a higher yield of CNTs [30]. This method entails several drawbacks, such as the complexity of scaling up, the need for increased temperature, reduced efficiency, and challenges in operational parameters [31,32].

3.3 CHARACTERIZATION OF CNTs

As the size of the nanoparticles (NPs) is in the range of 1–1,000 nm, it offers more surface area of contact, which results in the interaction of these particles with their surroundings [33]. Oral, rectal, ocular, parenteral, topical, transdermal, and pulmonary are some of the routes of administration of nanosized formulations; among all,

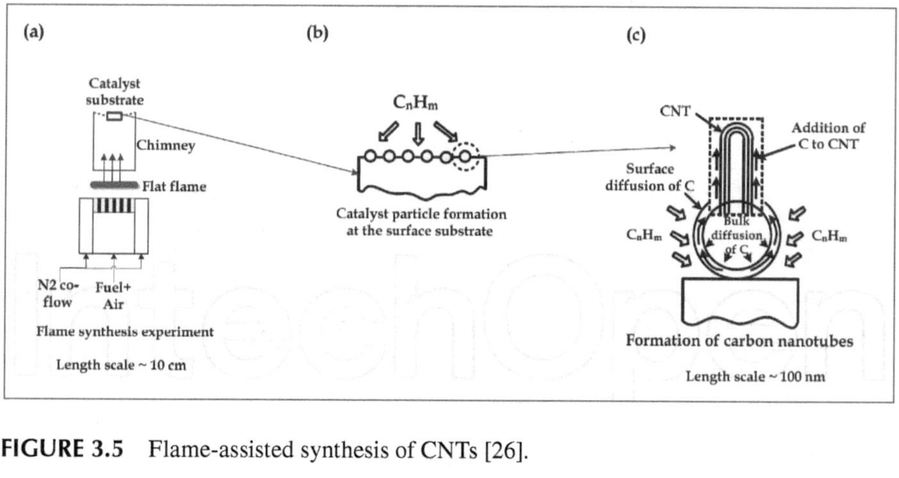

FIGURE 3.5 Flame-assisted synthesis of CNTs [26].

oral is a popular route of administration [34]. Properties such as size, surface charge, morphology, composition, and shape are some factors that determine *in vivo* fate of NPs [35]. Being NPs, CNTs also need to be accessed through these parameters to ensure effective drug delivery. Table 3.2 summarizes the principles and applications of each approach [36–41].

3.3.1 Particle Size

The diameter of the smallest capillary is 3–4 μm, so any delivery system must be smaller than this size to ensure irritation-free drug delivery [42]. Smaller particle size also improves penetration and retention [43]. CNTs below the size of 200 nm are better at crossing the blood–brain barrier [44]. Small-sized CNTs can also penetrate tumors passively [45]. The particle size of CNTs has been characterized by the following techniques.

3.3.1.1 Dynamic Light Scattering

Dynamic light scattering (DLS) or quasielastic scattering is a noninvasive method of size determination and size distribution. It can measure particles with a size of micron to nanometer range. Generally, DLS is used for particles that have been dissolved or suspended in liquid form [46]. Since DLS works with diffusion coefficient, it measures hydrodynamic radius, which is determined by the size and shape of the particle. Apart from size, DLS is also used for determining polydispersity index (PDI) [47]. Seyfoori et al. conducted the magnetic decoration of CNTs and to study the change in size of CNTs, the DLS method was adopted. The size of CNTs was observed to increase from 90 to 190 nm [48].

The Zetasizer uses a scattering detection angle of 173° for size measurements and 13° for zeta potential evaluation. Measurements are conducted at a temperature of 25°C with a precision of ±0.01°C. Backscattering was used as the detection mode with a scattering detection angle of 173° [49].

3.3.1.2 The Polydispersity Index

The PDI stands for the diversity and distribution of particles of NPs [50]. PDI of formulations should be as low as possible, which indicates the uniformity of NPs distribution [51]. Sharma et al. explored cancer targeting with doxorubicin (DOX)-loaded oxidized dextran conjugated polyethylene glycol (PEG)-MWCNTs with various methods. They reduced the toxicity of pristine nanotubes, and the final PDI of processed nanotubes was 0.233 ± 0.001 [52]. The decrease in PDI of docetaxel-loaded aspartic acid–conjugated hydroxylated CNTs was compared to pristine and partially processed CNTs, which denotes formulation devoid of any cluster [53]. Phospholipid 1,2-distearoyl-sn-glycero-3-phosphoethanolamine-Poly(ethylene glycol) (DSPE-PEG) carboxylic acid and ethylene diamine (EDA) were used to double functionalize SWCNTs, which were adopted for delivering amphotericin B. The PDI of these CNTs was noticed to be reduced as compared to SWCNTs resulting in dispersion with consistent size of CNTs [54].

TABLE 3.2
Various Characterization Techniques of CNTs

Sr. No	Characterization Parameters	Analytical Technique	Principle	Results Obtained
1	Size and polydispersity index (PDI)	Zetasizer	Dynamic light scattering	Size in nanometer and PDI
2	Surface charge	Zeta potential	Electrophoretic light scattering	Surface charge in kV
3	Spectroscopy	Fourier transfer infrared spectroscopy	Molecular vibrations at distinct frequencies when infrared light affects their dipole moments	Confirmation of chemical entities
		Raman spectroscopy	Evaluate molecular vibrations induced by monochromatic laser light	
		X-ray diffraction	Elastic scattering of X-rays	
4	Microscopic methods	SEM	An electron beam traverses the surface of a material, and the resultant secondary and backscattered electrons are utilized to generate a picture	Size, shape, and morphology
		TEM	The alteration in the scattering pattern of accelerated electrons resulting from change in surface dimensions produces a picture	
		AFM	It generates an image by quantifying the variations in forces across the pointed end of the cantilever and the material's surface	
5	Thermoanalytical techniques	DSC	Quantify the disparity in thermal flux between a specimen and a reference substance con relation to temperature	Melting point and glass transition temperature
		TGA	Determine the change in weight of the sample with respect to time or temperature when it is subjected to a regulated environment	

3.3.2 Surface Charge

Zeta potential is one of the factors that determine the stability of the formulation as it affects the flocculation of particles [55]. Zeta potential is a property of solid–liquid and liquid–liquid colloidal systems. When charged particles get surrounded by opposite charges of ions, these ions form a layer known as the Stern layer. Another layer of loosely clumped opposite charges to the surface's charge travels with particles in the medium. This motion of charge and particle is known as Brownian motion, and these layers are called double layers. Beyond this Stern layer, one more diffuse layer is present, which compromises both the same and oppositely charged ions. When an electric field is applied across the dispersant, charged particles travel to their respective opposite-charged electrode via a phenomenon called electrophoresis [56]. Surface charge is measured by dispersing CNTs in double distilled water, and to validate results, it is done three times [57]. The surface charge gets highly affected by the ligands attached to the CNTs. Nidhi et al. prepared MWCNTs, which were modified by attaching α tocopherol succinate and chondroitin sulfate (CHSL) where they observed that carboxylated CNTs (COOH-MWCNTs) had zeta potential of -15.1 ± 1.6 mV, which got shifted to $+40 \pm 3.2$ mV for aminated CNTs (NH_2-MWCNTs). After adding α tocopherol succinate and CHSL to the inherent property of these separate moieties for the final formulation of α tocopherol succinate and CHSL decorated DOX CNTs was 31.5 ± 2.1 mV [58].

3.3.3 Microscopic Analysis

Microscopic studies enable the examination of the surface characteristics of materials. It also assists in ascertaining its dimensions, form, and morphology.

3.3.3.1 Scanning Electron Microscope

Scanning electron microscope (SEM) is used widely for its magnification ability up to 10 nm, which can further go to 2.5 nm for its advanced version [59]. SEM is also associated with X-ray microanalysis machinery like energy dispersive X-ray analysis (EDX). For all elements except H_2, He and Li EDX can be recognized with EDX's help because every atom possesses its unique X-ray pattern [60]. Electrons in SEM are generated in lamps of tungsten filament or in a field emission gun. Electron's array gets its acceleration via a high-voltage system of 20 kilovolts (kV). Later, this gets narrowed through apertures and electromagnetic lenses. The scan coil scans the surface of a sample, which produces secondary electrons and backscattering electrons with unique X-rays. All these electrons get detected at the detector of SEM [59,61]. To study the morphological characteristics of CNTs, samples were stuck to SEM's stub using carbon tape. Then to get a conductive layer, samples were coated with gold and palladium by the sputtering system for at least 3 min. This 18 mm conductive layer helps to process images and magnification [62]. The shape of NPs is very crucial to penetrate into the cell [63]. Sharma et al. prepared functionalized CNTs, and SEM images of these CNTs showed that these CNTs were needle-shaped, and these CNTs can penetrate cell membranes [64] (Figure 3.6).

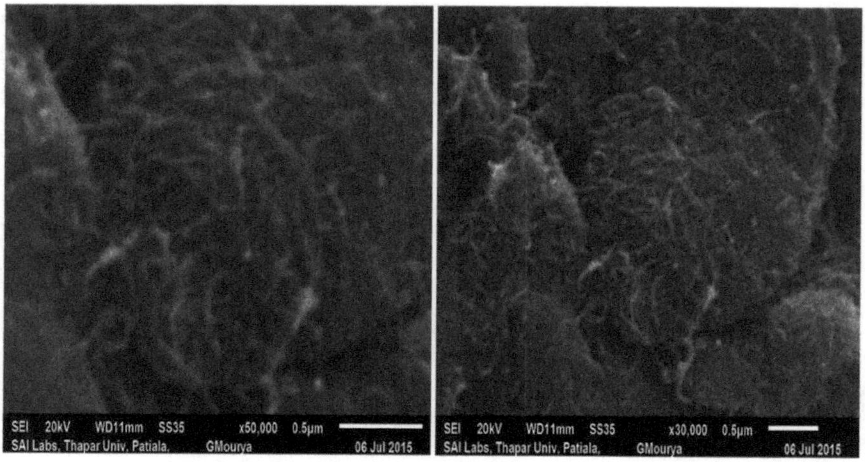

FIGURE 3.6 SEM images of needle-shaped CNTs [64].

The morphology of anchored chemicals can be seen because of the magnification of SEM. Azqhandi et al. prepared CNTs comprising PAA and PEG. He further added methotrexate and cyclophosphamide to these MWCNTs. They observed pictures of carboxylated CNTs (CNTs-COOH), CNTs with PEG (CNTs-PEG), and CNTs with PAA PEG (CNTs-PAA-PEG).

Unfunctionalized CNTs bearing carboxylic acid were seen to have the filament of COOH on a smooth surface as in Figure 3.7a. PEG has a characteristic long chain, which is visible in Figure 3.7b; PAA and PEG formed circular helical complexes, which are noticeable in Figure 3.7c and d [65].

Anchoring hydrophilic moieties such as PEG to the exterior of CNTs is a well-established method to address water insolubility issues. Moniriyan et al. selected magnetically active CNTs functionalized with the magnetic compound $FeCl_2 \cdot 4H_2O$ to deliver the chemotherapeutic agent cisplatin to the tumor. These PEGylated magnetic CNTS were then further coupled with 3-chloropropyltrimethoxysilane. Cisplatin (II) was finally coupled with these CNTs. They observed SEM photographs with different resolutions for morphologically analyzing these magnetic cisplatin-loaded CNTS. In contrast to the clean surface of pure CNTs, these modified CNTs exhibited a rough surface with an approximate diameter of 55 nm [66] (Figure 3.8).

3.3.3.2 Transmission Electron Microscope

Transmission electron microscopy (TEM) is commonly used to measure and analyze the size and distribution of particles [67]. TEM has a magnification power of 0.2 nm, allowing for the observation of particle morphology [68].

The transmission of highly energized electrons is focused on thin analytes by using electromagnetic lenses. The acceleration of electrons typically occurs at greater kilovolts (kV), leading to shorter wavelengths than visible light. Unlike an optical microscope, which relies on the wavelength of light to determine image magnification, a TEM utilizes the unique properties of electromagnetic lenses to achieve

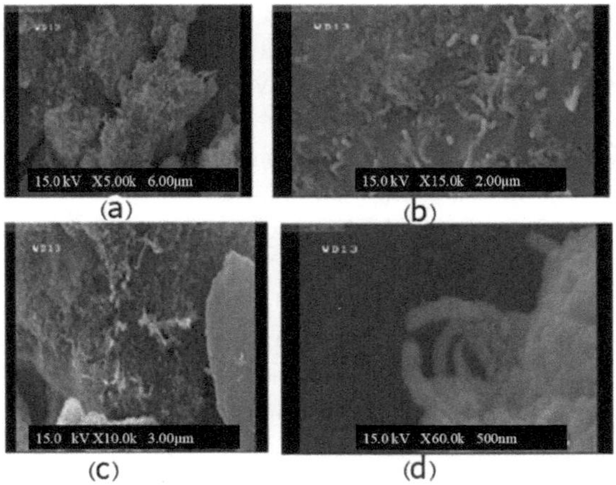

FIGURE 3.7 SEM analysis of (a) CNTs-COOH, (b) CNTs-PEG (c and d) CNTs-PAA-PEG [65].

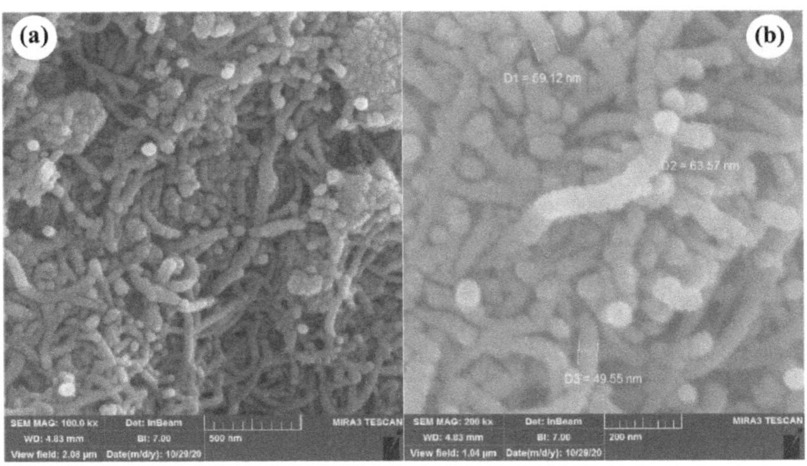

FIGURE 3.8 SEM images of magnetic CNTs nanocomposite at different magnifications [66].

magnification up to around 2,000 to one million folds. In TEM, our attention is not on the resolution of individual atoms, but rather on the collective behavior of groups of atoms and their resulting diffracted beams, which form an image on the screen. The crystal lattice of a specimen facilitates the identification of structural imperfections, such as dislocations, stacking faults, and interfaces between grains [39].

The TEM analysis of CNTs is done by dispersing them in water or a suitable solvent, followed by sonication of this solution. This solution is then spread on a carbon-coated copper grid of TEM [69–71]. TEM analysis helps to visualize the surface as well as the internal structures of CNTs. If CNTs are doped with metal particles, they can be seen as dark spots in the CNTs [72].

Cheng et al. examined the influence of PEGylation on the cellular uptake of CNTs and its consequences on the intracellular mechanisms of cells. MWCNTs were modified with PEG1500N after being synthesized using the vapor deposition process. The TEM pictures revealed a higher concentration of MWCNTs in the nucleus compared to the cytoplasm. Additionally, MWCNTs were observed in the cytoplasm, specifically in endosomes and near to the nucleus. The high-resolution TEM pictures revealed that the MWCNTs existed either as individual entities or in tiny clusters [73].

Mangifera indica, a plant, contains mangiferin, which is a phytoconstituent. To investigate the anticancer effects on brain cancer cells, Harsha et al. utilized CNTs to deliver this bioactive compound. In addition, they modified these CNTs with PEGylation to enhance their targeting efficiency and mitigate the harmful consequences associated with CNTs. The researchers chose TEM to examine any alterations in the morphology of CNTs following conjugation. However, the attachment of PEG and mangiferin did not result in any significant changes to the tubular structure of the CNTs [57].

The surface decoration of CNTs reduces the toxicity of pristine MWCNTs. In addition to increasing drug solubility, D-α-tocopheryl polyethylene glycol 100 succinate (TPGS) aids in the overcoming of multidrug resistance and triggers receptor-mediated endocytosis. Mehra et al. delivered DOX with TPGS-coupled MWCNTs (DOX/TPGS-MWCNTs). The combination of DOX and TPGS in MWCNTs results in DOX's sustained release and improved internalization in tumor tissue, both of which increase DOX's efficiency. Raw MWCNTs, oxidized MWCNTs, and TPGS-coupled MWCNTs were all compared in terms of their morphology. Size reduction occurs upon oxidation of MWCNTs. TEM confirmed that these DOX/TPGS-MWCNTs are nanoscale in size, exhibiting appropriate distribution and stability [74] (Figure 3.9).

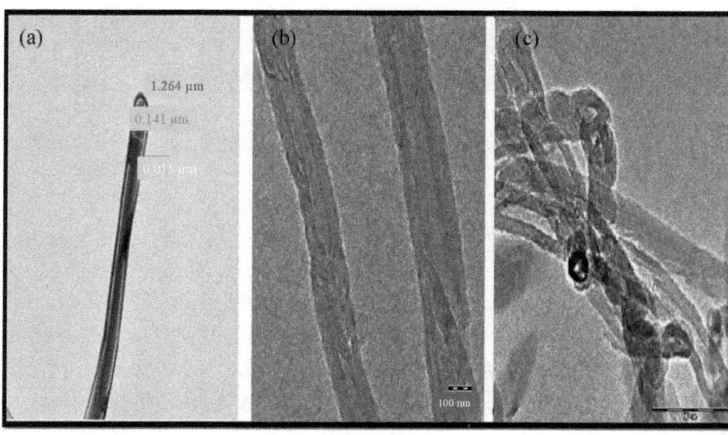

FIGURE 3.9 TEM images of (a) pristine MWCNTs, (b) oxidized MWCNTs, and (c) TPGS-coupled MWCNTs [74].

3.3.3.3 Atomic Force Microscopy

The field of scanning probe microscopy emerged in the early 1980s, after Gerd Binnig and Heinrich Rohrer developed the scanning tunneling microscope, for which they received the Nobel Prize for Physics in 1986. In the same year, Gerd Binning, Calvin Quate, and Christoph Gerber made a significant advancement by inventing the atomic force microscope (AFM). This discovery has had a lasting impact on nanometer characterization and evaluations, revolutionizing the field. Currently, AFM is the most widely recognized among all types of scanning probe microscopy [75].

AFM is a technique used to visualize and analyze the topography of a surface. The mechanism operates by sensing the attractive and repulsive interactions between a limited number of atoms, located at the tip of a cantilever and the surface under investigation. The three primary types of attractive forces that impact AFM imaging are the forces of van der Waals, electrostatic charges, and short-range chemical forces. Due to the unique effective range of each force, the dominance of each force fluctuates depending on the distance. AFM can be detected using two methods: static and dynamic. In static AFM, the force is quantified by observing the deflection of the cantilever resulting from the interaction between the tip and the sample. Dynamic AFM detects the force by monitoring alterations in the resonance characteristics of the cantilever, such as variations in frequency, amplitude, or phase, caused by the contact force. Static AFM provides a simple way to estimate the force, whereas dynamic AFM necessitates more intricate analysis to interpret the recorded amplitude, phase, or frequency data. Recent breakthroughs in ultra-high-resolution imaging have unveiled subatomic characteristics of specimens, facilitated by working in low-temperature, ultra-high vacuum conditions with a rigid cantilever. This section will examine the latest advancements in the theoretical and experimental aspects of imaging and force spectroscopy [76].

AFM is important in characterizing nanoscale drug delivery systems. It is used in high-resolution imaging to characterize surface properties and to measure mechanical properties like adhesion and elasticity. AFM can be coupled with other instruments like SEM, RAMAN, and confocal microscopy. Its real-time monitoring provides us with data useful for understanding various biological dynamic processes [77,78].

Curcumin is a proven anticancer medical product obtained from the plant *Curcuma longa*. It can stop the growth and spread of tumors and also helps to combat multi-target resistance in tumor tissue. Singh et al. loaded curcumin in chitosan (CS) and alginate functionalized SWCNTs. To functionalize CNTs further with alginate, CS oxidation of pristine CNTs is essential. Raw CNTS are hydrophobic metallic structures, which is responsible for their low water solubility and bioavailability. Before analysis, 20 μL samples were air-dried on a glass slide. AFM analysis was done to determine what changes had been brought up by oxidation on SWCNTs. Oxidized SWCNTs had high roughness, but functionalization with alginate and CS again made the surface smoother [79].

3.3.4 Spectroscopical Characterization

Spectroscopic techniques are employed in structural characterization, chemical identification, and impurity analysis. Fourier transform infrared spectroscopy (FTIR), Raman spectroscopy and X-ray diffraction (XRD) are commonly used spectroscopical methods.

3.3.4.1 Fourier Transform Infrared Spectroscopy

Fourier transform infrared spectroscopy (FTIR) characterization is advantageous for a comprehensive understanding of the molecular structure of a sample and its interactions. The primary distinguishing feature of FTIR, compared with other techniques, is its ability to scan all three phases of a substance. This method utilizes the analysis of molecular vibrations in compounds to identify and detect certain functional groups in materials. The ability of FTIR to discern chemical spectra is essential for comprehending the composition and behavior of NPs. FTIR spectroscopy measures the rotational and vibrational motions of molecules, which are affected by infrared radiation. This method is employed to identify modifications in molecular bonding structures and changes in functional groups. This property enables scientists to clarify the interactions between NPs and other substances, such as living beings or environmental components [80,81].

Infection by *Mycobacterium tuberculosis* leads to the onset of tuberculosis (TB). Isoniazid (INH) is utilized for managing TB. Zomorodbakhsh et al. employed MWCNTs-mediated medication delivery to minimize dosage and adverse effects. He attached INH to carboxylated MWCNTs. They evaluated carboxylated MWCNTs and INH–MWCNTs to verify the conjugation between the two entities. Carboxylated MWCNTs exhibited peaks at 1,527 and 3,304 cm^{-1}, corresponding to C=C and COOH groups, respectively. The amide bond formed between INH and MWCNTs exhibited a peak at 1,678.83, indicative of the amide bond's presence. The amide bond between the API and MWCNTs exhibited binding attributed to the pyrimidine ring, detected at 1,603 cm^{-1} [82].

Before the functionalization of CNTs, oxidation is a very crucial step. Ahmad Zawawi et al. compared three different oxidation steps for MWCNTs. These methods were (A) the addition of nitric and sulfuric acid assisted by stirring, (B) 2 hour of sonication, and (C) 6 hour of sonication. FTIR analysis involves collecting data on pure MWCNTs (p-MWCNT) before and after oxidation utilizing three distinct methods. All methods used to produce oxidized MWCNTs (Ox-MWCNTs) exhibited peaks at 3,400 cm^{-1} (indicating the presence of hydroxyl groups), 1,720 cm^{-1} (indicating the presence of carbonyl groups), 1,600 cm^{-1} (indicating the presence of carbon–carbon double bonds), and 1,200 cm^{-1} (indicating the presence of carbon–oxygen bonds). The presence of a carbon atom in the sp^2 hybridization state in the p-MWCNT is indicated by the peak at 1,623 cm^{-1}, suggesting that it is in an undamaged condition. FTIR analysis of batch A confirmed the initiation of the fault, as evidenced by the lack of CH_2 and CH_3 signals. Treatment A exhibited a lower carbonyl peak (1,714 cm^{-1}) compared with treatment B due to a higher degree of decarboxylation. This observation provides insight into the superiority of treatment B over treatment A, which was further validated by thermogravimetric analysis [83] (Figure 3.10).

Paclitaxel (PTX), being an antitumour agent for various cancers, is a drug of choice for many research projects. Magid and Al-Karam prepared CS functionalized MWCNTs (CS/MWCNTs) and conjoined this antitumour agent PTX to CS functionalized MWCNTs (CS/MWCNTs/PTX). FTIR analysis was done for the structural characterization of this complex. Sharp and larger peaks at 3,262 and 3,858.71 cm^{-1} are due to –OH in the carboxyl group, while 1,636.36 and 1,052.51 cm^{-1} relate to

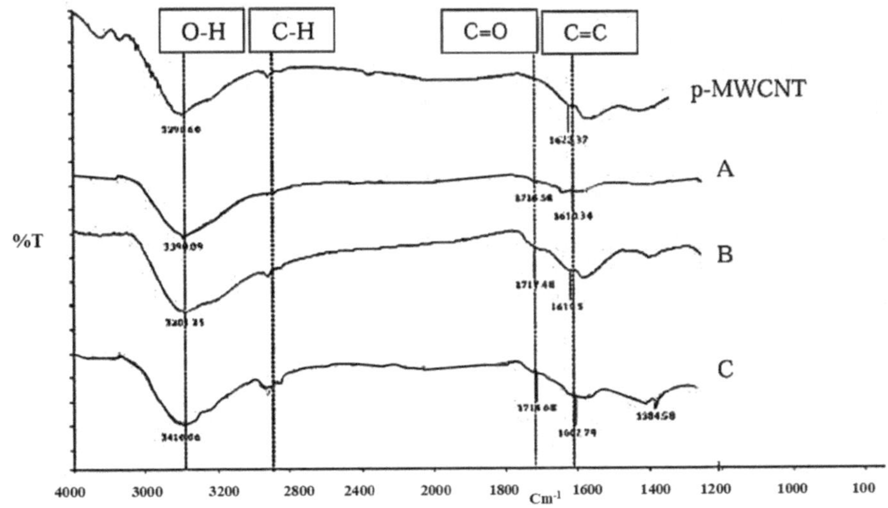

FIGURE 3.10 FTIR evaluation of MWCNTs and it is functionalized product prepared by (a) normal stirring, (b) sonication for 2 hours, and (c) sonication for 6 hours [83].

C=C and C–O. The secondary amine of CS and carboxyl group of MWCNTs fused to form an amide bond is confirmed by a peak at 1636 cm^{-1} [84] (Figure 3.11).

3.3.4.2 Raman Spectroscopy

Raman spectroscopy is a nondestructive technique for characterizing carbon distinguishing all allotropic forms of carbon. The location, intensity, and area of peaks vary according to the form of carbon [85]. This method works in two vibrational modes (i) radial breathing mode (RBM) and (ii) G-band and D-band. RBM is between 200 and 300 cm^{-1}, whereas G-band and D-band operate around 1,582 cm^{-1} [86].

To deliver semisynthetic anticancer agent docetaxel (DTX), DTX acts in tubulin's β part. The delivery via MWCNTs was adopted with attachments of PEG and folic acid (FA) to decrease toxicity and increase target ability. Data acquired from Raman spectroscopy elucidate CNTs' hybridization status and defects. DTX loading brought a shift of the G-band from 1,579.85 to 1,564 cm^{-1} and a movement of the D-band from 1,346.15 to 1,310 cm^{-1}. This alteration transpired due to an enhancement in single-bond attributes arising from FA functionalization [87].

Requardt et al. checked the influence of metal catalysts on their structure and cytotoxic action. Both iron-depleted CNTs (non-FE-CNT) and iron-containing CNTs (FE-CNT) led to a decrease in cell viability in A549 lung epithelial cells and HepG2 cells. In A549 cells, there was a significant drop in cell viability. The Raman spectra comparison of both CNTs revealed a G-band at 1,600 cm^{-1} and a D-band at 1,300 cm^{-1}; however, the D/G ratio for FE-CNT was approximately one, while for non-FE-CNT, it was below one. The D/G ratio suggests that non-Fe-CNT possesses fewer defects than the other types [88] (Figure 3.12).

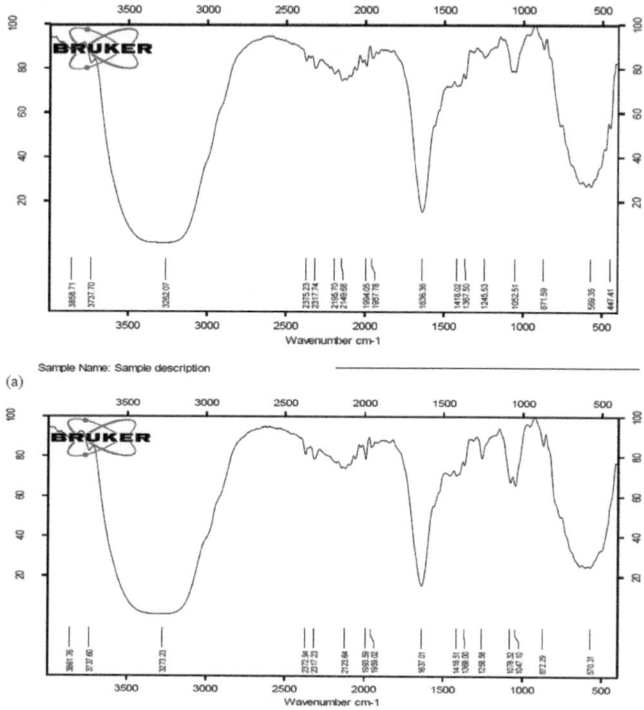

FIGURE 3.11 FTIR of (a) MWCNTs/CHI and (b) MWCNTs/CHI/PTX [84].

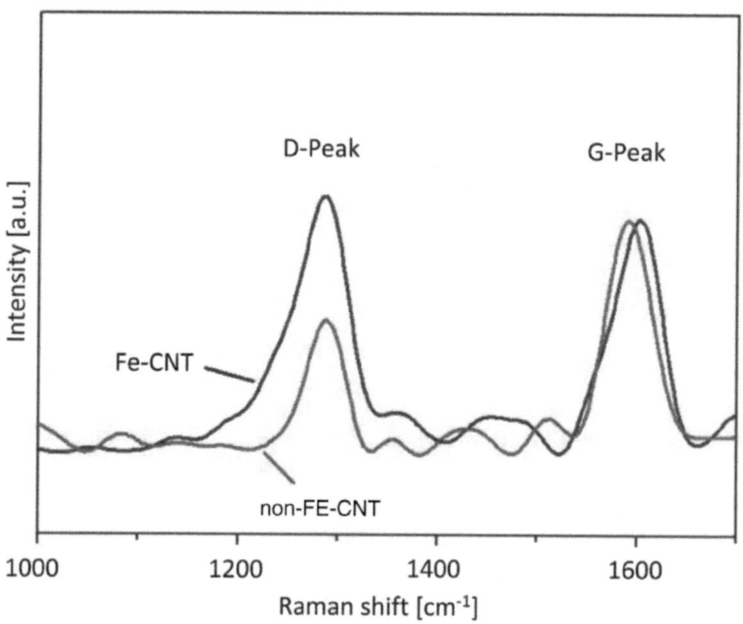

FIGURE 3.12 Raman spectra of Fe-CNT and non-FE-CNT [88].

3.3.4.3 X-Ray Diffraction

XRD is popular for predicting the crystallinity of a material. A cathode ray tube emits X-rays with different wavelengths, while a monochromator allows rays with a single wavelength. When these rays bombard on the atomic planes of a sample complying with Bragg's law, the rays are diffracted, transmitted, scattered, refracted, and adsorbed. Diffracted patterns help to predict the crystal lattice [89]. Single crystal XRD (SCXRD) and powder XRD (PXRD) are its two types. SCXRD describes elements and intermolecular interconnections. PXRD helps to discriminate the drug's different phases and the pureness of the material. PXRD is also used to look at the instability of a crystalline compound. Synchrotron sources aid in the characterization of amorphous materials. The configurational arrangement of atoms is studied with the help of the pair distribution function [90].

During formulation, XRD is used for checking on various steps for confirming the purity of a delivery system. Li et al. delivered DOX by using a system with a combination of magnetic CNTs and hydroxyapatite (CNT(Fe)/HA)). Further, they improved it with FA and CS attachments (CS-CNT(Fe)/HA). For CNT(Fe)/HA, signals at 26.4° and 44.6° relate with the planes of a graphite (002 and 013). Fe (110 and 101) showed peaks at 44.6° and 46.5°, while the rest were due to calcium hydroxyapatite's crystalline nature. When these CNTs were further modified with CS and FA (FA-CS-CNT(Fe)/HA), the absence of extra peaks confirmed that further synthesis was devoid of any extra-degraded product [91] (Figure 3.13).

Ahmad et al. utilized XRD analysis to ascertain the crystallinity of a composite material composed of polylactic acid (PLA) and PEG that was filled MWCNTs. The PLA/PEG spectrum exhibited a large attenuated peak at 16.99 and a peak centered at a 2θ magnitude of 32.30. These peaks indicate a decreased sharpness, a typical feature of the PLA polymer spectrum. The PLA polymer spectrum typically shows a broad band centered at a 2θ value of 16.99 with a considerable background signal. The absence of a crystalline peak in the PLA/PEG spectrum suggests an amorphous nature, as PEG does not disrupt the crystalline structure of the polyester. Characterization by XRD analysis showed an addition of MWCNTs composite of

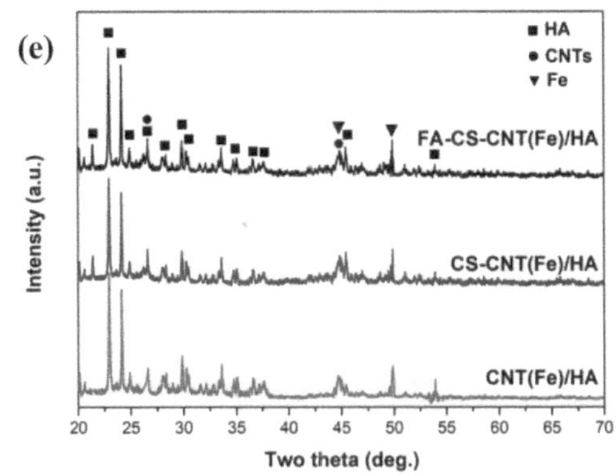

FIGURE 3.13 XRD diffraction patterns of CNT/Fe/HA, CS-CNT(Fe)/HA, and FA-CS-CNT(Fe)/HA [91].

amorphous PLA/PEG results in the progressive emergence of a separate diffraction peak. This implies that an increase in bond formation correlates with the amount of MWCNTs, which is reflected in the rise in diffraction peaks' intensity. Additionally, the XRD results suggest the transition of amorphous PLA/PEG composite to the semicrystalline mixture with the addition of MWCNTs [92] (Figure 3.14).

3.3.5 Thermoanalytical Techniques

Thermoanalytical tests like differential scanning calorimetry (DSC) and thermogravimetric analysis (TGA) are used to check the thermal behavior of materials.

3.3.5.1 Differential Scanning Calorimetry

DSC mainly studies thermal stabilities and properties. This analysis calculates heat transfer like absorbing and releasing under regulated conditions. This technique can detect minute thermal behaviors of material. TGA, XRD, and some microscopic devices are used to get details about the thermal nature of the sample [93]. Heat flux and power consumption are two types of DSC. Heat-flux DSC quantifies the disparity in thermal energy transfer between a specimen and a reference substance. In this technique, the sample and reference are positioned within a heating chamber outfitted with thermocouples to precisely measure their temperatures. The heating rate of the sample is influenced by the heat capacity as the temperature increases compared to the reference. During a thermal event, such as melting or crystallization, there is a shift in heat flow. This change produces a differential signal that shows the amount of energy absorbed or released during the transition. This approach is renowned for its simple and efficient design, along with its capability to deliver accurate thermal

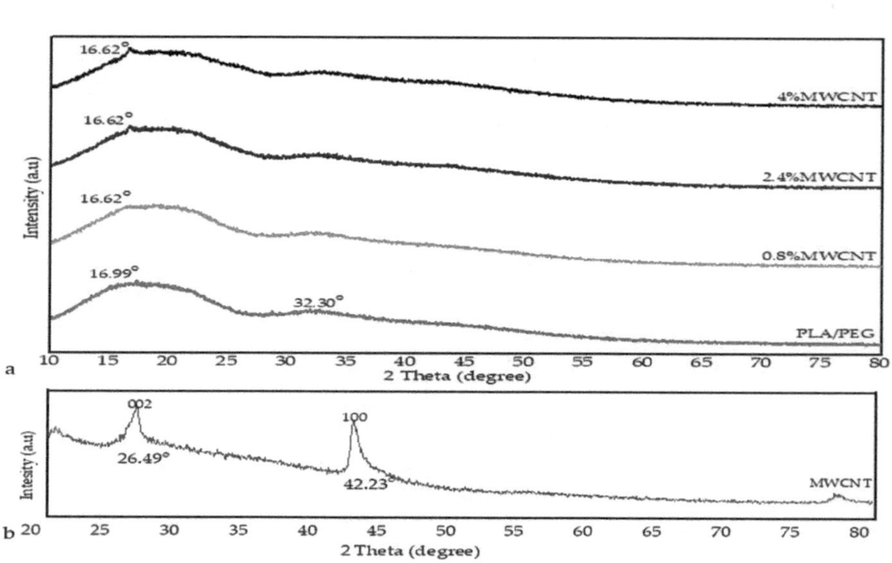

FIGURE 3.14 XRD spectra of (a) MWCNTs PLA/PEG and (b) PLA/PEG, 0.8% MWCNT, 2.4% MWCNT and 4% MWCNT [92].

measurements. Power-compensated DSC, which is also referred to as power differential DSC, operates distinctly. The procedure involves placing the sample and reference in thermally isolated chambers and then measuring the electrical power needed to keep both at an identical temperature. This method enables accurate temperature regulation and directly quantifies the energy required for thermal transitions, irrespective of the temperature discrepancy between the sample and reference [94].

DSC is also used to predict crystalline or amorphous material. Randive et al. prepared FA-decorated CS-coated SWCNTs. These CNTs have an energy-releasing peak at a temperature of 382.63°C. A dominant peak confirms the crystalline nature of drug-loaded SWCNTs [95] (Figure 3.15).

Similarly, Uttekar et al. formulated DOX containing MWCNTs decorated with FA and ethylene diamine (FA-EDA-MWCNTs) comparison between thermogram of drug and FA-EDA-MWCNTs-DOX. A peak at 222°C–224°C reflecting melting of an API. Another endothermic curve at 264°C–266°C. The additional endothermic peak detected at a temperature range of 264°C–266°C correlates to the melting point of the present conjugate. This outcome validates the absence of any indication of interaction or chemical incompatibility between the medication and the FA-EDA-MWCNTs-DOX nanoconjugate that was created. The decrease in crystalline nature was due to a reduction in the growth of the crystal [96] (Figure 3.16).

3.3.5.2 Thermogravimetric Analysis

Thermogravimetric analysis (TGA) measures the mass loss of a sample when it is heated under predetermined parameters and environments. The environment inside the furnace can be maintained as either inert or reactive, depending on the requirements of the sample. Nitrogen and argon are utilized to provide an inert atmosphere in the furnace, while air and oxygen are employed to maintain a reactive environment.

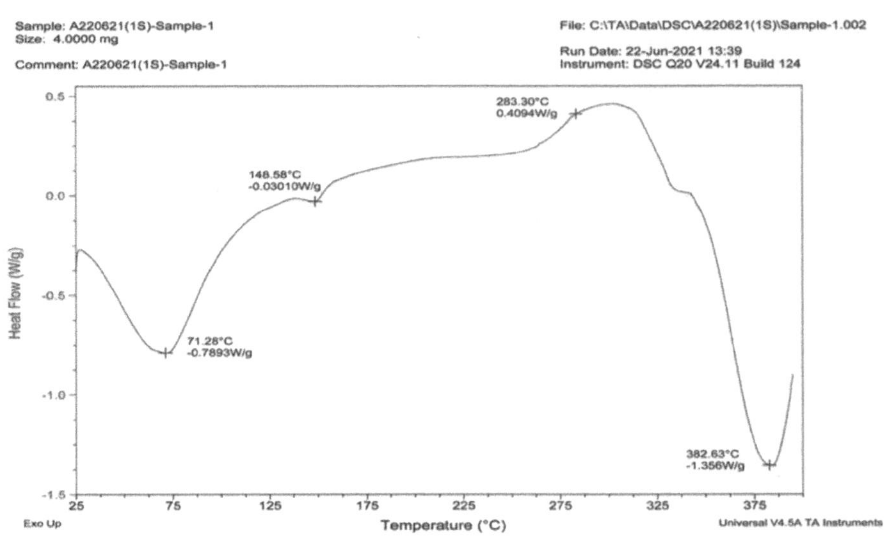

FIGURE 3.15 DSC thermogram of FA decorated CS coated SWCNTs [95].

The result of TGA is referred to as a thermogram, which is a graphical representation of the percentage of weight loss plotted against temperature or time [97].

The thermogravimetric approach has been employed due to its capacity to forecast the quantity of amorphous carbon, detect the presence of any metal, and differentiate between SWCNTs and MWCNTs. Carbon allotropes are evaluated using a semiquantitative method, while the quantity of metal is determined using a quantitative approach. All these values are in terms of percentage. Additional analytical

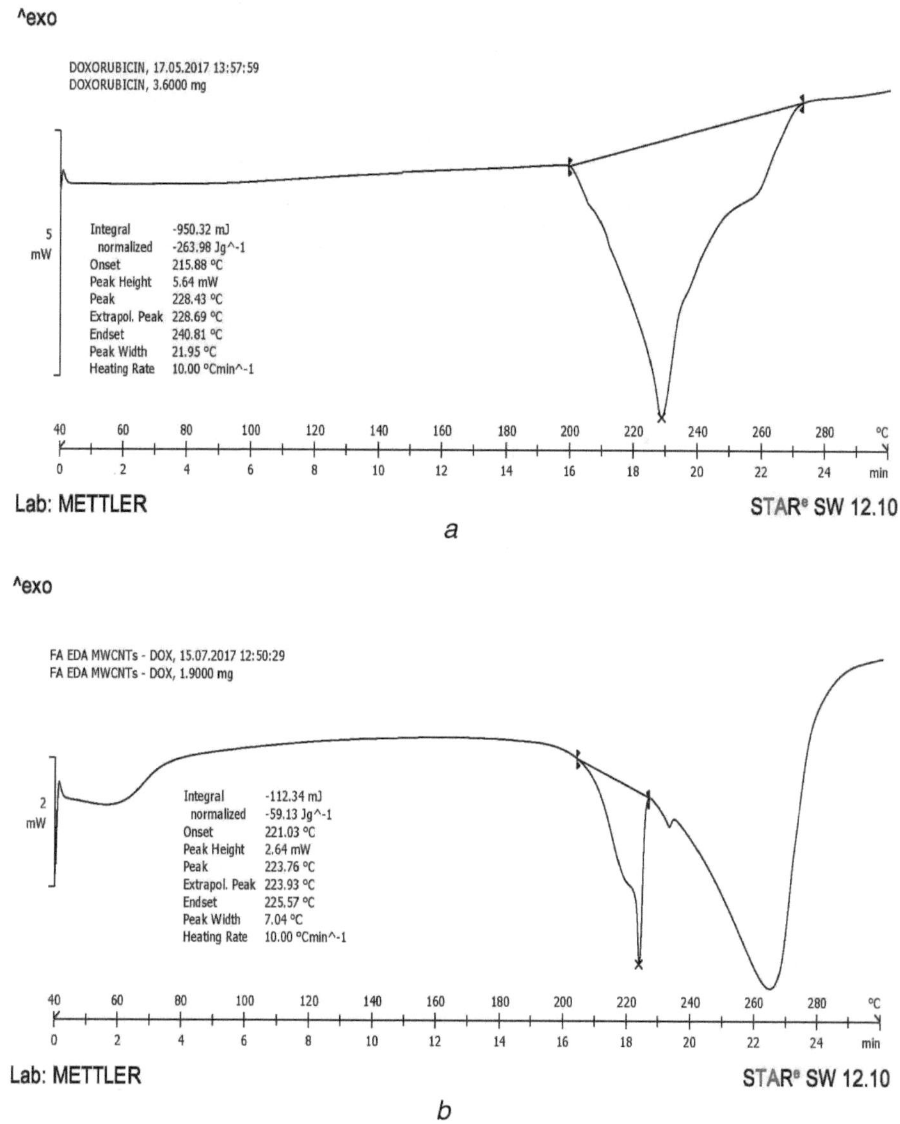

FIGURE 3.16 DSC spectra of (a) DOX and (b) FA-EDA-MWCNTs-DOX nanoconjugate [96].

techniques are employed to validate the TGA outcomes. The sample utilized in TGA is not retrievable or reusable since TGA measures the decrease in mass in proportion to heat. This equipment is capable of analyzing samples ranging from 1 to 10mg. Typically, this device breaks down all carbon-containing substances when exposed to temperatures over 600°C; this property of carbon can be used for quantifying metals in metal-incorporated CNTs. Likewise, amorphous carbon deteriorates within the temperature range of 200°C–400°C This temperature is below the breakdown temperature of CNTs, enabling the determination of the proportion of amorphous carbon [98].

Polyethylenimine (PEI) is a cationic polymer exhibiting a proton sponge effect; this property helps with its ability to endosomal escape. The needle shape of CNTs leads to a better penetration effect. Nia et al. tried to link these materials via a disulfide bond to get better gene delivery. Cystine (CYS)-functionalized CNTs were prepared with the help of linkers. They performed TGA to calculate parts of PEI on SWCNTs. For a better understanding of the behavior of these SWCNTs, they analyzed raw CNTs, carboxylated CNTs (SWCNTs-COOH), cysteine-functionalized SWCNTs (SWCNTs-CYS), and cysteine. Data obtained from TGA revealed that attached acidic moiety constituted 5% of SWCNTs-COOH, SWCNTs-CYS had 65% of SWCNTs and 35% of CNTs [99].

Liquid crystalline molecularly imprinted polymers (LC-MIP) is a new approach explored for drug delivery. Zhang et al. delivered levofloxacin (LVF) MWCNTs covered by liquid crystalline molecularly imprinted polymers (MWCNTs-LCMIP). MWCNTs-LCMIP was compared with other formulations such as the controlled group which was nonimprinted with MWCNTs (MWCNTs-LCNIP), MWCNTs which were imprinted without the liquid crystalline material (MWCNT-MIP), and MWCNTs without liquid crystalline material and did not get, and formulation where MWCNTs does not form perfect imprinting even absence of. The thermal behavior of particles was observed by using TGA. Reduction in mass was studied between 240°C and 450°C. MWCNTs with just molecular imprinted polymers (MWCNTs-MIP) degraded by 28% whereas MWCNTs-LCMIP degraded by 10% nearly at 330 °C. This data proved the stability of MWCNTs-LCMIP is more than that of its MWCNTs-MIP [100].

3.4 CONCLUSION

This chapter highlights the recent advancements in synthesizing and characterizing CNTs while addressing their critical aspects. CNTs are now rapidly transforming from laboratory research to more broad commercial applications. They are primarily utilized in the development of polymer nanocomposites, as components in electronic devices, and in other pharmaceutical and biomedical applications such as targeted drug delivery. However, despite their promising diversity, the large-scale production of CNTs remains challenging, and their tendency to aggregate due to strong van der Waals forces remains a significant hurdle for wider commercial adoption.

Addressing these challenges necessitates improvements in manufacturing processes to make the production of CNTs more scalable and cost-effective. Moreover, developing and improving the new techniques for synthesis and characterization of

CNTs will enhance the quality control essential for broader industrial applications. Advancements in surface modification like coating or conjugation can reduce aggregation and improve functionalities of CNTs and are anticipated to expand the range of their applications, potentially revolutionizing areas like aerospace, energy storage, pharmaceutical, and biomedical fields.

REFERENCES

1. Geim AK, Novoselov KS. The rise of graphene. *Nature Materials*. 2007;6(3):183–91. Available from: https://www.nature.com/articles/nmat1849
2. Iijima S. Helical microtubules of graphitic carbon. *Nature*. 1991;354:6348. Available from: https://www.nature.com/articles/354056a0
3. Habibi Y, Lucia LA, Rojas OJ. Cellulose nanocrystals: chemistry, self-assembly, and applications. *Chemical Reviews*. 2010;110:3479–500. Available from: https://pubs.acs.org/doi/full/10.1021/cr900339w
4. Anzar N, Hasan R, Tyagi M, Yadav N, Narang J. Carbon nanotube – a review on synthesis, properties and plethora of applications in the field of biomedical science. *Sensors International*. 2020;1:100003.
5. Rahamathulla M, Bhosale RR, Osmani RAM, Mahima KC, Johnson AP, Hani U, et al. Carbon nanotubes: current perspectives on diverse applications in targeted drug delivery and therapies. *Materials*. 2021;14:6707.
6. Szabó A, Perri C, Csató A, Giordano G, Vuono D, Nagy JB. Synthesis methods of carbon nanotubes and related materials. *Materials*. 2010;3:3092. Available from: https://pmc/articles/PMC5445923/
7. Lee C, Wei X, Kysar JW, Hone J. Measurement of the elastic properties and intrinsic strength of monolayer graphene. *Science*. 2008;321:385–8. Available from: https://pubmed.ncbi.nlm.nih.gov/18635798/
8. Tehrani M, Khanbolouki P. Carbon nanotubes: synthesis, characterization, and applications. *Advances in Nanomaterials: Fundamentals, Properties and Applications*. 2017;30:3–35. Available from: https://www.researchgate.net/publication/320595252_Carbon_Nanotubes_Synthesis_Characterization_and_Applications
9. De Volder MFL, Tawfick SH, Baughman RH, Hart AJ. Carbon nanotubes: present and future commercial applications. *Science*. 2013;339:535–9. Available from: https://pubmed.ncbi.nlm.nih.gov/23372006/
10. Eatemadi A, Daraee H, Karimkhanloo H, Kouhi M, Zarghami N, Akbarzadeh A, et al. Carbon nanotubes: properties, synthesis, purification, and medical applications. *Nanoscale Research Letters*. 2014;9:1–13. Available from: https://www.researchgate.net/publication/265174796_Carbon_nanotubes_Properties_synthesis_purification_and_medical_applications
11. Yahyazadeh A, Nanda S, Dalai AK. Carbon nanotubes: a review of synthesis methods and applications. *Reactions*. 2024;5:429–51.
12. Gacem A, Modi S, Yadav VK, Islam S, Patel A, Dawane V, et al. Recent advances in methods for synthesis of carbon nanotubes and carbon nanocomposite and their emerging applications: a descriptive review. *Journal of Nanomaterials*. 2022;2022:1–16.
13. Chen J, Wei S, Xie H. A brief introduction of carbon nanotubes: history, synthesis, and properties. *Journal of Physics: Conference Series*. 2021;1948:012184. Available from: https://iopscience.iop.org/article/10.1088/1742-6596/1948/1/012184
14. Shoukat R, Khan MI. Carbon nanotubes: a review on properties, synthesis methods and applications in micro and nanotechnology. *Microsystem Technologies*. 2021;27:4183–92.
15. Singh P, Tripathi RM, Saxena A. Synthesis of carbon nanotubes and their biomedical application. *Journal of Optoelectronics and Biomedical Materials*. 2010;2:91–8.

16. Naief MF, Mohammed SN, Mohammed AM. Carbon nanotubes: a review on synthesis and drug delivery for cancer treatment. *Inorganic Chemistry Communications.* 2024;159:111694.
17. Chernysheva M, Rozhin A, Fedotov Y, Mou C, Arif R, Kobtsev SM, et al. Carbon nanotubes for ultrafast fibre lasers. *Nanophotonics.* 2017;6:1–30.
18. Koziol K, Boskovic BO, Yahya N. Synthesis of carbon nanostructures by CVD method. *Advanced Structured Materials.* 2011;5:23–49. Available from: https://www.researchgate.net/publication/226054918_Synthesis_of_Carbon_Nanostructures_by_CVD_Method
19. Awasthi K, Srivastava A, Srivastava ON. Synthesis of carbon nanotubes. *Journal of Nanoscience and Nanotechnology.* 2005;5:1616–36. Available from: https://www.researchgate.net/publication/7520206_Synthesis_of_Carbon_Nanotubes
20. Singh R, Deshmukh R. Carbon nanotube as an emerging theranostic tool for oncology. *Journal of Drug Delivery Science and Technology.* 2022;74:1–18.
21. Jagadeesan AK, Thangavelu K, Dhananjeyan V, Jagadeesan AK, Thangavelu K, Dhananjeyan V. Carbon nanotubes: synthesis, properties and applications. In Pham P, Goel P, Kumar S, Yadav K (eds) *21st Century Surface Science – A Handbook.* 2020, IntechOpen, Rijeka. Available from: https://www.intechopen.com/chapters/72939
22. Bell MS, Teo KBK, Lacerda RG, Milne WI, Hash DB, Meyyappan M. Carbon nanotubes by plasma-enhanced chemical vapor deposition. *Pure and Applied Chemistry.* 2006;78:1117–25.
23. Yanase T, Miura T, Shiratori T, Weng M, Nagahama T, Shimada T. Synthesis of carbon nanotubes by plasma-enhanced chemical vapor deposition using $Fe_{1-x}Mn_xO$ nanoparticles as catalysts: how does the catalytic activity of graphitization affect the yields and morphology? *C (Basel).* 2019;5:1–9.
24. Zhang C, Tian B, Chong CT, Ding B, Fan L, Chang X, et al. Synthesis of single-walled carbon nanotubes in rich hydrogen/air flames. *Materials Chemistry and Physics.* 2020;254:123479.
25. Teoh WY, Amal R, Mädler L. Flame spray pyrolysis: an enabling technology for nanoparticles design and fabrication. *Nanoscale.* 2010;2:1–25.
26. Gore JP, Sane A, Gore JP, Sane A. Flame synthesis of carbon nanotubes. In Yellampalli S (eds) *Carbon Nanotubes – Synthesis, Characterization, Applications.* 2011, IntechOpen, Rijeka. Available from: https://www.intechopen.com/chapters/16801
27. Thakur CK, Karthikeyan C, Abou-Dahech MS, Altabakha MMAM, Al Shahwan MJS, Ashby CR, et al. Microwave-assisted functionalization of multi-walled carbon nanotubes for biosensor and drug delivery applications. *Pharmaceutics.* 2023;15:1–30.
28. Omoriyekomwan JE, Tahmasebi A, Dou J, Wang R, Yu J. A review on the recent advances in the production of carbon nanotubes and carbon nanofibers via microwave-assisted pyrolysis of biomass. *Fuel Processing Technology.* 2021;214:1–22.
29. Kure N, Hamidon MN, Azhari S, Mamat NS, Yusoff HM, Isa BM, et al. Simple microwave-assisted synthesis of carbon nanotubes using polyethylene as carbon precursor. *Journal of Nanomaterials.* 2017;2017:1–4.
30. Mubarak NM, Sahu JN, Abdullah EC, Jayakumar NS, Ganesan P. Single stage production of carbon nanotubes using microwave technology. *Diamond and Related Materials.* 2014;48:52–9.
31. Liu Y, Guo N, Yin P, Zhang C. Facile growth of carbon nanotubes using microwave ovens: the emerging application of highly efficient domestic plasma reactors. *Nanoscale Advances.* 2019;1:4546–59.
32. Adeola AO, Duarte MP, Naccache R. Microwave-assisted synthesis of carbon-based nanomaterials from biobased resources for water treatment applications: emerging trends and prospects. *Frontiers in Carbon.* 2023;2:1–29.
33. Adamo G, Campora S, Ghersi G. Functionalization of nanoparticles in specific targeting and mechanism release. *Nanostructures for Novel Therapy.* 2017;2017:57–80.

34. Nikolić V, Ilić-Stojanović S, Petrović S, Tačić A, Nikolić L. Administration routes for nano drugs and characterization of nano drug loading. *Characterization and Biology of Nanomaterials for Drug Delivery.* 2019;2019:587–625.
35. Agrahari V, Burnouf P-A, Burnouf T, Agrahari V. Nanoformulation properties, characterization, and behavior in complex biological matrices: challenges and opportunities for brain-targeted drug delivery applications and enhanced translational potential. *Advanced Drug Delivery Reviews.* 2019;148:146–80.
36. Bhattacharjee S. DLS and zeta potential – What they are and what they are not? *Journal of Controlled Release.* 2016;235:337–51.
37. Berthomieu C, Hienerwadel R. Fourier transform infrared (FTIR) spectroscopy. *Photosynthesis Research.* 2009;101:157–70.
38. Epp J. X-ray diffraction (XRD) techniques for materials characterization. *Materials Characterization Using Nondestructive Evaluation (NDE) Methods.* 2016;2016:81–124.
39. Inkson BJ. Scanning electron microscopy (SEM) and transmission electron microscopy (TEM) for materials characterization. *Materials Characterization Using Nondestructive Evaluation (NDE) Methods.* 2016;2016:17–43.
40. Kim Y, Kim W, Park JW. Principles and applications of force spectroscopy using atomic force microscopy. *Bulletin of the Korean Chemical Society.* 2016;37:1895–907.
41. Sakhare MA, Satpute RH, Thakur SV, Gadekar LS. XRD and TGA-DSC as an effective tools of research in physical science. *Journal of Basic and Applied Sciences.* 2014;01:1–4.
42. Circulatory System: The Histology Guide [Internet]. [cited 2024 Jan 4]. Available from: https://www.histology.leeds.ac.uk/circulatory/capillaries.php
43. Yokota J, Kyotani S. Influence of nanoparticle size on the skin penetration, skin retention and anti-inflammatory activity of non-steroidal anti-inflammatory drugs. *Journal of the Chinese Medical Association.* 2018;81:511–9.
44. Komane P, Kumar P, Choonara Y. Functionalised carbon nanotubes: promising drug delivery vehicles for neurovascular disorder intervention. *AAPS PharmSciTech.* 2023;24:201.
45. Kushwaha SKS, Ghoshal S, Rai AK, Singh S. Carbon nanotubes as a novel drug delivery system for anticancer therapy: a review. *Brazilian Journal of Pharmaceutical Sciences.* 2013;49:629–43. Available from: https://www.scielo.br/j/bjps/a/mRqMtpkd5MCxrLn99WZDJ9c/?lang=en
46. Malvern Panalytical | Malvern Panalytical [Internet]. [cited 2024 Jan 4]. Available from: https://www.malvernpanalytical.com/en/products/technology/light-scattering/dynamic-light-scattering
47. Stetefeld J, McKenna SA, Patel TR. Dynamic light scattering: a practical guide and applications in biomedical sciences. *Biophysical Reviews.* 2016;8:409–27.
48. Seyfoori A, Sarfarazijami S, Seyyed Ebrahimi SA. pH-responsive carbon nanotube-based hybrid nanogels as the smart anticancer drug carrier. *Artificial Cells, Nanomedicine, and Biotechnology.* 2019;47:1437–43.
49. Varenne F, Rustique E, Botton J, Coty J-B, Lanusse G, Ait Lahcen M, et al. Towards quality assessed characterization of nanomaterial: transfer of validated protocols for size measurement by dynamic light scattering and evaluation of zeta potential by electrophoretic light scattering. *International Journal of Pharmaceutics.* 2017;528:299–311.
50. Mourya A, Famta P, Shah S, Srinivasarao DA, Sharma A, Vambhurkar G, et al. Raloxifene loaded d-α-tocopherol polyethylene glycol 1000 succinate stabilized poly (ε-caprolactone) nanoparticles augmented drug delivery and apoptosis in breast cancer cells. *Journal of Drug Delivery Science and Technology.* 2024;92:105399.
51. Kolipaka T, Sen S, Mane SS, Bajad GD, Dengale SJ, Ranjan OP. Development of posaconazole nanosuspension for bioavailability enhancement: formulation optimization, in vitro characterization, and pharmacokinetic study. *Journal of Drug Delivery Science and Technology.* 2023;83:104434.

52. Sharma P, Jain K, Jain NK, Mehra NK. Ex vivo and in vivo performance of anti-cancer drug loaded carbon nanotubes. *Journal of Drug Delivery Science and Technology.* 2017;41:134–43.
53. Thotakura N, Sharma S, Khurana RK, Babu PV, Chitkara D, Kumar V, et al. Aspartic acid tagged carbon nanotubols as a tool to deliver docetaxel to breast cancer cells: reduced hemotoxicity with improved cytotoxicity. *Toxicology in Vitro.* 2019;59:126–34.
54. Yazdani S, Mozaffarian M, Pazuki G, Hadidi N, Gallego I, Puras G, et al. Design of double functionalized carbon nanotube for amphotericin B and genetic material delivery. *Scientific Reports.* 2022;12:21114.
55. Zeta Potential – An Introduction | Malvern Panalytical [Internet]. [cited 2024 Jan 25]. Available from: https://www.malvernpanalytical.com/en/products/measurement-type/zeta-potential
56. Zeta Potential Measurements – nanoComposix [Internet]. [cited 2024 Jan 25]. Available from: https://nanocomposix.com/pages/zeta-potential-measurements
57. Harsha PJ, Thotakura N, Kumar M, Sharma S, Mittal A, Khurana RK, et al. A novel PEGylated carbon nanotube conjugated mangiferin: an explorative nanomedicine for brain cancer cells. *Journal of Drug Delivery Science and Technology.* 2019;53:101186.
58. Singhai NJ, Maheshwari R, Jain NK, Ramteke S. Chondroitin sulphate and α-tocopheryl succinate tethered multiwalled carbon nanotubes for dual-action therapy of triple-negative breast cancer. *Journal of Drug Delivery Science and Technology.* 2020;60:102080.
59. Goldstein JI, Newbury DE, Michael JR, Ritchie NWM, Scott JHJ, Joy DC. *Scanning Electron Microscopy and X-Ray Microanalysis.* Springer, New York, 2018.
60. Scanning Electron Microscopy (SEM) with Energy Dispersive X-Ray Analysis (EDX) | Lucideon [Internet]. [cited 2024 Jun 21]. Available from: https://www.lucideon.com/testing-characterisation/techniques/sem-edx
61. SEM-EDX Analysis | Laboratory Services | Measurlabs [Internet]. [cited 2024 Jun 20]. Available from: https://measurlabs.com/methods/scanning-electron-microscopy-x-ray-spectroscopy-sem-edx/
62. Manoukian OS, Arul MR, Rudraiah S, Kalajzic I, Kumbar SG. Aligned microchannel polymer-nanotube composites for peripheral nerve regeneration: small molecule drug delivery. *Journal of Controlled Release.* 2019;296:54–67.
63. Augustine R, Hasan A, Primavera R, Wilson RJ, Thakor AS, Kevadiya BD. Cellular uptake and retention of nanoparticles: insights on particle properties and interaction with cellular components. *Materials Today Communications.* 2020;25:101692.
64. Sharma S, Mehra NK, Jain K, Jain NK. Effect of functionalization on drug delivery potential of carbon nanotubes. *Artificial Cells, Nanomedicine, and Biotechnology.* 2016;44:1851–60.
65. Azqhandi MHA, Farahani BV, Dehghani N. Encapsulation of methotrexate and cyclophosphamide in interpolymer complexes formed between poly acrylic acid and poly ethylene glycol on multi-walled carbon nanotubes as drug delivery systems. *Materials Science and Engineering: C.* 2017;79:841–7.
66. Moniriyan F, Jamal Tabatabaei Rezaei S, Javad Sabounchei S. Pegylated magnetic carbon nanotubes for efficient and safe delivery of poorly water-soluble platinum anticancer drugs. *Results in Chemistry.* 2024;7:1–12.
67. Rice SB, Chan C, Brown SC, Eschbach P, Han L, Ensor DS, et al. Particle size distributions by transmission electron microscopy: an interlaboratory comparison case study. *Metrologia.* 2013;50:663–78.
68. Borthakur P, Boruah PK, Sharma B, Das MR. Nanoemulsion: preparation and its application in food industry. *Emulsions.* 2016;2016:153–91. Available from: https://linkinghub.elsevier.com/retrieve/pii/B9780128043066000052

69. Yao Z, Braidy N, Botton GA, Adronov A. Polymerization from the surface of single-walled carbon nanotubes – preparation and characterization of nanocomposites. *Journal of the American Chemical Society*. 2003;125:16015–24.
70. David ME, Ion R-M, Grigorescu RM, Iancu L, Holban AM, Nicoara AI, et al. Hybrid materials based on multi-walled carbon nanotubes and nanoparticles with antimicrobial properties. *Nanomaterials*. 2021;11:1415.
71. Branca C, Frusteri F, Magazù V, Mangione A. Characterization of carbon nanotubes by TEM and infrared spectroscopy. *Journal of Physical Chemistry B*. 2004;108:3469–73.
72. Basheer A, Hanafiah M, Alsaadi M, Wan Yaacob W, Al-Douri Y. Synthesis, characterization, and analysis of hybrid carbon nanotubes by chemical vapor deposition: application for aluminum removal. *Polymers (Basel)*. 2020;12:1–18.
73. Cheng J, Meziani MJ, Sun Y-P, Cheng SH. Poly(ethylene glycol)-conjugated multi-walled carbon nanotubes as an efficient drug carrier for overcoming multidrug resistance. *Toxicology and Applied Pharmacology*. 2011;250:184–93.
74. Mehra NK, Verma AK, Mishra PR, Jain NK. The cancer targeting potential of d-α-tocopheryl polyethylene glycol 1000 succinate tethered multi walled carbon nanotubes. *Biomaterials*. 2014;35:4573–88.
75. AFM Modes and Theory — Introduction – Nanosurf [Internet]. [cited 2024 Jul 25]. Available from: https://www.nanosurf.com/en/support/history-and-background-of-afm
76. Seo Y, Jhe W. Atomic force microscopy and spectroscopy. Reports on Progress in Physics. 2007 [cited 2024 Jul 25];71:016101. Available from: https://iopscience.iop.org/article/10.1088/0034-4885/71/1/016101
77. Sitterberg J, Özcetin A, Ehrhardt C, Bakowsky U. Utilising atomic force microscopy for the characterisation of nanoscale drug delivery systems. *European Journal of Pharmaceutics and Biopharmaceutics*. 2010;74:2–13.
78. Smith JR, Olusanya TOB, Lamprou DA. Characterization of drug delivery vehicles using atomic force microscopy: current status. *Expert Opinion on Drug Delivery*. 2018;15:1211–21.
79. Singh N, Sachdev A, Gopinath P. Polysaccharide functionalized single walled carbon nanotubes as nanocarriers for delivery of curcumin in lung cancer cells. *Journal of Nanoscience and Nanotechnology*. 2018;18:1534–41.
80. Faghihzadeh F, Anaya NM, Schifman LA, Oyanedel-Craver V. Fourier transform infrared spectroscopy to assess molecular-level changes in microorganisms exposed to nanoparticles. *Nanotechnology for Environmental Engineering*. 2016;1:1–16.
81. Eid MM. *Characterization of Nanoparticles by FTIR and FTIR-Microscopy. Handbook of Consumer Nanoproducts*. Springer Nature, Singapore, 2022, p. 645–73.
82. Zomorodbakhsh S, Abbasian Y, Naghinejad M, Sheikhpour M. The effects study of isoniazid conjugated multi-wall carbon nanotubes nanofluid on *mycobacterium tuberculosis. International Journal of Nanomedicine*. 2020; 15:5901–9.
83. Ahmad Zawawi N, Abdul Majid Z, Aini Abdul Rashid N, Bahru J. Effect of acid oxidation methods on multi-walled carbon nanotubes (MWCNT) for drug delivery application. *International Journal of Advanced Scientific Research and Management*. 2016; 1:14–22. Available from: www.ijasrm.com
84. Magid M, Al-Karam LQ. Non-covalent functionalization of CNTs with chitosan for drug delivery system. *Journal of Physics: Conference Series*. 2021;2114:012038. Available from: https://iopscience.iop.org/article/10.1088/1742-6596/2114/1/012038
85. Kudelski A. Analytical applications of Raman spectroscopy. *Talanta*. 2008;76:1–8.
86. Dresselhaus MS, Dresselhaus G, Saito R, Jorio A. Raman spectroscopy of carbon nanotubes. *Physics Reports*. 2005;409:47–99.
87. Mehra NK, Jain NK. Cancer targeting propensity of folate conjugated surface engineered multi-walled carbon nanotubes. *Colloids and Surfaces B: Biointerfaces*. 2015;132:17–26.

88. Requardt H, Braun A, Steinberg P, Hampel S, Hansen T. Surface defects reduce carbon nanotube toxicity in vitro. *Toxicology in Vitro*. 2019;60:12–8.
89. Das R, Hamid S, Ali Md, Ramakrishna S, Yongzhi W. Carbon nanotubes characterization by X-ray powder diffraction – a review. *Current Nanoscience*. 2014;11:23–35.
90. Rodríguez I, Gautam R, Tinoco AD. Using X-ray diffraction techniques for biomimetic drug development, formulation, and polymorphic characterization. *Biomimetics*. 2021;6:1–23.
91. Li H, Sun X, Li Y, Li B, Liang C, Wang H. Preparation and properties of carbon nanotube (Fe)/hydroxyapatite composite as magnetic targeted drug delivery carrier. *Materials Science and Engineering: C*. 2019;97:222–9.
92. Ahmad AF, Aziz SA, Obaiys SJ, Zaid MHM, Matori KA, Samikannu K, et al. Biodegradable poly (lactic acid)/poly (ethylene glycol) reinforced multi-walled carbon nanotube nanocomposite fabrication, characterization, properties, and applications. *Polymers (Basel)*. 2020;12:1–22.
93. Jain AK, Thareja S. In vitro and in vivo characterization of pharmaceutical nanocarriers used for drug delivery. *Artificial Cells, Nanomedicine, and Biotechnology*. 2019;47:524–39.
94. Menczel JD, Kohl WS. Differential scanning calorimetry (DSC) in fiber research. *Thermal Analysis of Textiles and Fibers*. 2020;2020:17–69.
95. Randive DS, Shejawal KP, Bhinge SD, Bhutkar MA, Jadhav NR, Patil SB, et al. Efficient in vitro oxaliplatin delivery with functionalized single-walled carbon nanotube for enhanced colon cancer treatment. *Future Journal of Pharmaceutical Sciences*. 2023;9:91.
96. Uttekar PS, Lakade SH, Beldar VK, Harde MT. Facile synthesis of multi-walled carbon nanotube via folic acid grafted nanoparticle for precise delivery of doxorubicin. *IET Nanobiotechnology*. 2019;13:688–96.
97. Mansfield E, Banash M. Thermal analysis of nanoparticles: Methods, kinetics, and recent advances. *Modeling, Characterization, and Production of Nanomaterials*. 2023;2023:535–47.
98. Mansfield E, Kar A, Hooker SA. Applications of TGA in quality control of SWCNTs. *Analytical and Bioanalytical Chemistry*. 2010;396:1071–7.
99. Nia AH, Eshghi H, Abnous K, Ramezani M. The intracellular delivery of plasmid DNA using cationic reducible carbon nanotube — Disulfide conjugates of polyethylenimine. *European Journal of Pharmaceutical Sciences*. 2017;100:176–86.
100. Zhang L-P, Tan X-X, Huang Y-P, Liu Z-S. Floating liquid crystalline molecularly imprinted polymer coated carbon nanotubes for levofloxacin delivery. *European Journal of Pharmaceutics and Biopharmaceutics*. 2018;127:150–8.

4 Functionalization Approaches of Carbon Nanotubes

Shiv Kumar Prajapati, Sapna Joshi, Rajan Kumar Kurmi, Sweta Acharya, Pooja Mehra, Meenakshi Bajpai, and Ankit Jain

4.1 INTRODUCTION

Carbon is a multifunctional element exhibiting several allotropes and structures due to its sp, sp^2, or sp^3 hybridization. The hybridization enables the formation of CNTs with dimensions ranging from nanoscopic to millimeter scale. highlighting the diverse scale of structures. CNTs are promising for biomedical applications, especially in drug delivery systems [1]. CNTs are nanoscopic cylindrical formations engineered by wrapping graphene sheets into tubular configurations, possessing diameters spanning 1 and 100 nm. These tubes comprise layers of sp^2 in a hexagonal lattice-like arrangement with an interatomic distance of 1.4 Å [2].

The exclusive physicochemical properties of CNTs, including their excellent mechanical and thermal properties, make them highly suitable for targeted drug delivery. CNT's significant contribution to drug delivery lies in its ability to serve as an efficient drug carrier. The surface area of CNTs is large enough to allow various therapeutic moieties to be attached, facilitating localized drug delivery [3,4].

In addition to that, the encapsulation of various drugs onto their surfaces and cavities protects the drugs from getting degraded and enables drug release in a controlled manner [5]. Together, these properties improve the efficacy of drug delivery systems, highlighting the substantial capability of CNTs to advance biomedical applications. CNTs are categorized based on the number of layers of graphene present (Figure 4.1), with the primary categories being single-walled carbon nanotubes (SWCNTs), double-walled carbon nanotubes (DWCNTs), triple-walled carbon nanotubes (TWCNTs), and multiwalled carbon nanotubes (MWCNTs) [6–10]. SWCNTs consist of a cylindrical tube made from a single rolled graphene sheet, with diameters typically between 0.4 and 3 nm, often around 1 nm [11]. DWCNTs feature two concentric graphene cylinders, combining properties of both SWCNTs and MWCNTs, with the outer wall being modifiable while the inner wall retains its intrinsic characteristics [12,13]. MWCNTs, on the other hand, comprise multiple graphene layers, usually between 2 and 10, with an outer diameter exceeding 10 and 0.34 nm spacing within the internal layers [14].

DOI: 10.1201/9781003386537-4

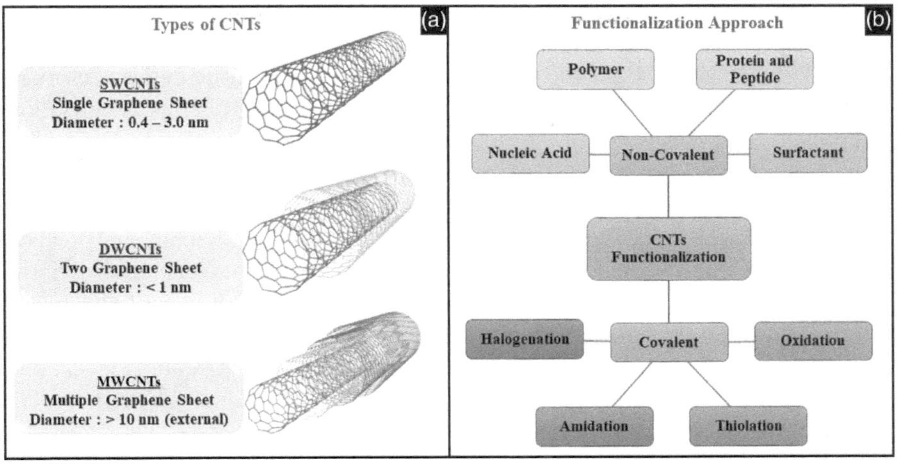

FIGURE 4.1 (a) Types of CNTs and (b) functionalization approaches.

Chemical modifications resulted in their enhanced solubility and biocompatibility and enabled targeted delivery. Because of their electrical conductivity, which results from the unique arrangement of carbon atoms and the delocalization of free electrons, CNTs show remarkable mechanical strength, which includes elevated tensile strength, Young's modulus, as well as elastic moduli, and a low density, helping to maintain structural integrity during transport. CNTs facilitate imaging and drug release in a controlled fashion [15,16]. Proper functionalization ensures good biocompatibility, minimizing adverse reactions. Additionally, CNTs can provide controlled and sustained drug release, penetrate cellular membranes, and serve as multifunctional platforms combining therapy and diagnostics [17]. Their high thermal conductivity can be exploited in hyperthermia treatments for cancer [18,19].

The drawbacks of CNTs for drug delivery include their toxicity, poor biodegradability, challenges in functionalization, potential immune responses, tendency to aggregate, high manufacturing costs, and regulatory hurdles. Surface modifications of CNTs can mitigate these shortcomings. Functionalizing CNTs enhances biocompatibility, reduces toxicity, improves dispersibility and stability in biological environments, facilitates targeted drug delivery, and minimizes immune responses [20]. This chapter describes various functionalization approaches for CNTs that aim to overcome their inherent limitations for drug delivery applications. It covers covalent and noncovalent methods and highlights strategies to improve biocompatibility, alleviate toxicity, improve stability and dispersibility in biological environments, and enable targeted drug delivery.

4.2 FUNCTIONALIZATION/SURFACE MODIFICATION OF CNTs

Although the use of CNTs in drug delivery is promising, it presents numerous challenges. The inherent limited miscibility of CNTs greatly complicates the effective development of drug delivery systems. Strong stacking interactions and van der Waals attractions cause CNTs to aggregate and clump [21]. This tendency to aggregate leads to poor dispersion of CNTs in aqueous media and similar solvents and polymer matrices, mainly due to insufficient interfacial contact [22–24]. Addressing these dispersion challenges is crucial for further developing CNT-based drug delivery carriers. One of the primary strategies for overcoming these limitations is functionalization. Functionalization or surface modifications involve not only chemically modifying the surface of CNTs to disrupt their aggregation and enhance their solubility, but also reduces bundling and enhances cellular internalization. Functionalized CNTs exhibit a greater affinity for pairing with various chemical moieties and ligands, thereby improving their compatibility with biological environments and optimizing their efficacy as drug carriers. These advancements can significantly impact the effectiveness of targeted drug delivery, advancing the pharmaceutical sciences and offering new possibilities for therapeutic interventions. For therapeutic applications, the functionalization of CNTs involves the conjugation of several polymers, nucleic acids (DNA, RNA), proteins, peptides, and various ligands, such as drug molecules, to the native CNT scaffold. Immaculate CNTs exhibit inherent aqueous insolubility, cytotoxicity, and a propensity for aggregation within biological tissues, necessitating functionalization for their increase in solubility and biocompatibility. The primary methodologies for functionalization are bifurcated into covalent and noncovalent approaches. Covalent modifications involve the formation of stable covalent bonds between the CNT surface and functional groups, ensuring robust attachment. Conversely, noncovalent functionalization relies on weaker intermolecular forces, such as π–π stacking and hydrophobic interactions, to anchor functional molecules onto the CNTs. Additionally, surface functionalization with bioactive or biological molecules constitutes a critical strategy for enhancing the interaction of CNTs with biological systems. This modification significantly improves CNT dispersion in physiological environments and augments their specificity for cellular targets, thereby optimizing the therapeutic efficacy of the conjugated agents.

4.2.1 Covalent Functionalization

Covalent functionalization is a crucial approach for surface engineering CNTs in drug delivery. The process involves the formation of chemical bonds with the carbon atoms on the sidewalls or ends through various reactions such as amidation, halogenation, oxidation, thiolation, and the inclusion of carbenes, nitrenes, and radicals [25–27]. Covalent functionalization increases the functionality of CNTs, improves dispersion stability in aqueous and organic solvents, and facilitates their processing and integration into composite materials [28]. This technique can be tailored to enhance biocompatibility and targeting efficiency, making CNTs effective carriers for therapeutic agents [29]. However, the process poses challenges, such as transforming sp^2-hybridized carbon atoms into sp^3-hybridized ones [30], disrupting the π-electron system, and affecting electrical and mechanical properties. Functional group introduction might also lead to structural defects, compromising CNT strength and integrity. Covalent surface modifications of CNTs via oxidation are a critical method for enhancing their properties for drug delivery. Using this method, CNT surfaces are modified into oxygen-containing functional groups, such as hydroxyl and carbonyl groups [31,32]. Oxidation typically employs oxidizing solid agents like nitric acid, sulfuric acid, or a combination. These effectively open the CNT ends and create defect sites along the sidewalls where functional groups can attach. Introducing these functional groups significantly enhances the hydrophilicity of CNTs, resulting in more dispersibility in aqueous environments. The presence of carboxyl groups improves solubility and provides reactive sites for further chemical modifications. It also allows the attachment of several therapeutic agents, targeting ligands, or imaging molecules [33,34]. Modification of the CNTs via thiolation is crucial for improving its effectiveness in drug delivery. This process involves the incorporation of thiol (–SH) groups into the CNT structure [35]. Thiolation can be performed through various chemical reactions involving the attachment of thiol-containing molecules to preoxidized CNTs or through direct chemical reactions. Thiol groups provide reactive sites for further conjugation with numerous biomolecules, allowing targeted delivery of therapeutic agents. Moreover, thiol groups on CNTs facilitate the binding of metal nanoparticles, such as gold, which can be used in photothermal therapy and as contrasting agents in imaging [36,37]. This dual functionality enhances the theranostic capabilities of CNTs, enabling simultaneous treatment and diagnostic imaging. Amidation involves forming amide bonds between carboxylated CNTs and amines. The process begins with CNTs oxidizing to introduce carboxyl groups (–COOH) on their surface [38,39]. These carboxylated CNTs are then reacted with amines to form amide bonds (–CONH–), building a stable and robust link that facilitates the attachment of various bioactive molecules. Amidation significantly upgrades the aqueous solubility and dispersibility of CNTs. The enhanced dispersibility ensures that functionalized CNTs can be easily integrated into biological systems without forming aggregates [40].

Badea et al. developed a nanoconjugate by functionalizing SWCNTs with carboxyl groups and loading them with cisplatin (CDDP) to enhance anticancer efficacy against triple-negative breast cancer cells. Figure 4.2a illustrates that these SWCNT–COOH–CDDP nanoconjugates showed a significant decline in cell sustainability (MDA-MB-231) in a dose- and time-dependent manner compared to free CDDP or SWCNT–COOH alone. Figure 4.2b presents fluorescence microscopy images showing

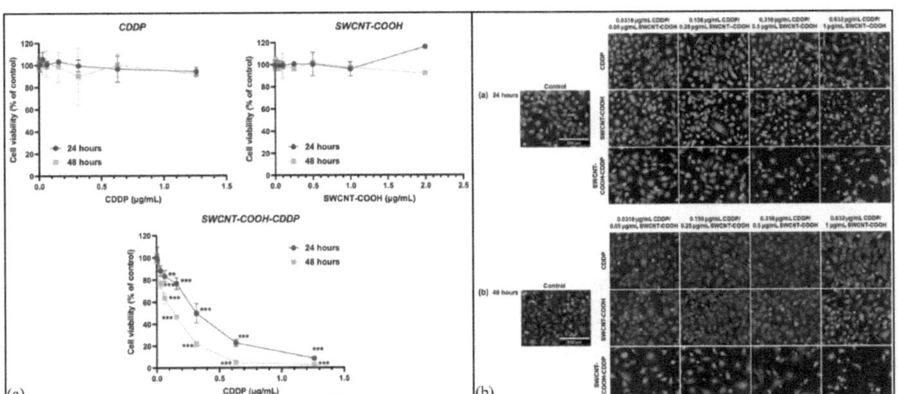

FIGURE 4.2 Cell viability and cellular uptake study of functionalized SWCNT–COOH. (Adapted From Ref. [41], CC-BY 4.0 license).

increased cell death posttreatment, marked by black fluorescence, especially at higher nanoconjugate doses. The study highlighted the potential of COOH–SWCNTs as an efficient delivery system for CDDP, providing improved cytotoxic effects and the possibility to inhibit cancer cell migration and the PI3K/Akt signaling pathway [41].

The stability of the amide bond ensures that the drug remains attached to the CNTs under physiological conditions, preventing premature release. Halogenation involves the incorporation of halogen atoms, such as bromine, chlorine, or fluorine, into the CNT structure, which can occur either on the sidewalls or the ends of the nanotubes [42,43]. The halogenation of CNTs significantly improves their solubility in various solvents, including organic ones, thereby facilitating their dispersibility and processability, which are crucial for their integration into pharmaceutical formulations. The introduction of halogen atoms creates active sites that can further react with other functional groups, enabling the conjugation of a wide range of therapeutic agents. Figure 4.3 describes different covalent functionalization approaches.

4.2.2 Noncovalent Functionalization

Surface modifications via CNT noncovalent interactions of CNTs involve wrapping various molecules around their surfaces, like amphiphilic surfactants, polymers, or other biomolecules. This enhances the miscibility and dispersibility of CNTs without

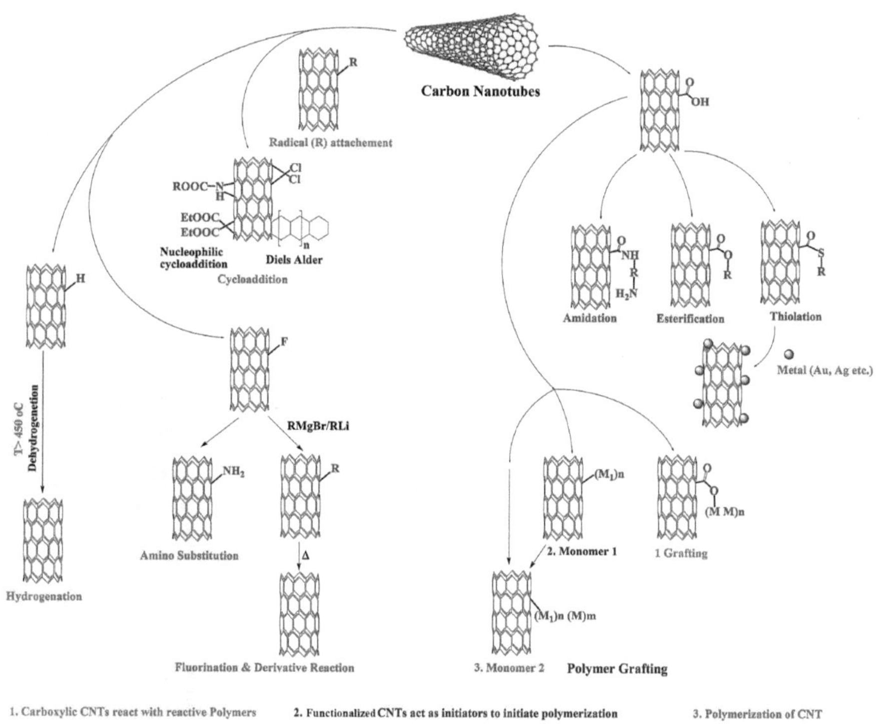

FIGURE 4.3 Strategies for the covalent surface modifications of carbon nanotubes (CNTs) encompass (a) defect-group and (b) sidewall functionalization.

disrupting their inherent sp^2-hybridized sidewall architecture, which is crucial for maintaining their structural integrity. Utilizing weak interactions like van der Waals forces, hydrophobic interactions, π-stacking, ionic bonding, and hydrogen bonding, noncovalent functionalization can effectively modify CNTs [44,45]. These methods fall within supramolecular chemistry, enabling the attachment of aromatic compounds, polymers, biopolymers, and surfactants to CNT surfaces. Noncovalent functionalization offers several advantages in drug delivery applications. It allows CNTs to retain their desirable physicochemical properties while significantly improving their solubility in aqueous environments, which is critical for biomedical applications. By enhancing dispersibility, noncovalent modifications enable better interaction with biological systems, potentially improving cellular uptake and bioavailability of therapeutic payloads carried by CNTs [34,46]. Moreover, the versatility of noncovalent functionalization facilitates the loading of a wide range of molecules onto CNTs, including drugs, nucleic acids, and imaging agents, thereby expanding their utility in targeted or localized drug delivery systems [47,48]. However, the weak bonds of noncovalent functionalization can destabilize CNT complexes in biological fluids.

4.2.2.1 Biomolecule Conjugation

4.2.2.1.1 Nucleic Acid Functionalization

CNTs can noncovalently bind with nucleic acids like DNA, RNA, and siRNA to improve their delivery into cells [49–51]. By penetrating cell membranes, CNTs can release these biomolecules into the nucleus, aiding in treating gene-related disorders [7]. Fujigaya et al. reported a novel SWCNT hybrid functionalized noncovalently with DNA and polyethylene glycol-grafted poly(L-lysine) (PLL-g-PEG), demonstrating significantly enhanced cellular uptake. The SWCNT/DNA/PLL-g-PEG hybrid shows improved biocompatibility, avoiding nonspecific protein binding and aggregation while maintaining stability and dispersion in aqueous and biological media. The hybrid's effectiveness in entering cells is due to the electrostatic interaction between PLL, which is positively charged, and the cell membrane, which is negatively charged [52]. Siu et al. reported on developing noncovalently functionalized SWCNTs with succinate polyethyleneimine (PEI-SA) for efficient topical siRNA delivery into melanoma cells. The study demonstrated that these functionalized SWCNTs (PEI-SA/CNT) significantly enhance siRNA's cellular consumption and gene silencing in both *in vitro* and *in vivo* models. Specifically, the PEI-SA/CNT–siRNA complexes could penetrate the stratum corneum and deliver siRNA into melanoma cells in a murine model, resulting in significant gene silencing of Braf and reduced tumor growth over 25 days [53]. Varkouhi et al. reported on applying functionalized CNTs for siRNA delivery, examining two variants: CNT-PEI (covalently functionalized with polyethyleneimine) and CNT-pyridinium (noncovalently functionalized). CNT forms successfully complexed with siRNA and enhanced cellular uptake, achieving 10%–30% gene silencing rates. However, the study found no significant advantage of these CNT-based systems over traditional transfection methods like PEI and lipofectamine in silencing efficiency and cytotoxicity [54].

4.2.2.1.2 Polymer Functionalization

The polymer functionalization of CNTs can be done via chemical and physical methodologies. Further, the physical approach is subdivided into three primary mechanisms: π–π stacking interactions, electrostatic interactions, and hydrogen bonding. In the π–π stacking interactions, the π-donor benzene rings of the CNTs form stable complexes with polymers that contain π-acceptor aromatic rings, enhancing the CNTs' compatibility and functionality [55]. The surface introduction of carboxylic acid groups is essential for CNTs to undergo polymer functionalization through hydrogen bonding [56]. The surface of CNTs is charged by the electrostatic interaction mechanism, allowing for functionalization with polymers that carry opposite electrical charges [57]. The functionalization of CNT through some chemical modifications can be achieved through grafting methods, such as "grafting to," "grafting from," and "grafting through" [38,58,59]. "Grafting to" involves the reaction of end-functional polymer chains with functional groups on the CNT surface. This method is also acceptable for some coupling reactions, radical addition, and click chemistry, which typically results in low grafting density due to steric hindrance [60]. In "grafting from," the initiator is functionalized on the surface of CNT and initiates polymerization directly from the CNT surface, yielding high grafting density and thickness [61]. "Grafting through" involves functionalizing the CNT surface with double bonds, allowing polymer chains to anchor to the surface during polymerization. This method, though less common, also results in high grafting density and thickness [62]. The "grafting from" and "grafting through" methods are *in situ* techniques of polymerization, leading to denser and thicker polymer brushes compared to the "grafting to" method. The commonly used polymers for the functionalization in drug delivery include polyethylene glycol (PEG), poly(lactic-co-glycolic acid) (PLGA), chitosan, polycaprolactone (PCL), poly(vinyl alcohol) (PVA), polyethyleneimine (PEI), polyglycolide (PGA), poly(methyl methacrylate) (PMMA), polyanhydrides, polydioxanone (PDO), and hyaluronic acid (HA) [9,63,64]. Cheng et al. reported developing PLGA–CNT conjugates for delivering caspase-3 (CP3) into osteosarcoma cells. The study showed that PLGA-modified CNTs are effectively attached to CP3, forming CNT–PLGA–CP3 conjugates. These conjugates efficiently penetrated cancer cells, releasing CP3 through PLGA degradation, which induced apoptosis and suppressed cell proliferation. This highlights the potential of CNTs functionalized with PLGA as a nonviral, controlled-release gene delivery system with reduced cytotoxicity for cancer therapy [65]. Zheng et al. reported using chitosan (CHI)-functionalized SWCNTs–CHI to target SNX-2112, a novel Hsp90 inhibitor, in cancer cells. The study demonstrated that SWCNTs–CHI significantly improved the solubility and biocompatibility of SNX-2112. The drug-loaded SWCNTs–CHI exhibited pH-responsive and time-dependent release, with enhanced reactive oxygen species production and higher apoptosis induction in K562 cancer cells compared to unmodified SWCNTs. The results suggested that SWCNTs–CHI/SNX-2112 could effectively inhibit cancer cell proliferation, making it a promising drug delivery carrier for cancer therapy [66]. Patel et al. investigated using hydroxyl-functionalized MWCNTs combined with polymethyl methacrylate (PMMA) to treat oral squamous carcinoma (KB) cells. MWCNTs improved their PMMA dispersion, enhancing mechanical properties and reducing cytotoxicity. This approach offered significant benefits, such as increased tensile and flexural strength and enhanced biocompatibility, which enabled targeted cancer cell treatment while minimizing harm to healthy cells. Figure 4.4 shows the dose-dependent cytotoxic effects of these composites, demonstrating a notable reduction in cancer cell viability and the induction of apoptosis with increasing concentrations [67].

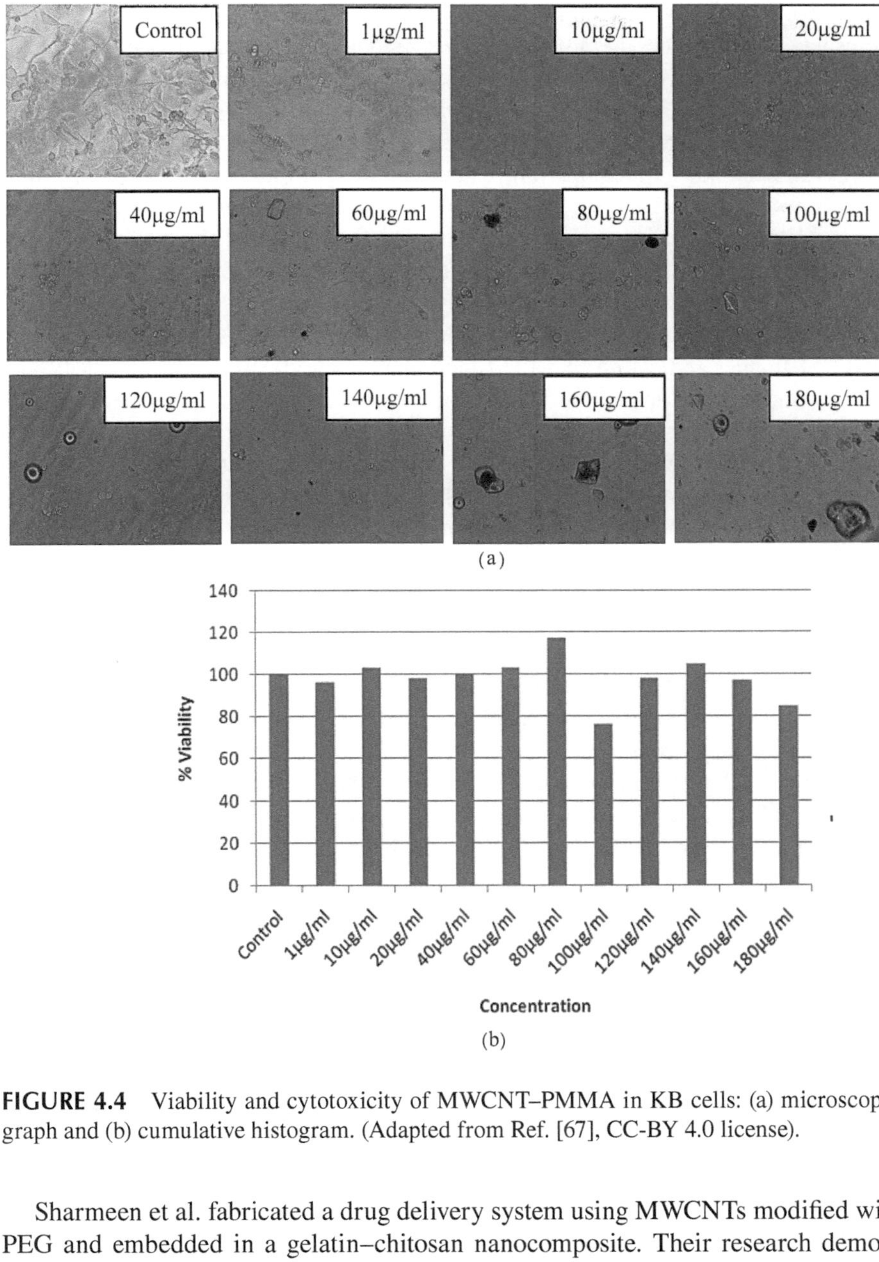

FIGURE 4.4 Viability and cytotoxicity of MWCNT–PMMA in KB cells: (a) microscopic graph and (b) cumulative histogram. (Adapted from Ref. [67], CC-BY 4.0 license).

Sharmeen et al. fabricated a drug delivery system using MWCNTs modified with PEG and embedded in a gelatin–chitosan nanocomposite. Their research demonstrated that this nanocomposite could achieve a stable dispersion of MWCNTs, confirmed through FTIR and TGA analyses. The MWCNTs/gelatin–chitosan matrix exhibited improved mechanical and thermal properties and enhanced water uptake capabilities. When loaded with the antibiotic ciprofloxacin, the nanocomposite showed rapid burst release and sustained drug release, resulting in significantly enhanced antibacterial activity compared to gelatin–chitosan composites without MWCNTs. This indicates effective drug delivery and tissue engineering applications of the above carrier system [68]. Prajapati et al. developed HA conjugated MWCNTs for targeted colon cancer therapy through functionalization steps. Initially, MWCNTs were purified and oxidized to introduce carboxylic functional groups. These carboxylated MWCNTs underwent thionyl chloride acylation, followed by amination with ethylenediamine (EDA). Subsequently, PEG was conjugated to the aminated MWCNTs to enhance biocompatibility and circulation time. Finally, HA was conjugated to both aminated and PEGylated MWCNTs to target CD44 receptors on colon cancer cells, thereby improving drug loading efficiency and therapeutic efficacy against colon cancer [69]. Pennetta et al. developed a drug carrier system by functionalizing CNTs with pyrrole-derived polypropylene glycol (PPGP), forming supramolecular (CNT/PPGPs) and covalent (CNT/PPGPc) adducts. Figure 4.5a revealed HR-TEM images indicating that the functionalization enhanced dispersibility and added a polymer coating to the CNTs while maintaining their structural integrity. Figure 4.5b illustrates molecular simulations of doxorubicin (DOX) loading, which showed efficient drug encapsulation via π–π stacking interactions. This functionalization significantly improved CNT solubility and enabled controlled DOX release at acidic pH, highlighting its potential as a targeted cancer therapy delivery system [70].

4.2.2.1.3 Proteins and Peptide Functionalization

Protein-functionalized CNTs have a very high surface area and cell-penetrating ability, allowing them to have a significant role in the delivery of therapeutics, biosensing, and gene therapy applications [71]. Proteins can be attached to functionalized CNTs through various reactive groups, including hydroxyls, amines, thiols, and carboxylic acids. Studies have explored the covalent bonding of proteins like chymotrypsin, hemoglobin, and streptavidin to CNTs. The surface modification typically involving the covalent bonds employs a cross-linking agent such as 1-ethyl-3-(3-dimethyl aminopropyl) carbodiimide [72]. Peptide-functionalized CNTs can be utilized in various delivery systems, including sensors, biosensors, drug delivery, and tissue engineering [73]. The exceptional loading capacity of functionalized CNTs makes them ideal for delivering peptide antigens. Consequently, synthetic peptides were attached to CNTs and rigorously tested for vaccine efficacy by analyzing their immunological responses [74]. Zhang et al. found that functionalized MWCNTs can specifically bind to and inhibit α-chymotrypsin (ChT) at its catalytic site. Using a diverse f-MWCNT library, they identified nanotubes that competitively inhibit ChT's enzymatic activity without altering its secondary structure. This precise binding and inhibition suggest that f-MWCNTs could be helpful for targeted enzyme inhibition in diagnostic and therapeutic applications [75]. Karousis et al. reported the synthesis of MWCNTs functionalized with peptidomimetics, designed to mimic metabolites of antagonist G of substance P, which were soluble in aqueous environments. These peptidomimetic-functionalized MWCNTs demonstrated enhanced antitrypsin activity. The functionalization involved the attachment of peptidomimetics to the CNTs, resulting in a high loading of bioactive molecules on the nanotube surface. The conjugates showed improved solubility, stability, and bioactivity, effectively inhibiting trypsin and suggesting potential applications in anti-inflammatory and anticancer therapies [76]. Li et al. developed a pH-responsive drug carrier system using CNTs functionalized with polydopamine and gelatin (CNTs/Gel), loaded with cisplatin (Cp).

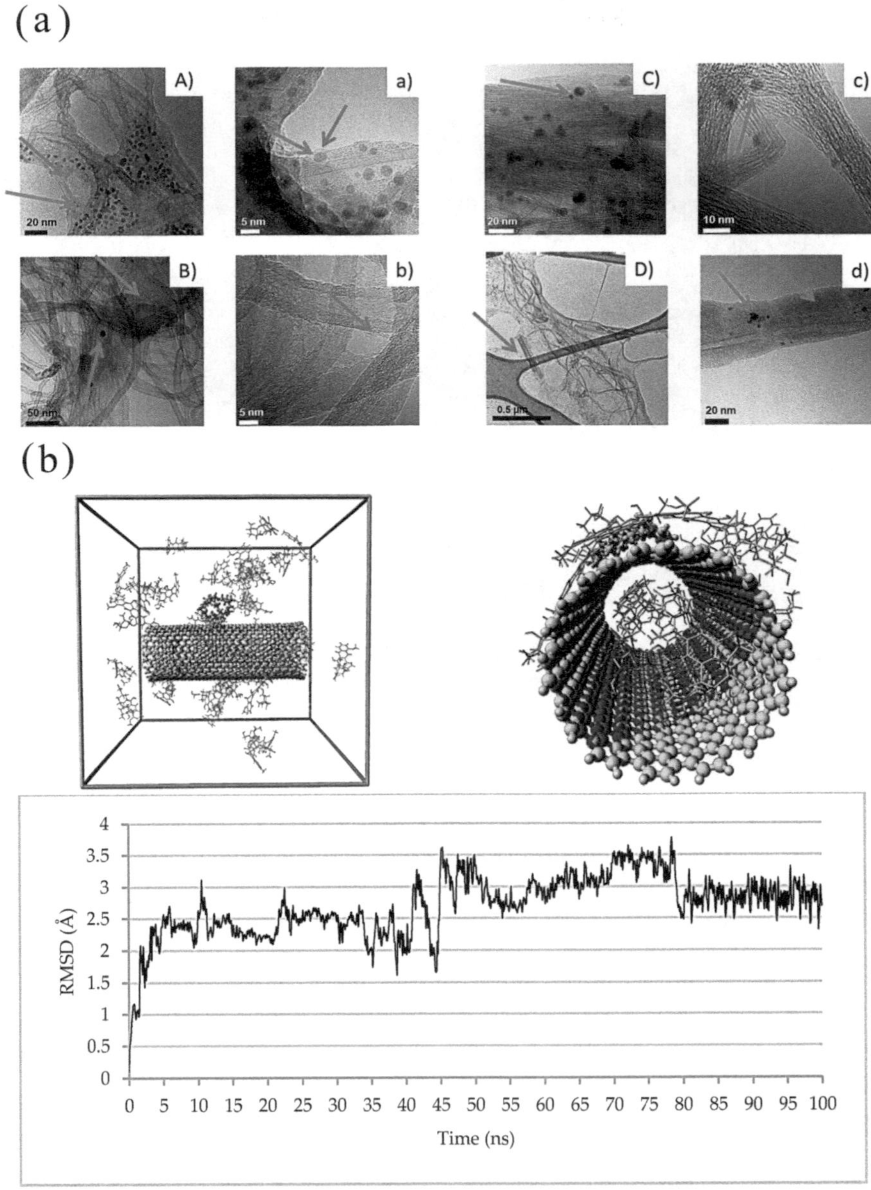

FIGURE 4.5 (a) HR-TEM images of CNT/PPGP adducts and (b) molecular simulations illustrating efficient DOX. (Adapted from Ref. [70], CC-BY 4.0 license).

Figure 4.6 demonstrated the TEM images of CNTs/Gel and CNTs/Gel/Cp, showing improved dispersion and successful Cp loading on CNTs, evidenced by an uneven or rough surface and uniform adsorption of materials. The study found that the CNTs/Gel/Cp system effectively facilitated controlled Cp release under acidic conditions, maintained high drug concentrations in tumors, and exhibited significant antitumor activity comparable to free Cp [77].

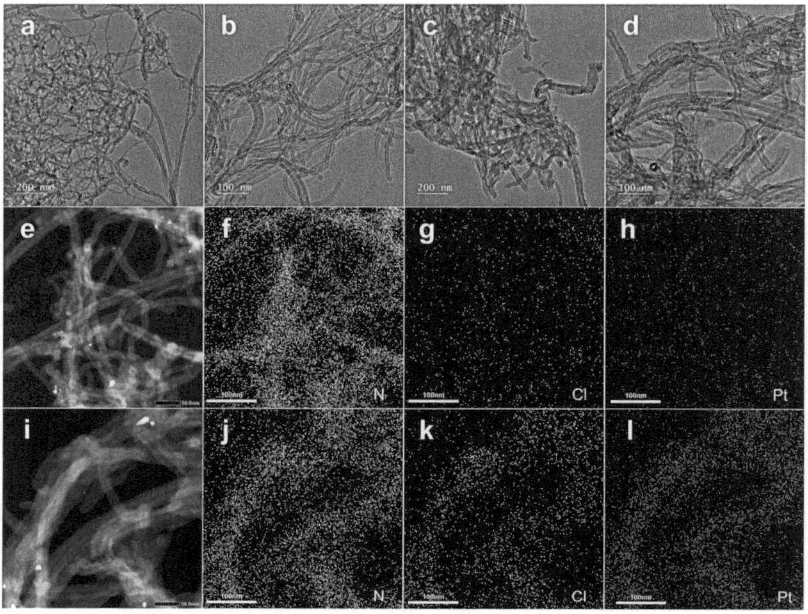

FIGURE 4.6 TEM images of CNTs/Gel and CNTs/Gel/Cp showing enhanced dispersion and cisplatin adsorption. (Adapted from Ref. [77], CC-BY 4.0 license).

4.2.2.1.4 Antibodies Functionalization

The antibody functionalization of CNTs enhances localized drug delivery by attaching antibodies to CNT surfaces. The antibody-functionalized CNTs improve targeted delivery, controlled release, and treatment outcomes in biomedical and drug applications. Antibodies (Abs) are conjugated to CNTs through direct or indirect methods. Direct conjugation involves the adsorption of Abs onto CNTs, but this weak interaction often results in the loss of the ability to target the antibodies. Indirect conjugation requires an additional step, such as utilizing the interaction of streptavidin and biotin to form an antibody–biotin conjugate [78]. Wu et al. developed MWCNTs grafted with PEI in conjugation with fluorescein isothiocyanate (FITC) and prostate stem cell antigen (PSCA) monoclonal antibody (mAb). The resulting CNT–PEI(FITC)–mAb had good biocompatibility. Specifically, it targeted PSCA-overexpressing cancer cells, proven as an influential targeted ultrasound contrast agent in delivering drugs to tumors, suppressing growth in PC-3 tumor-bearing mice. The study highlights CNT–PEI(FITC)–mAb as a promising platform for ultrasound imaging and drug delivery [79]. Gavanji et al. assessed carboxylated MWCNTs for immunochromatographic assays to detect morphine. They characterized carboxylated MWCNTs with FTIR, Raman spectroscopy, Field Emission Scanning Electron Microscopy (FESEM), and Dynamic Light Scattering (DLS). Using 0.5 mg/mL polyvinyl pyrrolidone (PVP) improved the zeta potential from −41.9 to −54.1 mV. Antimorphine antibodies showed high binding efficiency (91.1%–95.5%) to MWCNTs, creating test strips with a visual detection limit of 1.0 and a 10.0 ng/mL cutoff, enhancing assay sensitivity [80].

4.2.2.1.5 Aptamer Functionalization

Aptamers are oligonucleic acid molecules with high affinity and peculiarity for target proteins and are advantageous for CNT functionalization due to their flexibility, reversible denaturation, and the ability to function under nonphysiological conditions compared to antibodies [81,82]. Mohammadi et al. have developed an innovative gene delivery system using SWCNTs functionalized with (PEI), and piperazine conjugated with an EpCAM RNA aptamer. This system targets EpCAM-positive cancer cells associated with metastasis and poor survival rates. The functionalized SWCNTs facilitate the delivery of siRNA against BCL9l, a critical gene in the signaling pathway linked to cancer progression. This method significantly enhances transfection efficiency and selectively induces apoptosis in EpCAM-positive cells [83]. Barbosa et al. elaborated on developing a nanoscale platform achieved through the noncovalent interactions of MWCNTs with a DNA aptamer designed to target tumor cells. Successive washes removed excess aptamer, exposing the nanocomplex surface. The π-stacking interaction ensured a stable aqueous configuration with good cell viability. It also resulted in the intact nucleotide sequence of the aptamer even after functionalization [84]. Blondeau et al. detailed the process of functionalizing SWCNTs with aptamers by covalently attaching adenosine monophosphate (AMP) through carbodiimide and acylation methods. The synthesis of SWCNT–AMP hybrids involves heating in HNO_3, treating with EDC/NHS and various adenosine salts, followed by steps with 1,3-diaminopropane, AMP, and thionyl chloride in water and DMF over specific time intervals. This modification improved the specificity and binding affinity of SWCNTs for target molecules, making them promising for biosensing and therapeutic applications [48].

4.2.2.1.6 Miscellaneous

4.2.2.1.6.1 Ionic Liquid Functionalization Ionic liquids (ILs) are functional materials of large organic cations and small inorganic or organic anions. They have received notable interest because of their distinctive features, such as highly customizable properties, minimal vapor pressure, and thermal solid stability [85,86]. Wei et al. reported that imidazolium ILs modified on MWCNTs' surface to improve their interfacial charge transfer capabilities and prevent aggregation. The IL-functionalized MWCNTs exhibited enhanced dispersion and hydrophilicity, facilitating their application in electrochemical sensors to determine dihydroxybenzene isomers simultaneously. The sensors showed excellent sensitivity and selectivity for hydroquinone, catechol, and resorcinol [87]. The Table 4.1 below summarizes key findings from recent studies on CNTs, highlighting their functionalization and drug delivery efficiency.

4.3 CONCLUSION

The remarkable mechanical and thermal properties of CNTs allow for functionalization, which augments their capability in biomedical applications, especially in drug carrier systems, by improving solubility, biocompatibility, and targeting capabilities. Both covalent and noncovalent approaches enable the attachment of various ligands and drug molecules to the CNT surface, thereby eliminating limitations such as poor

TABLE 4.1
Overview of Surface Modified Carbon Nanotube-Based Nanocarriers for Drug Delivery

Nanocarrier	Drug	Key Outcome	References
pH-responsive gelatin functionalized CNTs	Cisplatin	• Polydopamine and gelatin functionalization improved dispersion and stability. • Drug loading capacity was significantly increased. • Biocompatibility was enhanced.	[77]
Dual-functionalized MWCNTs with ethylenediamine and phenylboronic acid	Paclitaxel	• Noncovalent π–π stacking resulted in high drug loading. • Release was controlled effectively. • Biocompatibility was improved.	[88]
PEGylated magnetic CNTs	Platinum anticancer drug	• Solubility and bioavailability were augmented. • Covalent bonding resulted in more significant drug loading. • The conjugate demonstrated pH-responsive behavior.	[89]
Chitosan or chitosan–stearic acid-coated CNTs functionalized with carboxylic acid or amine	Ferulic acid and diosgenin	• Zero-order release transitioned to a sustained release. • Improved targeting of HepG2 cells induced apoptosis.	[90]
Folic acid-coated poly(ethylene glycol) SWCNTs	Doxorubicin	• Maintained stability under neutral pH. • Enhanced entry into cancer cell lysosomes via clathrin-mediated endocytosis. • Minimal cytotoxic to standard 3T3 cells.	[91]
Diphenylalanine peptide CNTs	–	• The conjugate showed no toxicity against glioblastoma cells. • Following the Weibull model, over 85% of sodium fluorescein-loaded peptide nanotubes exhibited burst release.	[92]

(Continued)

TABLE 4.1 (*Continued*)
Overview of Surface Modified Carbon Nanotube-Based Nanocarriers for Drug Delivery

Nanocarrier	Drug	Key Outcome	References
Carboxylic acid-functionalized SWCNTs	Betulinic acid	• Betulinic acid was conjugated via π–π stacking. • The drug exhibited controlled and prolonged release. • The conjugate was nontoxic in fibroblast cells. • It showed increased anticancer activity in A549 cells.	[93]
Carboxylic acid-functionalized SWCNTs	Droxidopa	• The robust chemisorption process enabled efficient drug loading. • The dispersibility of single-walled CNTs increased. • Bioavailability was improved while reducing systemic toxicity.	[94]
Oxidized SWCNTs and MWCNTs	Biochanin A and formononetin	• Acidic treatment enhanced the water dispersibility of SWCNTs and MWCNTs.	[95]
Biotinylated chitosan MWCNTs	Neratinib	• Noncovalent binding with biotin–chitosan was achieved. • Higher drug release was noted under acidic pH conditions. • Dose-dependent cytotoxic effects were demonstrated against the SkBr3 cell line. • Cellular uptake was improved due to surface functionalization.	[96]
PEGylated functionalization on carboxylated MWCNTs	Doxorubicin	• Folic acid was bonded to MWCNTs as a ligand. • Drug loading was efficient, with targeted, pH-dependent release in cancer cells. • High drug release occurred at acidic pH.	[97]

dispersion, toxicity, and aggregation. Further investigations should focus on optimizing these processes to maintain the intrinsic properties of CNTs while improving their specificity and stability in biological environments. Exploring novel functionalization techniques, combining CNTs with other nanomaterials, and developing advanced drug delivery strategies will be crucial. Furthermore, it will be essential to address regulatory considerations and expand functionalization methods for clinical use to exploit the potential of CNTs for innovative therapeutic interventions fully.

REFERENCES

1. Murjani, B.O., et al., Carbon nanotubes in biomedical applications: current status, promises, and challenges. *Carbon Letters*, 2022. **32**(5): pp. 1207–1226.
2. Agnihotri, T., et al., Chemically modified carbon nanotubes in cancer therapy, in *Chemically Modified Carbon Nanotubes for Commercial Applications*, J. Aslam, C. M. Hussain, and R. Aslam, Editors. 2023, Wiley-VCH GmbH, Berlin and Weinheim, pp. 299–330.
3. Mahor, A., et al., Carbon-based nanomaterials for delivery of biologicals and therapeutics: a cutting-edge technology. *C-Journal of Carbon Research*, 2021. **7**(1): p. 19.
4. Zare, H., et al., Carbon nanotubes: smart drug/gene delivery carriers. *International Journal of Nanomedicine*, 2021. **16**: pp. 1681–1706.
5. Rahamathulla, M., et al., Carbon nanotubes: current perspectives on diverse applications in targeted drug delivery and therapies. *Materials (Basel)*, 2021. **14**(21): p. 6707.
6. Hasnain, M.S. and A.K. Nayak, Classification of carbon nanotubes, in *Carbon Nanotubes for Targeted Drug Delivery*, J. Aslam, C. M. Hussain, and R. Aslam, Editors. 2019, Springer, Singapore, pp. 11–15.
7. Debnath, S.K. and R. Srivastava, Drug delivery with carbon-based nanomaterials as versatile nanocarriers: progress and prospects. *Frontiers in Nanotechnology*, 2021. **3**: p. 644564.
8. Mehra, N.K., V. Mishra, and N.K. Jain, A review of ligand tethered surface engineered carbon nanotubes. *Biomaterials*, 2014. **35**(4): pp. 1267–1283.
9. Prajapati, S.K., et al., Biomedical applications and toxicities of carbon nanotubes. *Drug and Chemical Toxicology*, 2022. **45**(1): pp. 435–450.
10. Mehra, N.K. and S. Palakurthi, Interactions between carbon nanotubes and bioactives: a drug delivery perspective. *Drug Discovery Today*, 2016. **21**(4): pp. 585–597.
11. Karimi, M., et al., Carbon nanotubes: properties and classification, in *Carbon Nanotubes in Drug and Gene Delivery*, J. Aslam, C. M. Hussain, and R. Aslam, Editors. 2017, Morgan & Claypool Publishers, San Rafael, CA, pp. 2-1–2-9.
12. Barrejon, M., et al., TEGylated double-walled carbon nanotubes as platforms to engineer neuronal networks. *ACS Applied Bio Materials*, 2023. **15**(1): pp. 77–90.
13. Moore, K.E., D.D. Tune, and B.S. Flavel, Double-walled carbon nanotube processing. *Advanced Materials*, 2015. **27**(20): pp. 3105–3137.
14. Sakharova, N.A., et al., Mechanical characterization of multiwalled carbon nanotubes: numerical simulation study. *Materials (Basel)*, 2020. **13**(19): p. 4283.
15. Kaur, J., G.S. Gill, and K. Jeet, Applications of carbon nanotubes in drug delivery, in *Characterization and Biology of Nanomaterials for Drug Delivery*, S. S. Mohapatra, et al., Editors. 2019, Elsevier, Amsterdam, The Netherlands, pp. 113–135.
16. Saifuddin, N., A.Z. Raziah, and A.R. Junizah, Carbon nanotubes: a review on structure and their interaction with proteins. *Journal of Chemistry*, 2013. **2013**(1): p. 676815.
17. de Almeida Barcelos, K., et al., Recent advances in the applications of CNT-based nanomaterials in pharmaceutical nanotechnology and biomedical engineering. *Journal of Drug Delivery Science and Technology*, 2023. **87**: p. 104834.

18. Sanginario, A., B. Miccoli, and D. Demarchi, Carbon nanotubes as an effective opportunity for cancer diagnosis and treatment. *Biosensors (Basel)*, 2017. **7**(1): p. 9.
19. Singh, R. and S.V. Torti, Carbon nanotubes in hyperthermia therapy. *Advanced Drug Delivery Reviews*, 2013. **65**(15): pp. 2045–2060.
20. Maghimaa M., et al., Enhancing biocompatibility and functionality: carbon nanotube-polymer nanocomposites for improved biomedical applications. *Journal of Drug Delivery Science and Technology*, 2024. **99**: p. 105958.
21. Krasley, A.T., et al., Carbon nanomaterial fluorescent probes and their biological applications. *Chemical Reviews*, 2024. **124**(6): pp. 3085–3185.
22. Naqvi, S.T.R., et al., Modification strategies for improving the solubility/dispersion of carbon nanotubes. *Journal of Molecular Liquids*, 2020. **297**: p. 111919.
23. Jogi, B.F., et al., Dispersion and performance properties of carbon nanotubes (CNTs) based polymer composites: a review. *Journal of Encapsulation and Adsorption Sciences*, 2012. **02**(04): pp. 69–78.
24. Fatemi, S.M. and M. Foroutan, Recent developments concerning the dispersion of carbon nanotubes in surfactant/polymer systems by MD simulation. *Journal of Nanostructure in Chemistry*, 2015. **6**(1): pp. 29–40.
25. Holzinger, M., et al., Sidewall functionalization of carbon nanotubes. *Angewandte Chemie-International Edition*, 2001. **40**(21): p. 4002.
26. Kharlamova, M.V., M. Paukov, and M.G. Burdanova, Nanotube functionalization: investigation, methods and demonstrated applications. *Materials (Basel)*, 2022. **15**(15): p. 5386.
27. Prajapati, S.K., et al., Surface modification strategies for the carbon nanotubes, in *Chemically Modified Carbon Nanotubes for Commercial Applications*, J. Abraham, S. Thomas, and N. Kalarikkal, Editors. 2023, Wiley-VCH GmbH, Berlin and Weinheim, pp. 51–74.
28. Yadav, R., K. Kumar, and P. Venkatesu, Covalent functionalization of carbon nanotube, in *Handbook of Carbon Nanotubes*, J. Abraham, S. Thomas, and N. Kalarikkal, Editors. 2022, Springer International Publishing, Cham, pp. 393–420.
29. Mahajan, S., et al., Functionalized carbon nanotubes as emerging delivery system for the treatment of cancer. *International Journal of Pharmaceutics*, 2018. **548**(1): pp. 540–558.
30. Jeon, I.-Y., et al., Functionalization of carbon nanotubes, in *Carbon Nanotubes – Polymer Nanocomposites*, Y. Siva, Editor. 2011, IntechOpen, Rijeka, p. 5.
31. Khan, A. and K.A. Alamry, Surface modified carbon nanotubes: an introduction, in *Surface Modified Carbon Nanotubes Volume 1: Fundamentals, Synthesis and Recent Trends*, J. Aslam, C.M. Hussain, and J. Aslam, Editors. 2022, American Chemical Society, Washington, DC, pp. 1–25.
32. Wepasnick, K.A., et al., Surface and structural characterization of multi-walled carbon nanotubes following different oxidative treatments. *Carbon*, 2011. **49**(1): pp. 24–36.
33. Zhang, W., Z. Zhang, and Y. Zhang, The application of carbon nanotubes in target drug delivery systems for cancer therapies. *Nanoscale Research Letters*, 2011. **6**(1): p. 555.
34. Costa, P.M., et al., Functionalised carbon nanotubes: from intracellular uptake and cell-related toxicity to systemic brain delivery. *Journal of Controlled Release*, 2016. **241**: pp. 200–219.
35. Curran, S.A., et al., Thiolation of carbon nanotubes and sidewall functionalization. *Journal of Materials Research*, 2006. **21**(4): pp. 1012–1018.
36. Ferrag, C., M. Noroozifar, and K. Kerman, Thiol functionalized carbon ceramic electrode modified with multi-walled carbon nanotubes and gold nanoparticles for simultaneous determination of purine derivatives. *Materials Science & Engineering C-Materials for Biological Applications*, 2020. **110**: pp. 110568.
37. Yola, M.L. and N. Atar, Novel voltammetric tumor necrosis factor-alpha (TNF-alpha) immunosensor based on gold nanoparticles involved in thiol-functionalized multi-walled carbon nanotubes and bimetallic Ni/Cu-MOFs. *Analytical and Bioanalytical Chemistry*, 2021. **413**(9): pp. 2481–2492.

38. Dehkharghani, R.A., Covalent functionalization of carbon nanostructures, in *Handbook of Functionalized Carbon Nanostructures: From Synthesis Methods to Applications*, A. Barhoum and K. Deshmukh, Editors. 2023, Springer, Cham, pp. 1–43.
39. Salah, L.S., et al., Carbon nanotubes (CNTs) from synthesis to functionalized (CNTs) using conventional and new chemical approaches. *Journal of Nanomaterials*, 2021. **2021**(1): p. 4972770.
40. Kharisov, B.I., O.V. Kharissova, and A.V. Dimas, The dispersion, solubilization and stabilization in "solution" of single-walled carbon nanotubes. *RSC Advances*, 2016. **6**(73): pp. 68760–68787.
41. Badea, M.A., et al., Carboxyl-functionalized carbon nanotubes loaded with cisplatin promote the inhibition of PI3K/Akt pathway and suppress the migration of breast cancer cells. *Pharmaceutics*, 2022. **14**(2): p. 469.
42. Qian, Z.S., et al., Facile synthesis of halogenated multi-walled carbon nanotubes and their unusual photoluminescence. *Journal of Materials Chemistry*, 2012. **22**(41): pp. 22113–22119.
43. Taborowska, P., et al., Doping of carbon nanotubes by halogenated solvents. *Scientific Reports*, 2022. **12**(1): p. 7004.
44. Li, H.F., et al., Functionalization of carbon nanotubes in polystyrene and properties of their composites: a review. *Polymers*, 2024. **16**(6): p. 770.
45. Matiyani, M., et al., Noncovalent functionalization of carbon nanotubes, in *Handbook of Carbon Nanotubes*, J. Abraham, S. Thomas, and N. Kalarikkal, Editors. 2022, Springer International Publishing, Cham, pp. 421–448.
46. Bhattacharya, K., et al., Biological interactions of carbon-based nanomaterials: from coronation to degradation. *Nanomedicine*, 2016. **12**(2): pp. 333–51.
47. Beg, S., et al., Emergence in the functionalized carbon nanotubes as smart nanocarriers for drug delivery applications, in *Fullerens, Graphenes and Nanotubes*, A. M. Grumezescu, Editor. 2018, William Andrew Publishing, Norwich, NY, pp. 105–133.
48. Burdanova, M.G., et al., Applications of pristine and functionalized carbon nanotubes, graphene, and graphene nanoribbons in biomedicine. *Nanomaterials (Basel)*, 2021. **11**(11): p. 3020.
49. Umemura, K., Hybrids of nucleic acids and carbon nanotubes for nanobiotechnology. *Nanomaterials (Basel)*, 2015. **5**(1): pp. 321–350.
50. Alidori, S., et al., Deploying RNA and DNA with functionalized carbon nanotubes. *Journal of Physical Chemistry C, Nanomaterials and Interfaces*, 2013. **117**(11): pp. 5982–5992.
51. Karimi, M., et al., Carbon nanotubes in gene delivery, in *Carbon Nanotubes in Drug and Gene Delivery*. 2017, Morgan & Claypool Publishers, San Rafael, CA, pp. 7-1–7-9.
52. Fujigaya, T., et al., Enhanced cell uptake via non-covalent decollation of a single-walled carbon nanotube-DNA hybrid with polyethylene glycol-grafted poly(l-lysine) labeled with an Alexa-dye and its efficient uptake in a cancer cell. *Nanoscale*, 2011. **3**(10): pp. 4352–8.
53. Siu, K.S., et al., Non-covalently functionalized single-walled carbon nanotube for topical siRNA delivery into melanoma. *Biomaterials*, 2014. **35**(10): pp. 3435–42.
54. Varkouhi, A.K., et al., SiRNA delivery with functionalized carbon nanotubes. *International Journal of Pharmaceutics*, 2011. **416**(2): pp. 419–425.
55. Zhao, D., et al., Construction of a different polymer chain structure to study pi-pi interaction between polymer and reduced graphene oxide. *Polymers (Basel)*, 2018. **10**(7): p. 716.
56. Abel, S.B., et al., Functionalization of conductive polymers through covalent postmodification. *Polymers*, 2023. **15**(1): p. 205.
57. Eskandari, P., et al., Polymer-functionalization of carbon nanotube by in situ conventional and controlled radical polymerizations. *Advances in Colloid and Interface Science*, 2021. **294**: p. 102471.

58. Qin, S.H., et al., Functionalization of single-walled carbon nanotubes with polystyrene via grafting to and grafting from methods. *Macromolecules*, 2004. **37**(3): pp. 752–757.
59. Gao, C., et al., Polymer-grafted carbon nanotubes via "grafting from" approach, in *Carbon Nanotube-Polymer Composites*, D. Tasis, Editor. 2013, The Royal Society of Chemistry, Washington, DC, pp. 120–181.
60. Tsubokawa, N., Preparation and properties of polymer-grafted carbon nanotubes and nanofibers. *Polymer Journal*, 2005. **37**(9): pp. 637–655.
61. Hu, Z., et al., Surface initiated grafting of polymer chains on carbon nanotubes via one-step cycloaddition of diarylcarbene. *Composites Science and Technology*, 2017. **142**: pp. 294–301.
62. Salami-Kalajahi, M., et al., Properties of PMMA/carbon nanotubes nanocomposites prepared by "grafting through" method. *Polymer Composites*, 2012. **33**(2): pp. 215–224.
63. Prajapati, S.K., et al., Biodegradable polymers and constructs: a novel approach in drug delivery. *European Polymer Journal*, 2019. **120**: p. 109191.
64. Kesharwani, P., et al., A glimpse of biomedical application potential of biodegradable polymers for anticancer drug delivery, in *Polymeric Biomaterials for Healthcare Applications*, K. Varaprasad, Editor. 2022, Woodhead Publishing, Cambridge, pp. 211–234.
65. Cheng, Q.S., et al., PLGA-carbon nanotube conjugates for intercellular delivery of caspase-3 into osteosarcoma cells. *PLoS One*, 2013. **8**(12): p. e81947.
66. Zheng, L., et al., Chitosan-functionalised single-walled carbon nanotube-mediated drug delivery of SNX-2112 in cancer cells. *Journal of Biomaterials Applications*, 2016. **31**(3): pp. 379–386.
67. Patel, V., et al., Cytotoxicity analysis for the hydroxyl functionalized MWCNT reinforced PMMA nanocomposites in oral squamous carcinoma (KB) cells. *Polymers (Basel)*, 2023. **15**(5): p. 1192.
68. Sharmeen, S., et al., Polyethylene glycol functionalized carbon nanotubes/gelatin-chitosan nanocomposite: an approach for significant drug release. *Bioactive Materials*, 2018. **3**(3): pp. 236–244.
69. Prajapati, S.K., et al., Hyaluronic acid conjugated multi-walled carbon nanotubes for colon cancer targeting. *International Journal of Biological Macromolecules*, 2019. **123**: pp. 691–703.
70. Pennetta, C., et al., Functionalization of single and multi-walled carbon nanotubes with polypropylene glycol decorated pyrrole for the development of doxorubicin nano-conveyors for cancer drug delivery. *Nanomaterials (Basel)*, 2020. **10**(6): p. 1073.
71. Nagaraju, K., R. Reddy, and N. Reddy, A review on protein functionalized carbon nanotubes. *Journal of Applied Biomaterials & Functional Materials*, 2015. **13**(4): pp. e301–e312.
72. Maheshwari, N., et al., Functionalized carbon nanotubes for protein, peptide, and gene delivery, in *Biomaterials and Bionanotechnology*, R. K. Tekade, Editor. 2019, Academic Press, New York, pp. 613–637.
73. Hassan, H.A., et al., Carbon nanotubes' surface chemistry determines their potency as vaccine nanocarriers in vitro and in vivo. *Journal of Controlled Release*, 2016. **225**: pp. 205–216.
74. Zhao, L., et al., Nanoparticle vaccines. *Vaccine*, 2014. **32**(3): pp. 327–337.
75. Zhang, B., et al., Functionalized carbon nanotubes specifically bind to alpha-chymotrypsin's catalytic site and regulate its enzymatic function. *Nano Letters*, 2009. **9**(6): pp. 2280–2284.
76. Karousis, N., et al., Peptidomimetic-functionalized carbon nanotubes with antitrypsin activity. *Carbon*, 2009. **47**(15): pp. 3550–3558.
77. Li, R., et al., Gelatin-functionalized carbon nanotubes loaded with cisplatin for anti-cancer therapy. *Polymers (Basel)*, 2023. **15**(16): p. 3333.
78. Xiao, Y., et al., Anti-HER2 IgY antibody-functionalized single-walled carbon nanotubes for detection and selective destruction of breast cancer cells. *BMC Cancer*, 2009. **9**(1): p. 351.

79. Wu, H., et al., Prostate stem cell antigen antibody-conjugated multiwalled carbon nanotubes for targeted ultrasound imaging and drug delivery. *Biomaterials*, 2014. **35**(20): pp. 5369–5380.
80. Gavanji, K., et al., Determination of morphine using carboxylated multiwalled carbon nanotubes (MWCNTs) as the label for an immunochromatographic assay (ICA). *Analytical Letters*, 2024. **57**(17): pp. 1–11.
81. Blondeau, P., et al., Covalent functionalization of single-walled carbon nanotubes with adenosine monophosphate: towards the synthesis of SWCNT-Aptamer hybrids. *Materials Science & Engineering C-Materials for Biological Applications*, 2011. **31**(7): pp. 1363–1368.
82. Barani, M., et al., Aptamer-conjugated carbon nanotubes or graphene for targeted cancer therapy and diagnosis, in *Aptamers Engineered Nanocarriers for Cancer Therapy*, P. Kesharwani, Editor. 2023, Woodhead Publishing, Cambridge, pp. 277–294.
83. Mohammadi, M., et al., Single-walled carbon nanotubes functionalized with aptamer and piperazine-polyethylenimine derivative for targeted siRNA delivery into breast cancer cells. *International Journal of Pharmaceutics*, 2015. **485**(1–2): pp. 50–60.
84. Barbosa, M.B., et al., A carefully designed nanoplatform based on multi walled carbon nanotube wrapped with aptamers. *Colloids and Surfaces B: Biointerfaces*, 2019. **175**: pp. 175–183.
85. Khan, O., et al., Ionic liquid-based formulation approaches for enhanced transmucosal drug delivery. *Drug Discovery Today*, 2024. **29**(9): p. 104109.
86. Jain, A., et al., An Insight into pharmaceutical challenges with ionic liquids: where do we stand in transdermal delivery? *Frontiers in Bioengineering and Biotechnology*, 2024. **12**: p. 1454247.
87. Wei, H., et al., Imidazolium ionic liquid functionalized carbon nanotubes for improved interfacial charge transfer and simultaneous determination of dihydroxybenzene isomers. *Molecules*, 2016. **21**(5): p. 617.
88. Rathod, V., et al., Paclitaxel encapsulation into dual-functionalized multi-walled carbon nanotubes. *AAPS PharmSciTech*, 2019. **20**(2): p. 51.
89. Moniriyan, F., S. Jamal Tabatabaei Rezaei, and S. Javad Sabounchei, Pegylated magnetic carbon nanotubes for efficient and safe delivery of poorly water-soluble platinum anticancer drugs. *Results in Chemistry*, 2024. **7**: p. 101275.
90. AbouAitah, K., et al., Functionalized carbon nanotubes for delivery of ferulic acid and diosgenin anticancer natural agents. *ACS Applied Bio Materials*, 2024. **7**(2): pp. 791–811.
91. Niu, L., L. Meng, and Q. Lu, Folate-conjugated PEG on single walled carbon nanotubes for targeting delivery of doxorubicin to cancer cells. *Macromol Biosciences*, 2013. **13**(6): pp. 735–744.
92. Porter, S.L., et al., Pharmaceutical formulation and characterization of dipeptide nanotubes for drug delivery applications. *Macromol Biosciences*, 2020. **20**(7): p. e2000115.
93. Tan, J.M., et al., Sustained release and cytotoxicity evaluation of carbon nanotube-mediated drug delivery system for betulinic acid. *Journal of Nanomaterials*, 2014. **2014**(1): p. 862148.
94. Yoosefian, M. and M. Jahani, A molecular study on drug delivery system based on carbon nanotube for the novel norepinephrine prodrug, Droxidopa. *Journal of Molecular Liquids*, 2019. **284**: pp. 258–264.
95. Novais, G.B., et al., Isoflavones-functionalized single-walled and multi-walled carbon nanotubes: synthesis and characterization of new nanoarchitetonics for biomedical uses. *Journal of Molecular Structure*, 2023. **1294**: p. 136351.
96. Abu Lila, A.S., et al., Dual targeting multiwalled carbon nanotubes for improved neratinib delivery in breast cancer. *RSC Advances*, 2023. **13**(35): pp. 24309–24318.
97. Zhou, Y.D., et al., Hyper-branched multifunctional carbon nanotubes carrier for targeted liver cancer therapy. *Arabian Journal of Chemistry*, 2022. **15**(3): p. 103649.

5 Hybrid Carbon Nanotubes

A New Approach in Biopharmaceuticals

Nalla Usha kumari, Sri Pada Datta Chigurupati, Naveen Rajana, Jitender Madan, and Neelesh Kumar Mehra

ABBREVIATIONS

18FDG	^{18}F-fluorodeoxyglucose
CNT	carbon nanotubes
DOPC	1,2-dioleoyl-sn-glycero-3-phosphocholine
DOTAP	2-dioleoyl-3-trimethylammonium propane
DSPE-mPEG	1,2-distearoyl-sn-glycero-3-phosphoethanolamine N-[amino (polyethylene glycol)-5000]
ECM	extracellular matrix
EDC	1-Ethyl-3-(3-dimethylaminopropyl)carbodiimide
GLUT	glucose transporters
IC50	inhibitory concentration 50
Lys	lysine
MDR	multidrug resistance
MWCNTs	multi-walled CNTs
NIR irradiation	near-infrared irradiation
NHS	N-Hydroxysuccinimide
PEG	polyethylene glycol
PLA-co-mPEG	poly(lactide)eco-methoxy polyethylene glycol
PMSA	prostate-specific membrane antigen
PNIPAAm	poly(N-isopropyl acrylamide)
siRNA	small interfering RNA
SWCNTs	single-walled CNTs
TEM	transmission electron microscopy

DOI: 10.1201/9781003386537-5

5.1 INTRODUCTION

Nanotechnology-aided drug delivery systems are gaining momentum as a new way of delivering chemotherapeutics to the target site. It is a fascinating emerging science that aids in the improved efficacy of therapeutics while reducing the off-target effects. The various limitations of conventional chemotherapy, like toxicity, inadequate absorption, rapid clearance, etc, can be overcome by using nanocarriers. Nanocarriers are self-assembled colloidal dispersions in the nano-size range with properties like small size, high surface-to-volume ratio, tunable drug kinetics, etc. These nanocarriers have garnered superior interest in therapeutics and imaging applications [1]. They improve the stability, pharmacokinetics, prolonged circulation time, and codelivery of drugs. Surface modification of nanocarriers with ligands and antibodies improves the internalization of the nanocarriers into the cells. Nanocarriers are classified into various types, such as vesicular, non-vesicular, carbon based, particulate, non-particulate, emulsion, and miscellaneous systems. Vesicular systems include liposomes, niosomes, transferosomes, etc., whereas the particulate system includes polymeric nanoparticles, solid lipid nanoparticles, carbon nanotubes, gold nanoparticles, etc [2,3].

Carbon-based nanomaterials exhibited significant potential in various areas due to their unique physicochemical properties and structural dimensions. These are structurally categorized into graphenes, fullerenes, nanodiamonds, mesoporous carbon, carbon quantum dots, carbon nanohorns, and carbon nanotubes. Nanotubes are an exclusive class of nanomaterials made of carbon, which are condensed benzene rings rolled up into tubular structures. Carbon nanotubes (CNTs) are distinctive nanodrug delivery systems fabricated exclusively of carbon. Since its discovery in 1991, CNTs have been used in various industries, from aerospace to energy, due to their high aspect ratio with intriguing electrochemical, optical, mechanical, and surface properties. CNTs are applied in energy harvesting, targeted delivery, chemical sensing, catalysis, and therapeutics [4]. They have an excellent mechanical and chemical stabilities owing to the strong network of C–C covalent bonds. It is being used as a nanocarrier for different diseases, including cancer, due to its cell membrane crossing ability and tumor accumulation properties, which, in turn, enhances the stability and therapeutic efficacy of the nanocarrier [5]. We discussed the structure, and functionalization of CNTs along with their types and limitations.

5.2 STRUCTURE AND FUNCTIONALIZATION OF CNTs

CNTs comprise sheets of graphene, a monolayer of carbon atoms integrated into a two-dimensional honeycomb lattice. It is made of sp^2 hybridized carbon atoms arranged hexagonally into a cylinder (tube-like hollow fullerenes). CNTs are considered to be hybrids of fullerene and carbon fibers. These well-ordered hollow graphitic materials are of various lengths with high aspect ratios, allowing them to be tailored for multiple applications [6]. CNTs are classified into single-walled CNTs (SWCNTs) and multi-walled CNTs (MWCNTs). The SWCNTs have diameters between 0.4 and 2 nm, whereas the MWCNTs have diameters between 1 and 3 nm. CNTs can be made using laser ablation, chemical vapor deposition, and arc discharge methods [7]. The chemical deposition method utilizes methane, ethane, and other hydrocarbon sources. The arc discharge method utilizes high temperatures (>3,000°C) required for converting carbon atoms into plasma to form SWCNTs and MWCNTs. The MWCNTs can be prepared without the catalyst and obtained with considerable purity. In contrast, the SWCNTs use cobalt, iron, nickel, and yttrium as catalytic agents for synthesis and production, and they are bulky and have low purity. The laser ablation method includes the graphite vaporization in an electrical furnace operated at 1,200°C, and the CNT purity depends on the graphite quality. Macroscopic processing is utilized for biomedical applications to obtain CNTs with specific characteristics like lengths, alignment, etc [8]. *In situ* emulsion polymerization techniques were also exercised to synthesize CNT composites in a colloidal system with polystyrene to form a nanostructured brush [9].

CNTs show exceptional physical and chemical properties like fast electron transfer kinetics, unique electronic structures, ultralight weight, high tensile strength, and high thermal and chemical stability. These outstanding properties made CNTs the most exploited carbon-based nanomaterials. CNTs have an intrinsic limitation of solubility and dispersibility in aqueous and organic solvents, which, in turn, limits the conjunction of CNTs with the bioactive. Functionalization of CNTs improves their properties by raising their solubility in biological fluids, endowing them to deliver biomolecules, chemotherapeutics, etc., along with improving bioavailability. They can readily cross biological barriers through tiny nanoneedle mechanisms and vehiculate hydrophilic and lipophilic drugs. A recent investigation disclosed that CNTs have higher drug entrapment capacity than dendrimers, nanoparticles, and liposomes, showing their high potential in drug delivery systems [10]. Apart from this, they're being explored in diagnostics as a component of biosensors and as a scaffold for neuronal regeneration and tissue culture.

5.3 TYPES OF FUNCTIONALIZATION OF CNTs

Functionalization of CNTs can be classified into covalent, noncovalent, and endohedral filling inside the CNTs. Direct modifications of the CNTs attenuate the optical and physical properties despite exhibiting low yields. To tackle these shortcomings, the CNTs were coated with polymer, lipid, targeting ligands, antibodies, nucleic acid, aptamers, biomaterials, etc. Hybrid carbon nanotubes are a promising drug delivery system that reduces the toxicity and improves the biocompatibility. The carbon scaffold forms a complex interface with polymers, lipids, inorganic and biological materials, improving the electrical, thermal, vehiculating, and targeting properties. The broad classification of CNTs and their applications are illustrated in Figure 5.1. Covalent and noncovalent binding has been utilized to bind peptides, polymers, enzymes, bacterial cells, proteins, and nucleic acids to fullerenes for imaging, therapy, labeling, targeted delivery, etc. [11]. Covalent functionalization involves stable chemical linkage formation onto the carbon with functional groups, which can stabilize over a wide pH range. Due to this alteration, the *p*-conjugation system on the GR layer is lost, changing the carbon hybridization from sp^2 to sp^3 [12]. Depending on the properties, the reactions to graft molecules are categorized into *grafting from* reactions, where the monomers are polymerized from surface-derived initiators on

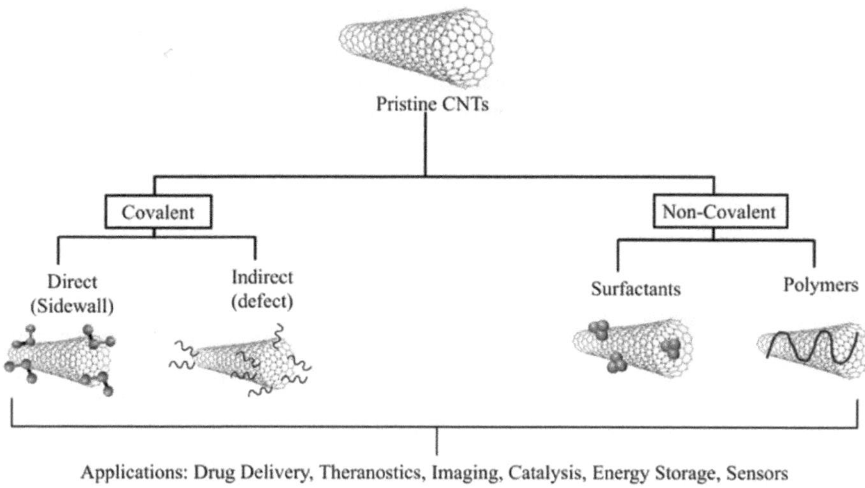

FIGURE 5.1 Classification and applications of functionalized CNTs.

CNTs, and *grafting to* reactions, where preformed polymer chains are added to the CNTs [13]. The molecules are attached to the CNTs through the reaction on the surface of oxidized, functionalized, or pristine CNTs, where oxidized CNTs are mostly used for the fabrication of hybrid CNTs. Covalent modifications affect the intrinsic properties like Raman scattering and photoluminescence, which have impact on their application in imaging and photothermal ablation purposes. Noncovalent binding of amphiphilic molecules or polymers on CNTs is another approach for functionalization. The physical properties of CNTs are not altered; hence, they are useful for applications like imaging, photothermal ablation, etc. [14].

5.4 NEED FOR HYBRID CARBON NANOTUBES

Pristine CNTs have few safety concerns as they can potentially release toxic ions into the biological milieu. Other limitations are high hydrophobicity, formulation issues, and increased viscosity at low concentrations [4]. The naked CNTs are toxic and need surface-tailored functionalization to reduce their toxicity and immunogenicity. Surface functionalization improves biocompatibility for conjugation with biomolecules or favorable candidates for biomedical applications. These hybrid systems reduce the CNT cytotoxicity without affecting other optical, electronic, and mechanical properties. Grafting amends CNTs for targeting functions, theranostic, and multifunctional drug delivery systems, which imparts stimuli responsiveness and site-specific delivery [15]. The CNT hybrids show exclusive properties like stability, reusability, high activity, lower mass transfer resistance, etc., which endows solubility and immobilization of enzymes, showing its applications in diagnosis, biocatalysis, biosensors, biochips, etc [16]. The advantages of the hybrid CNTs are illustrated in Figure 5.2.

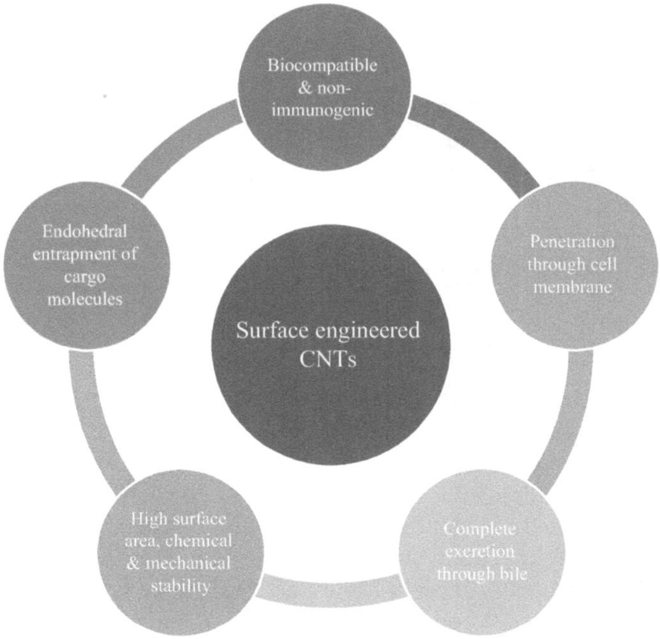

FIGURE 5.2 Advantages of surface-engineered CNTs.

5.4.1 Advantages

- It exhibits high loading efficiency, improved cellular internalization compared to pristine CNTs.
- It provides codelivery of bioactives with controlled release kinetics.
- Coating of CNTs reduces the cytotoxicity by rendering them biocompatible and reduces the ROS generation by the CNTs.
- Improved biocompatibility and biodegradable nature.
- Improved pharmacokinetics parameters.
- Minimization of the toxicities.
- Ease in internalization.
- Improvement in the solubility.
- Improvement in the stability.
- More precise and selective targeting.
- Chances of increased endohedral filling and loading capacity of the therapeutics.
- Dual benefits of hybrid CNTs in biomedical application.

5.5 Types of Hybrid CNTs

Investigations have reported the superior advantages of hybrid systems while masking the limitations of the individual systems [17,18]. For example, polymer-lipid hybrid nanocarriers exhibit the advantages of both polymeric and lipid nanoparticles while overpowering the shortcomings of polymers and lipids. This new system emerges as the promising nanosystem for the treatment of various ailments, including cancer [19], rheumatoid arthritis [20], and respiratory disorders [21]. As discussed earlier, the limitations of the CNTs, like toxicity, aggregation, etc., can be overcome by the use of hybrid systems. Here, we devised the CNT-hybrid carriers into six main classes: lipid-CNT hybrids, polymer-CNT hybrids, protein-CNT hybrids, inorganic-CNTs hybrids, carbohydrate-CNT hybrids, and cell membrane-CNT hybrids. The various types of these hybrid CNTs are discussed below.

5.6 LIPID-COATED CNTs

Fahrenholtz fabricated 1,2-distearoyl-sn-glycero-3-phosphoethanolamine N-[amino (polyethylene glycol)-5000] (DSPE-mPEG)-coated MWCNTs to deliver platinum chemotherapeutic agent for breast cancer therapy. The DSPE-mPEG was added to enhance the biocompatibility and dispersibility of the MWCNTs, and the fabricated hybrid system exhibited colloidal stability in physiological solution and delivered high amounts of the drug. This hybrid system induced S-phase arrest with non-apoptotic cell death in the breast cancer cell lines, whereas the free platinum agent exhibited progression to the G2/M phase with induction of apoptosis. The photothermal activation of hybrid MWCNTs synergized the cytotoxicity by generating mild hyperthermia [22]. Due to the limitation of drug conjugation with the CNTs, a novel strategy was used by Shao et al. for the targeted delivery of paclitaxel (PTX) to the human breast cancer xenograft animal model. They conjugated docosanol to the PTX and folic acid (FA) to the phospholipid for targeted delivery. Targeted CNTs with lipid–drug conjugate exhibited 74% arrest in the G2/M cell cycle phase in cell cycle analysis similar to that of Taxol. *In vivo* studies showed significant inhibition of tumors with no detectable changes in the hematological and histological parameters. This approach indicates that the CNT-coating polymers can be used for multifunctional targeted drug delivery systems [23].

Zhao and colleagues delivered anti-survivin siRNA using CNTs coated with peptide lipid and sucrose laurate for tumor therapy. The schematic representation of the process is shown in Figure 5.3. The peptide lipid and sucrose laurate phase transition facilitated siRNA release from the CNTs. The temperature sensitivity of the lipids exhibited synergy with the photothermal and gene therapy. NIR irradiation increases the temperature at the tumor vicinity, which disorders the lipids and triggers siRNA release. *In vivo* studies on nude mice with HeLa tumor xenografts exhibited more significant antitumor activity, resulting in complete tumor inhibition. Caspase-8, caspase-9, and cleaved caspase-3 levels were increased after combining photothermal and gene therapy. *In vivo* toxicity studies showed no change in the liver and renal enzymes and no damage to the organs, showing temperature-sensitive CNTs as promising nanocarriers for stimuli-responsive gene therapy [24].

Liposomes are colloidal dispersions of phospholipid bilayers with biodegradation and biocompatibility but short circulation time. The PEGylation approaches improved the circulation time, but steric interference is limiting the internalization of liposomes. CNTs are being used to overcome this hurdle due to their high

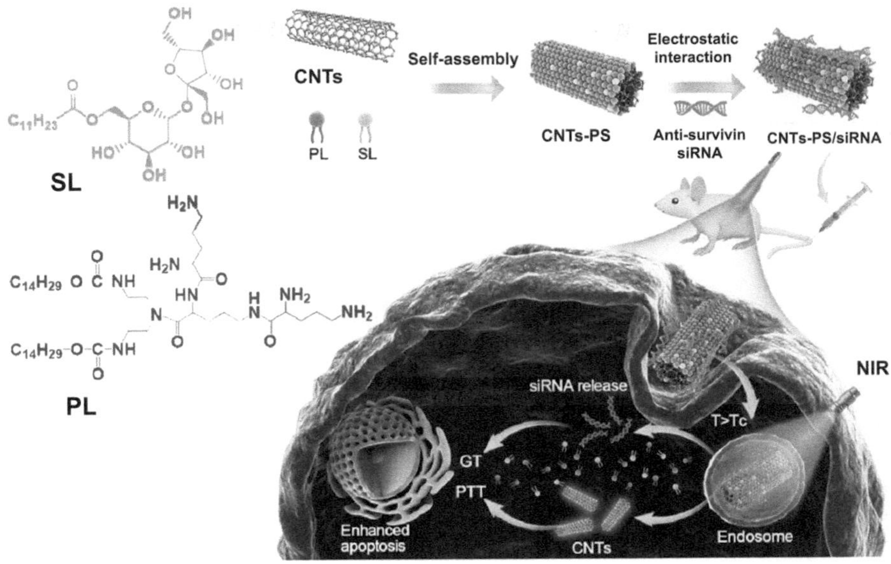

FIGURE 5.3 Schematic representation of temperature-sensitive CNT-sucrose-laurate and peptide-lipid/siRNA nanoparticles for synergistic photothermal therapy and gene therapy for cancer cells. Reproduced with copyright permission from Ref [24]. Copyright 2021, American Chemical Society.

internalization efficiency. Karchemski and coworkers covalently attached on to the liposomes to the partially carboxylated MWCNTs using EDC chemistry to deliver enormous amounts of drug into the cells. The cryo-TEM images revealed the binding of 2–6 liposomes per CNTs, indicating 50,000 drug molecules compared to the direct covalent attachment of the drug to the CNTs (80 molecules per CNTs). This shows that the low dose of CNTs could deliver high amounts of drug when conjugated with liposomes, which can reduce CNT-mediated toxicity. The authors suggested that combination therapy by loading different drugs and active targeting with CNTs/liposomes can significantly increase the drug delivery to target cells [25].

Zhu et al. reported that the hybrid system of thermosensitive liposomes loaded with doxorubicin (DOX) and lysine-modified CNTs reduces the side effects and enhances the antitumor effect of DOX. The concept of the hybrid system is shown in Figure 5.4. The thermosensitive liposomes were prepared by the reverse phase evaporation, which resulted in a particle size of 232 ± 5.6 nm and $86.5\% \pm 3.7\%$ drug entrapment with quicker drug release at 42°C compared to 37°C. The fabricated system exhibited excellent tumor growth suppression in human hepatic carcinoma cells and S180-bearing mice, which may be attributed to the combined chemo-photodynamic therapy. The results confirmed that the controlled drug release, passive tumor targeting, and synergistic effect of chemo-photodynamic therapy of fabricated hybrid systems can be a promising strategy for tumor treatment [26].

In a recent study, the incorporation of MWCNTs into liposomes was successfully performed and analyzed using thermal and spectroscopic techniques. The

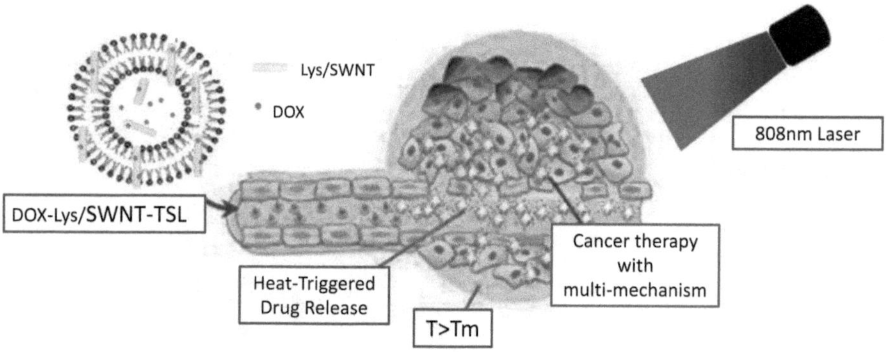

FIGURE 5.4 Concept of thermosensitive liposomes loaded with DOX and Lys-modified SWCNTs (DOX, doxorubicin; Lys, lysine; SWCNT, single-walled carbon nanotubes). (Reproduced with copyright permission from

zeta potential values are near zero, which confirms the successful incorporation of MWCNTs into liposomal bilayers. The *in vitro* cell line studies exhibited low cytotoxicity at high concentrations (500 μM), indicating its safety and suitability as nanocarriers for vehiculating bioactives [27]. Miyako and colleagues have effectively created self-assembling CNT–liposomes supramolecular nanotrains, which serve as sophisticated molecular transport mechanisms. By employing the temperature-responsive liposomes and photothermal CNTs, the controlled release was accomplished by laser light, which destroyed the liposomes and allowed the release of its payload [28].

Madani and coworkers fabricated near-infrared responsive 3D alginate hydrogels loaded with DNA-wrapped SWCNT–liposome complexes for triggered drug release. Cationic liposomes made of 1,2-dioleoyl-sn-glycero-3-phosphocholine (DOPC) and 2-dioleoyl-3-trimethylammonium propane (DOTAP) self-assemble with the negatively charged DNA-wrapped SWCNT giving a net positive charge to the complex, which attributes to the prolonged encapsulation of complex in the negatively charged alginate hydrogel. Cryo-TEM imaging visualized that the liposomal integrity was preserved even after the incorporation of SWCNTs. NIR light was used to trigger the cargo release from the hydrogel scaffold, demonstrating that modulation of stimulation time alters the cargo release from the matrix. The *in vitro* studies exhibited minimal cytotoxicity, suggesting its potential as a carrier for the delivery of various chemotherapeutics [29]. Cationic liposomes–CNT hybrids were used to deliver siRNA and DOX to A-549 cells. Pristine MWCNTs were functionalized using ammonia, followed by coating with liposomes by hydrating the thin film of DOTAP and cholesterol. The fabricated complexes loaded with siRNA and DOX were studied for stability, siRNA binding ability, cellular internalization, and *in vitro* gene silencing efficiency. The hybrid complexes exhibited enhanced gene silencing compared to the cationic liposomes, which may be attributed to the enhanced cellular internalization. These findings show the CNT–liposome as a promising carrier for the codelivery of siRNA with chemotherapeutics [30].

The self-assembly of colloidal particles at the interface of emulsion droplets forms solid microcapsules known as colloidosomes. These hollow shell microstructures control the permeability of the entrapped components. In 2005, the new term "carbon nanotubosomes" was introduced by Marc and Paunov. They utilized the self-assembly of amine-functionalized MWCNTs into shells around the water-in-oil emulsion, followed by an emulsion inversion to produce the novel carbon nanotubosomes, as illustrated in Figure 5.5 [31].

5.7 CARBOHYDRATE-COATED CNTs

Tumor cells show high glucose uptake due to alteration in normal glucose metabolism, further increasing the expression of the glucose transporters. Carbohydrate-coated CNTs have proven their efficiency in specific molecular binding, bacterial interaction, biosensing, and low *in vitro* and *in vivo* toxicity. ^{18}F-fluorodeoxyglucose (^{18}FDG) is used clinically to identify high glucose uptake tumors using positron emission tomography (PET) [32]. Fahrenholtz et al. fabricated sugar-coated CNTs for targeting breast cancer. They used glucosamine, a derivative of ^{18}FDG, for functionalizing CNTs to improve the circulation time, reduce clearance, more significant accumulation at the tumor site, and more excellent signal-to-noise (S/N) ratio in detecting the tumor [33].

Chitosan is a biocompatible polysaccharide made from deacetylated chitin and has wide applications in pharmaceuticals, 3D printing, tissue engineering, cosmetics, etc. Sobh et al. fabricated pH-sensitive nanocarriers for delivering 5-flurouracil, curcumin, and its derivative using a chitosan and CNTs-based system. The CNTs improved the thermal stability and drug loading with selective toxicity to cancer cells. Drug release was dependent on the pH as the chitosan is relatively insoluble at pH 7.4 and soluble at pH 1.2, and the pH affected the molecular features and swelling nature of chitosan and chitosan–CNTs [34]. In another study, SWCNTs were modified with chitosan and epidermal growth factor to deliver etoposide to lung cancer. The fabricated CNTs exhibited excellent solubility and dispersibility with high drug loading and 2.7 times greater cytotoxicity than the etoposide. The authors concluded that the drug release occurs at low pH-dependent and uptake by the tumor cells through adenosine triphosphate-dependent endocytosis [35].

Folic acid is used extensively to target folate receptors, which are overexpressed in certain tumors. Singh and coworkers conjugated folic acid (FA) with chitosan

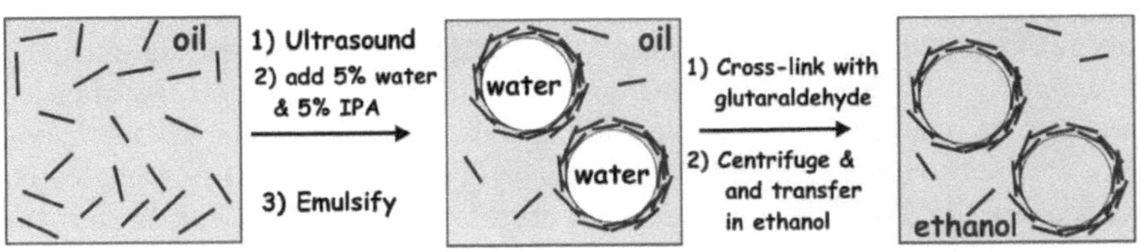

FIGURE 5.5 Schematic presentation of novel carbon nanotubosomes by emulsion inversion technique. (Reproduced with copyright permission from Ref [31]. Copyright 2005 Royal Society of Chemistry.)

(CHI) using EDC chemistry, which is further coated onto the MWCNTs to deliver docetaxel (DTX) to lung cancer cells. The cell uptake studies showed that the coated MWCNTs had higher internalization through folate receptor-mediated endocytosis (RME). The IC_{50} values demonstrated that coated MWCNTs are 89 times more effective than the marketed formulation of docetaxel [36]. Diverse stimuli-responsive nanocarriers are designed to provide "on-demand" drug release to treat various ailments. Different stimuli like light, magnetic fields, electric fields, ultrasound, pH, and enzymes are being utilized to trigger cargo release from the delivery systems. In a recent study, chitosan-coated CNTs was loaded into a thermosensitive hydrogel for photothermal therapy of tumors. The CNTs exhibited the gel-sol-gel transition similar to the unloaded gel and did not show a noticeable change in the gelation temperature. Upon irradiation, the CNT hydrogel undergoes photothermal conversion, which loosens the hydrogel, accelerating drug release. This system can potentially fulfill the spatial and temporal drug release requirements for chemotherapy [37].

In another study, Seyfoori et al. developed chitosan-coated iron nanoparticles conjugated with MWCNTs for multifunctional magneto/pH-responsive hydrogels prepared to deliver DOX to tumor cells. The hybrid systems exhibited more excellent pH responsiveness and drug release compared to just functionalized CNTs, which could be attributed to the (i) weakened hydrogen bond among CNTs and doxorubicin due to the competition role of protons, and (ii) electrostatic repulsion among the protonated amines of chitosan that permit more incredible drug release from swollen nanogel. All these features make this a promising system for innovative, targeted, and controlled drug delivery for cancer [38]. Das et al. fabricated curcumin and DOX-loaded CNT-embedded cyclodextrin polymer-derived nanocarrier for MDR reversal. Incorporating CNTs into cyclodextrin polymer exhibited photothermal therapy, anti-angiogenesis, and synergistic therapy for tumor localization and multidrug resistance with lower cytotoxicity on normal cells. These multiple approaches exhibited sustained release of the two drugs for more than 30 hours with superior antitumor effects *in vitro* cell line studies and *in vivo* studies using Balb/c mice, which shows that novel nanocarrier can be a promising therapy for MDR reversal [39].

Investigations have shown that sugar-functionalized CNTs exhibit greater uptake by the tumor cells *via* glucose transporters (GLUTs). GLUT5 is responsible for the absorption and metabolism of fructose in the cells. Various fructosylated CNTs are used in the treatment of cancers. For example, carboxylated CNTs are modified with fructosylated glycoblock copolymers and folic acid to improve the water solubility, compatibility, biointeractions, and targetability to breast cancer cells. The fructose and folic-acid-tethered CNTs enhanced the stability and facilitated the intracellular entrance of the loaded drug (DOX), thereby improving the apoptosis of cancer cells [40]. In another study, carbohydrate ligands, galactose, mannose, and lactose were tethered to carboxylated MWCNTs using a lysine linker. They found carbohydrate-coated MWCNTs exhibited high drug loading capacity with higher drug release at pH 5, making them suitable carriers for tumor targeting compared to carboxylated CNTs. Among the carbohydrate-modified CNTs, the galactosylated and mannosylated MWCNTs exhibited greater cellular uptake and cytotoxicity than lactosylated MWCNTs [41]. Further, hybrid CNTs were also utilized in immunotherapy.

Dong and the research group fabricated mannose-modified MWCNTs to deliver ovalbumin antigen to dendritic cells. The mannose-modified MWCNTs exhibited high drug loading of ovalbumin, and the CNT–antigen complex exhibited higher uptake by the dendritic cells to induce maturation and cytokine release *in vitro* [42]. In another study, mannose-modified CNT complexes enhanced the antigen internalization and the secretion of interleukin-1β and interleukin-6 from macrophages. This complex induced long-term antigen-specific IgG antibodies, which may be attributed to the ovalbumin release from mannose-modified MWCNTS. The mannose–CNTs complex showed lower cytotoxicity, indicating it could be a potential nano-vector for antigen presentation of immune cells [43].

Glycans, proteoglycans, glycoproteins, and glycolipids are forms of carbohydrates that are essential parts of the cell membrane and significant factors influencing molecular recognition processes on the cell surface. They are engaged in various biological processes, including endocytosis, cell trafficking, immune responses, cell signaling, proliferation, metastasis, etc. Abnormal glycosylation involves multiple autoimmune diseases, cancers, and viral and bacterial infections. Gu et al. fabricated multivalent ligand-conjugated SWCNTs for capturing pathogens. Galactose-functionalized SWCNTs exhibited binding to pathogenic *Escherichia coli* strains under physiological conditions, which may be attributed to galactose binding proteins on the surface of *E. coli*. The greater surface area and aspect ratio of SWCNTs bound more galactose molecules, and the semiflexible nature of SWCNTs facilitates *E. coli* binding [44]. Galactose and mannose functionalized SWCNTs exhibited aggregation and reduction of colony-forming units in the presence of divalent calcium ions due to divalent cation-mediated multivalent carbohydrate–carbohydrate interactions [45].

Many natural and synthetic polymers are used for electrospun scaffolds for bone tissue regeneration. Electrospun scaffolds provide optimum porosity, strong biomechanical properties, and the encapsulation of bioactives in osteogenic environments. Shresta and coworkers fabricated a fibrous scaffold by amalgamation of chitosan and zein and polyurethane-associated functionalized MWCNTs as an artificial bone extracellular matrix (ECM). The scaffold exhibited potent synergistic effects with antibacterial effects, hydrophilicity, and strong mechanical strength, allowing biomineralization and biodegradation. Biological evaluation with alizarin red staining, Western blotting analysis, and alkaline phosphatase activity confirmed the hydroxyapatite nanocrystal nucleation and excellent osteoinductive properties. All these results show that the fibrous scaffold is a promising artificial ECM for bone cell regeneration [46]. It is also possible to electrospun hyaluronic acid into nanofibers using extremely tiny amounts of carboxylated MWCNTs. To safely and effectively deliver current for stimulating neurons, the MWCNTs are inserted into the HA fibers to lower their impedance and improve their charge capacity. When subjected to 200 mV/mm electrical stimulation for a minimum of half an hour, dorsal root ganglion cells growing on these substrates were found to have longer neurite lengths than unstimulated controls [47].

CNTs conjugated with polysaccharides have exhibited immunogenicity by potentiating cellular and humoral immunity. Lentinan is an immunostimulatory herbal polysaccharide that potentiates the response of standard antigenic protein (OVA) by conjugating with MWCNTs. The lentinan improved the solubility of CNTs, whereas the CNTs provided the depot effect and biostability to the lentinan, making them an excellent combination for vaccine delivery. *In vitro* analysis demonstrated that the lentinan-coated MWCNTs rapidly internalized into dendritic cells, improving antigen delivery and retention in the cells. *In vivo* studies in healthy mice did not show any adverse reactions or toxic effects in the organs, confirming the potential of this combination as a promising vaccine delivery system [48].

5.8 PROTEIN-COATED CNTs

Proteins are heterogeneous molecules with multiple functional groups that provide diverse chemical derivatization. They are extensively investigated for understanding the signaling pathways involved in physiological and pathological processes. Additionally, because of their quick adaptability to complex 3D structures with fine specificity, they have found a broad array of applications in the domains of biology and sensing. Their capacity to respond rapidly to environmental changes is an intriguing area of research because they form complexes, catalyze chemical changes, and alter biological cascades through quick chemical changes (such as the acquisition and loss of phosphate groups). The CNTs and protein hybrids are attracting increased attention from researchers due to their great potential in medicine, diagnosis, smart materials, nanotechnology, and biology. Since the crystallization of streptavidin (biotin-binding protein) on the MWCNTs in 1999, the interactions between CNTs (carbon-based materials) and proteins were intensively investigated. The CNT-protein hybrids can effectively reduce the toxicity of the CNTs without affecting the optical, mechanical, and electronic properties of the CNTs. As it holds the potential of proteins and CNTs, the hybrid systems have significant applications in medicine and biology. The effective interactions due to the high surface area of the CNTs provide unique properties like stability, high activity, and low mass transfer resistance, enabling the immobilization and solubility of the enzymes for diagnosis, biochip, and biosensor applications [16]. The van der Waals interactions, electrostatic interactions, and hydrophobic interactions, along with amphiphilicity, are the main driving forces for the adsorption of proteins on the CNTs. Each protein has a "unique molecular personality," and to maximize assembly on the SWCNT surface, many factors must be considered based on the protein structure and field of application. Proteins bind to SWCNTs spontaneously when the Gibbs binding free energy is negative ($\Delta G < 0$) [49]. Polyethyleneimine-functionalized MWCNTs were used to conjugate PTX for the treatment of prostate cancer. The polyethyleneimine functionalization improved the dispersibility by reducing their ability to agglomerate, which was confirmed by TEM imaging. These MWCNTs were coated with prostate-specific membrane antigen (PMSA) antibodies, which enhanced the internalization of modified CNTs into the cancer cells, further improving the toxicity. *In vitro* cytotoxicity analysis of modified CNTs exhibited cytotoxicity in $PMSA^+$ prostate cancer cells and PMSA-colorectal cancer (HCT-116 and CaCo-2) but not in human peripheral monocytes and lymphocytes. However, the anti-PMSA-coated PTX-loaded CNTs exhibited improved cytotoxicity in $PMSA^+$ cells but not in PMSA cells, indicating the specificity of the nanocarriers. All these findings conclude that the antibody-coated MWCNTs can be a promising carrier for targeted cancer therapy [50].

Namgung et al. investigated fibronectin and CNTs to support the adhesion and growth of cells. Force spectroscopy and immunofluorescence demonstrated that the fibronectin has greater affinity for CNTs. Human mesenchymal stem cells exhibited more excellent cell adhesion on the CNT–fibronectin hybrid compared to CNTs, fibronectin, and glass. This shows that ECM molecules have a greater affinity for the CNTs and support selective proliferation of cells [51]. In line with this, Vidal and colleagues observed that CNT papers adsorbed with fibronectin triggered cell adhesion and proliferation on both hydrophilic and hydrophobic CNPs. In contrast, significant proliferation occurred only on hydrophilic CNT papers [52]. Pantarotto et al. fabricated peptide-functionalized SWCNTs conjugates to enhance virus-neutralizing antibodies without any detectable immunogenicity. Successful presentation of attached peptides *in vivo* may be responsible for the induction of peptide-specific antibodies, which shows the potential of CNTs for the delivery of peptides [53].

5.9 POLYMER-COATED CNTs

Polymers are widely used to improve the limitations of CNTs, such as low functionality, large size of pure CNTs, high hydrophobicity, and so on, through covalent and noncovalent functionalization of CNT surfaces. Polyethylene glycol (PEG) is widely used, which renders hydrophilicity and stealth characteristics for RES escape. Liu and colleagues fabricated PEGylated SWCNTs to deliver the doxorubicin to the tumor. This supramolecular chemistry of SWCNTs exhibited 50%–60% wt of DOX compared to the 8% in the conventional liposomes, and also, the system exhibited pH-dependent drug release. The authors suggested that surface functionalization can be utilized for active targeting of the CNTs [54]. In a recent investigation, Zhao studied the effect of length and PEGylation degree on the cytotoxicity and DOX loading capacity. They observed that PEG-MWCNTs less than 300 nm exhibited superior cytocompatibility among the 800, 500, 300, and 100 nm CNTs. The length of the CNTs can be regulated through oxidation time and sonication. The CNTs with 300 nm exhibited a drug loading of 0.55 mg/mg, which might be due to the π–π stacking of the drug with the MWCNTs, and the system showed the pH-triggered drug release with up to 57% release in the first 24 hours compared to 10% release in the physiological pH. These findings support using the PEGylated CNTs as a promising carrier for the stimuli-responsive release of the encapsulated drugs at the tumor site [55]. In another study, Liu and colleagues covalently conjugated PTX to PEG segments, which were then adsorbed onto the CNTs' sidewalls, as shown in the Figure 5.6a. Prolonged circulation time and 10 times higher uptake of conjugate might be responsible for the more significant tumor suppression in the murine 4T1 breast cancer model. The CNT–PTX conjugate did not exhibit detectable toxic effects, showing its efficacy as a promising drug delivery system with minimum side effects [56].

Sobhani and colleagues utilized polycitric acid (PCA) to improve the functionality and hydrophilicity of MWCNTs. They conjugated PTX to the PCA through a cleavable ester bond, which exhibited a particle size of 200 nm and a drug loading of 38% w/w. The hydrophobicity of MWCNTs and hydrophilicity of the PCA

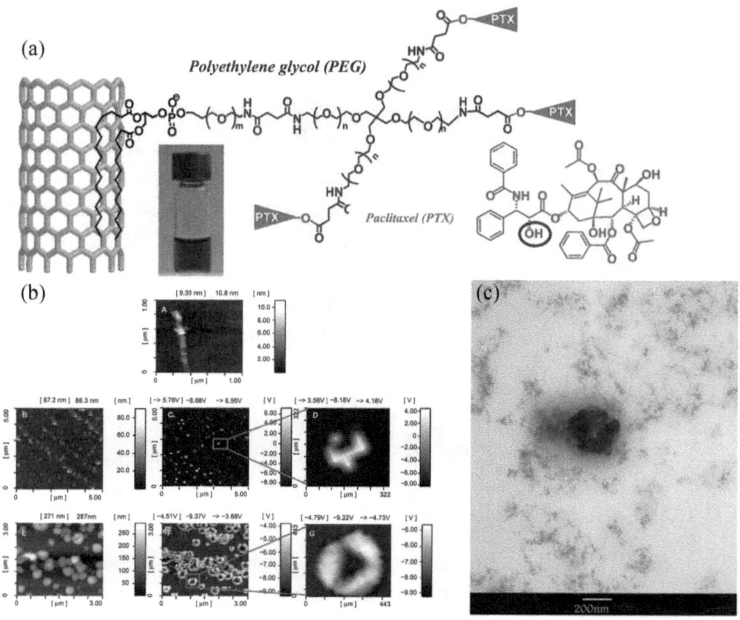

FIGURE 5.6 Polymer–CNT hybrids. (a) PTX-PEG-CNT conjugate. (Reproduced with copyright permission from Ref [56].) (b) AFM images of oxidized MWCNT. MWCNT-PCA, MWCNT- PCA-PTX conjugate. (c) TEM image of MWCNT- PCA-PTX conjugate. (Reproduced with copyright permission from Ref [57]. Copyright 2011 Dovepress.)

exhibited conformational changes in the conjugate, which might be responsible for the curved shape in the AFM and TEM images, as shown in Figure 5.6b and c. The PTX-conjugated PCA MWCNTs exhibited pH-dependent drug release and significant cytotoxic effects in the A549 and SKOV3 cell lines, which infers improved cell penetration in the system [57].

Cao et al utilized pH-responsive surface-modified SWCNTs for delivering siRNA and DOX to A549 cells for enhanced endosomal escape. The schematic representation of the nanocomplex activity in A549 cells is illustrated in Figure 5.7. The oxidized SWCNTs were tethered with betaine–polyethyleneimine conjugate, which renders pH-responsive lysosomal escape. The functionalized CNTs were coated with targeting peptide BR2 to improve the cellular internalization. Confocal laser microscopy images revealed endosomal escape of siRNA, which might be attributed to the pH sponge effect of the betaine–polyethyleneimine conjugate. The subcellular localization of the DOX around the nucleus and FAM-siRNA in the cytoplasm demonstrates the effectiveness of the nanosystem in facilitating the endosomal escape of the siRNA and DOX penetration into the nucleus. The codelivery siRNA and DOX showed a significant reduction in the tumor volume in A549-bearing mice without affecting the normal tissues. All these results demonstrate the effectiveness of fabricated CNT complex for the codelivery of chemotherapy and gene therapy [58].

Cheng et al. fabricated PLGA-CNTs conjugates for delivering caspase-3 into osteosarcoma cells. The caspase-3 is a pro-apoptotic enzyme involved in cellular

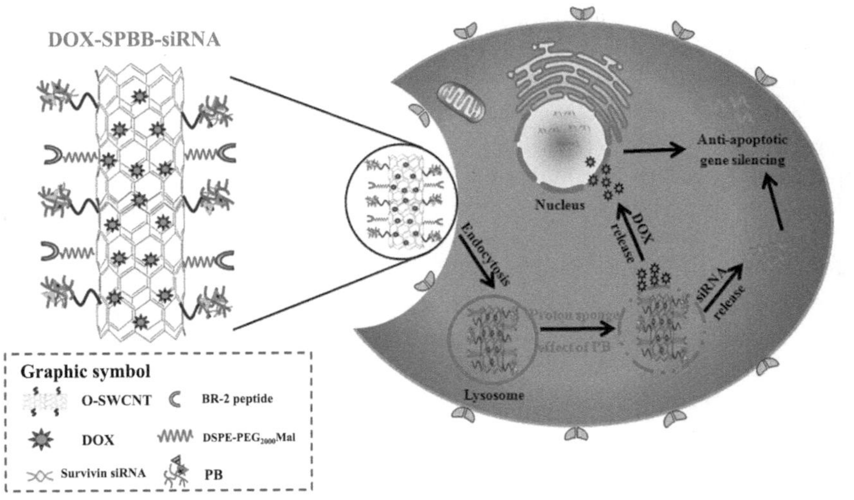

FIGURE 5.7 Schematic representation of pH-responsive surface-modified SWCNT for codelivery of siRNA and doxorubicin in A549 cells. (Reproduced with copyright permission from Ref [58]. Copyright 2019, American Chemical Society.)

apoptosis. The attachment of the protein to the CNT was observed using fluorescent-tagged BSA and FTIR, and cellular transfection rates of the prepared nanocarrier were 30%–40% greater than that achieved by the liposomes or polymeric nanoparticles. These PLGA-CNTs-caspase 3 conjugates exhibited a time-dependent release profile and high transfection rates, indicating that the PLGA-CNTs are a promising carrier for protein delivery [59].

Hydrogels are three-dimensional polymeric networks that can imbibe large quantities of water inside them. The hydrogel can be formed by hydrogen bonding or electrostatic interactions or by covalent bonding using cross-linkers. Mechanical strength, porosity, biocompatibility, stability, and biodegradability can be modified by changing the polymerization technique or altering the polymeric composition. The limitations of hydrogels, like increased toxicity, low mechanical strength, low biodegradability, etc., can be addressed by hybrid hydrogels. They are being applied in imaging, biosensing, microfluidics, scaffolds, biorobotics, gene targeting, targeted drug delivery, etc, because of their desired properties such as high electrical conductivity, effective surface area, and high mechanical strength [60]. The interactions between the drugs and CNT hybrid hydrogels are affected by the electrical properties of the drugs and CNT counterparts. The electromagnetic behavior, high surface area, drug loading, excellent stability, and cellular uptake of CNT allow them to be nanofillers for various applications. The combination of CNTs and hydrogels has demonstrated potential for application in chondrogenic differentiation of stem cells, near IR responsive systems, pulsatile drug delivery, tissue engineering, biosensing, etc. The hybrid hydrogels can be synthesized using *in situ* polymerization techniques, physical crosslinking, *ex situ* polymerization techniques, grafting techniques, etc. A thermosensitive poly(N-isopropyl acrylamide) PNIPAAm/CNT composite hydrogel shows a volume transition upon electrical stimulation. The electrothermal conversion of the hybrid gel triggers the material's shape-modifying ability, resulting in shrinking behavior. This electro-sensitive nature of the hybrid can have promising functions in sensors [61]. CNT-doped acrylate hydrogels are synthesized by Cirillo and coworkers for the electro-induced release of curcumin for wound healing treatments. The drug release from the hydrogel was modulated by the strength of the applied voltage and the amount of CNT. The creation of highly mouldable releasing profiles appropriate for wound healing treatment is made possible by the intricate interactions between the drug and hybrid hydrogel, as well as the system's polarization as a result of electric stimulation [62]. In another study, Servant et al. combined poly(methacrylic acid) hydrogel matrix with pristine MWCNTs to improve its electro-responsiveness. The electro-responsiveness was investigated using ^{14}C-sucrose release from the matrix *in vitro* and *in vivo*, which showed increased release and sharper response upon electrical stimulation. These hybrid implants demonstrated excellent biocompatibility in the absence of electrical simulation. However, the electrical simulation exhibited inflammation and necrosis at the application site with no adverse reactions, which is attributed to the oxidation of the electrodes upon skin contact. The developed hybrid hydrogels for electro-responsive drug delivery have the potency to improve personalized medicine using short electric impulses [63]. Collagen is a structural protein present in the ECM in connective tissues of animals. Type I collagen can be used as an ECM for cardiac tissue. It is combined with CNTs to improve electrical performance and mechanical strength while preserving the microporous nature of hydrogels. Cardiac myocytes cultured in hybrid CNT–collagen hydrogel scaffolds improved the cardiac cell function, as indicated by the increased rhythmic contraction area. This shows that hybrid CNT–collagen hydrogel scaffolds can be a potential strategy for enhancing the systolic contraction of cardiac tissue [64].

5.10 INORGANIC NANOPARTICLES WITH CARBON NANOTUBES

Singh et al. developed a multifunctional hybrid nanocarrier of magnetic CNTs layered with mesoporous silica nanoparticles for theranostics. The magnetic and mesoporous hybrid systems showed greater encapsulation of loaded cargo, i.e., gentamicin and cytochrome C, due to mesopore structure, whereas siRNA was loaded by charge–charge interaction on aminated mesoporous silica of magnetic CNTs. The hybrid system also exhibited excellent magnetic properties *in vitro* and *in vivo*, and high cellular internalization via magnetofection while showing minimal cytotoxicity to the cells [65]. The MWCNTs were decorated with other nanoparticles like zinc oxide, hydroxyapatite, and silver nanoparticles and were tested for antimicrobial properties against opportunistic pathogens. The presence of inorganic nanoparticles on MWCNT surface improved their antimicrobial properties [66]. A CNT ring was used as a template for coating gold nanoparticles to enhance the surface-enhanced Raman scattering property and detect cancer cells. The resonance of the CNT ring coated with gold nanoparticles increased 110 times compared to the CNT ring, which may be attributed to the strong electromagnetic field of the nanoshell of gold nanoparticles, which enhanced the resonance of the CNT. Numerical simulations and perturbation theory analysis also supported the experimental results. This modified

CNT platform exhibited more excellent photoacoustic signals and photothermal therapy, indicating its potential in various cancer therapies [67]. In another study, iron nanoparticle-doped CNTs were coated with poly(lactide)eco-methoxy polyethylene glycol (PLA-co-mPEG) micelles for controlled and smart release of docetaxel were investigated. The coating of biocompatible micelles was used to circumvent the hydrophobicity of the CNTs, and the high surface area of the CNTs allowed more significant drug loading and cellular internalization. The nanocomposite exhibited no detectable cellular toxicity on MCF-7 at 1,000 μg/mL, even after 72 hours, showing its potential for controlled and targeted delivery of docetaxel [68].

5.11 CELL-MEMBRANE-COATED CNTs

Nanoparticles coated with cell membranes are widely explored for their inherent abundant source cell relevant functions, specific molecular recognition, and enhanced bio-interfacing capabilities [69]. Different types of cells, like stem cells, RBC, WBC, platelets, etc, are being utilized to coat the nanoparticles. Ultrasonication, microfluidic extrapolation, and membrane extrusion are used to fuse the cell membranes with the synthetic nanoparticles [70,71]. The first CMNs are fabricated using the coextrusion of RBC membrane on PLGA cores. They mimic the surface properties of the source cells and exhibit exclusive properties like decreased RES clearance, enhanced biocompatibility, tumor penetration, etc [72]. The limitations of pristine CNTs can be overcome by utilizing cell membrane coating on the CNTs. Cell membrane-cloaked magnetic CNTs were used to screen the drug leads from traditional Chinese medicine. The cell membrane was obtained from hypotonic lysis of high-expression ephrinb2/HEK293 cells, which were then coated to the surface of EDC-NHS-modified magnetic CNTs. The resultant membranes were used for screening of leads from *Aconitum Carmichaeli Dedx* with better targetability and selectivity. The hybrid system exhibited better stability of the cell membrane while preserving the cell membrane features [73]. CMN-coated CNTs were also utilized in selective solid-phase extraction of bioactive components from natural compounds [74]. The cell-membrane-coated CNTs are still needed to be explored for the delivery of chemotherapeutics.

5.12 CHALLENGES OF HYBRID CNTs

Hybrid CNTs are potential delivery systems for the controlled delivery of various therapeutic agents. The multi-targetability, tunable particle size, controlled release, etc., are needed in various clinical settings such as gene therapy, tissue engineering, drug delivery, imaging, and so on. Nevertheless, designing such a system with desirable characters for targeted applications is still a challenge that hinders clinical translation. From the manufacturing point, the scalability challenges from the lab scale to the industrial scale of the CNTs are also a major concern. The multistep fabrication process is one of the objectionable financial standpoints for the developed hybrid CNTs as they increase manufacturing costs [75]. The architectural integrity of the CNTs hybrid still needs to be addressed. The nano–bio interactions may alter the hybrid CNT properties and may show varied responses *in vitro* due to the complex nature of the biological system. Another hurdle that has impacts on the clinical translation is the adverse effects of the hybrid CNTs and regulatory concerns regarding the novel formulations. The hybrid CNTs are still in their infancy and require collaborative efforts from multidisciplinary scientists for clinical translation.

5.13 CONCLUSION AND FUTURE PERSPECTIVE

Though CNTs have emerged as a novel material for drug delivery, much progress has not been achieved. They are unable to substitute the other polymeric materials used for drug delivery due to their toxicity concerns and are expensive to synthesize, large-scale manufacturing, etc. The biomedical sector has shown significant progress in the development of biocompatible hybrid nanocarriers. Functionalization of the CNTs is considered a promising approach for improving the properties and overcoming the downsides of the CNTs. These nanocarriers are able to reach the target site, and the inclusion of smart polymers in the hybrid systems will render stimuli-responsive properties to the system. These CNT hybrids are capable of exhibiting controlled drug delivery, targeting, codelivery, etc. Besides the advantages, further investigations are required to fully understand the mechanics behind the CNT hybrids. The underlying mechanisms of biointeractions will provide a path for a better understanding and development of CNT hybrids, which enable accurate delivery at the targeted site.

ACKNOWLEDGMENT

The authors thank the National Institute of Pharmaceutical Education & Research (NIPER), Hyderabad, for their support during the writing of this chapter.

REFERENCES

1. Fraguas-Sánchez AI, Martín-Sabroso C, Fernández-Carballido A, Torres-Suárez AI. Current status of nanomedicine in the chemotherapy of breast cancer. *Cancer Chemother. Pharmacol.* 84(4), 689–706 (2019).
2. Kumari NU, Pardhi E, Chary PS, Mehra NK. Exploring contemporary breakthroughs in utilizing vesicular nanocarriers for breast cancer therapy. *Ther. Deliv.* 15, 279–303 (2024) https://doi.org/10.4155/tde-2023-0092.
3. Kumari NU, Pardhi E, Madan J, Singh PK, Mehra NK. Non-vesicular nanocarriers for breast cancer management: current advances, challenges, and future perspective. *Int. J. Polym. Mater. Polym. Biomater.*74, 424–444 (2024).
4. Simon J, Flahaut E, Golzio M. Overview of carbon nanotubes for biomedical applications. *Materials (Basel).* 12(4), 1–21 (2019).
5. Mehra NK, Palakurthi S. Interactions between carbon nanotubes and bioactives: a drug delivery perspective. *Drug Discov. Today.* 21(4), 585–597 (2016).
6. Sajid MI, Jamshaid U, Jamshaid T, Zafar N, Fessi H, Elaissari A. Carbon nanotubes from synthesis to in vivo biomedical applications. *Int. J. Pharm.* 501(1–2), 278–299 (2016).
7. Chary PS, Bhawale R, Vasave R, et al. A review on emerging role of multifunctional carbon nanotubes as an armament in cancer therapy, imaging and biosensing. *J. Drug Deliv. Sci. Technol.* 85, 104588 (2023).

8. Anzar N, Hasan R, Tyagi M, Yadav N, Narang J. Carbon nanotube - a review on synthesis, properties and plethora of applications in the field of biomedical science. *Sensors Int.* 1(December 2019), 100003 (2020).
9. Khan MU, Reddy KR, Snguanwongchai T, Haque E, Gomes VG. Polymer brush synthesis on surface modified carbon nanotubes via in situ emulsion polymerization. *Colloid Polym. Sci.* 294(10), 1599–1610 (2016).
10. Mody N, Tekade RK, Mehra NK, Chopdey P, Jain NK. Dendrimer, liposomes, carbon nanotubes and PLGA nanoparticles: one platform assessment of drug delivery potential. *AAPS PharmSciTech.* 15(2), 388–399 (2014).
11. Naief MF, Mohammed SN, Mohammed AM. Carbon nanotubes: a review on synthesis and drug delivery for cancer treatment. *Inorg. Chem. Commun.* 159(May 2023), 111694 (2024).
12. Naebe M, Wang J, Amini A, et al. Mechanical property and structure of covalent functionalised graphene/epoxy nanocomposites. *Sci. Rep.* 4, 4375(2014).
13. Spitalsky Z, Tasis D, Papagelis K, Galiotis C. Carbon nanotube–polymer composites: chemistry, processing, mechanical and electrical properties. *Prog. Polym. Sci.* 35(3), 357–401 (2010).
14. Kaur J, Gill GS, Jeet K. Applications of carbon nanotubes in drug delivery: a comprehensive review. *Charact. Biol. Nanomater. Drug Deliv. Nanosci. Nanotechnol. Drug Deliv.* 2019, 113–135 (2019).
15. Vardharajula S, Ali SZ, Tiwari PM, et al. Functionalized carbon nanotubes: biomedical applications. *Int. J. Nanomed.* 7, 5361–5374 (2012).
16. Zhou X, Su J, Wang C, et al. Design, preparation and measurement of protein/CNTs hybrids: a concise review. *J. Mater. Sci. Technol.* 46, 74–87 (2020).
17. Mohammadi M, Taghavi S, Abnous K, Taghdisi SM, Ramezani M, Alibolandi M. Hybrid vesicular drug delivery systems for cancer therapeutics. *Adv. Funct. Mater.* 28(36), 1802136 (2018).
18. Pandita D, Kumar S, Lather V. Hybrid poly(lactic-co-glycolic acid) nanoparticles: design and delivery prospectives. *Drug Discov. Today.* 20(1), 95–104 (2015).
19. Rajana N, Mounika A, Chary PS, et al. Multifunctional hybrid nanoparticles in diagnosis and therapy of breast cancer. *J. Control. Release.* 352(August), 1024–1047 (2022).
20. Brzezicka KA, Arlian BM, Wang S, Olmer M, Lotz M, Paulson JC. Suppression of autoimmune rheumatoid arthritis with hybrid nanoparticles that induce B and T cell tolerance to self-antigen. *ACS Nano.* 16(12), 20206–20221 (2022).
21. Maurya P, Singh S, Saraf SA. Inhalable hybrid nanocarriers for respiratory disorders. In: Dua K, Hansbro PM, Wadhwa R, et al. (eds.) *Targeting Chronic Inflammatory Lung Diseases Using Advanced Drug Delivery Systems*, Academic Press, New York, 281–302 (2020).
22. Fahrenholtz CD, Ding S, Bernish BW, et al. Design and cellular studies of a carbon nanotube-based delivery system for a hybrid platinum-acridine anticancer agent. *J. Inorg. Biochem.* 165, 170–180 (2016).
23. Shao W, Paul A, Zhao B, Lee C, Rodes L, Prakash S. Carbon nanotube lipid drug approach for targeted delivery of a chemotherapy drug in a human breast cancer xenograft animal model. *Biomaterials.* 34(38), 10109–10119 (2013).
24. Zhao Y, Zhao T, Cao Y, et al. Temperature-sensitive lipid-coated carbon nanotubes for synergistic photothermal therapy and gene therapy. *ACS Nano.* 15(4), 6517–6529 (2021).
25. Karchemski F, Zucker D, Barenholz Y, Regev O. Carbon nanotubes-liposomes conjugate as a platform for drug delivery into cells. *J. Control. Release.* 160(2), 339–345 (2012).
26. Zhu X, Xie Y, Zhang Y, et al. Thermo-sensitive liposomes loaded with doxorubicin and lysine modified single-walled carbon nanotubes as tumor-targeting drug delivery system. *J. Biomater. Appl.* 29(5), 769–779 (2014) https://doi.org/10.1177/0885328214543211.

27. Pippa N, Chronopoulos DD, Stellas D, et al. Design and development of multi-walled carbon nanotube-liposome drug delivery platforms. *Int. J. Pharm.* 528(1–2), 429–439 (2017).
28. Miyako E, Kono K, Yuba E, Hosokawa C, Nagai H, Hagihara Y. Carbon nanotube–liposome supramolecular nanotrains for intelligent molecular-transport systems. *Nat. Commun.* 3(1), 1–9 (2012).
29. Madani SZM, Safaee MM, Gravely M, et al. Carbon nanotube-liposome complexes in hydrogels for controlled drug delivery via near-infrared laser stimulation. *ACS Appl. Nano Mater.* 4(1), 331–342 (2021).
30. Pereira S, Lee J, Rubio N, et al. Cationic liposome- multi-walled carbon nanotubes hybrids for dual siPLK1 and doxorubicin delivery in vitro. *Pharm. Res.* 32(10), 3293–3308 (2015).
31. In Het Panhuis M, Paunov VN. Assembling carbon nanotubosomes using an emulsion-inversion technique. *Chem. Commun.* 13, 1726–1728 (2005).
32. Ono K, Ochiai R, Yoshida T, et al. The detection rates and tumor clinical/pathological stages of whole-body FDG-PET cancer screening. *Ann. Nucl. Med.* 21(1), 65–72 (2007).
33. Fahrenholtz CD, Hadimani M, King SB, Torti SV, Singh R. Targeting breast cancer with sugar-coated carbon nanotubes. *Nanomedicine (Lond).* 10(16), 2481–2497 (2015) https://doi.org/10.2217/nnm.15.90.
34. Sobh RA, Nasr HES, Mohamed WS. Formulation and in vitro characterization of anti-cancer drugs encapsulated chitosan/multi-walled carbon nanotube nanocomposites. *J. Appl. Pharm. Sci.* 9(8), 032–040 (2019).
35. Chen C, Xie XX, Zhou Q, et al. EGF-functionalized single-walled carbon nanotubes for targeting delivery of etoposide. *Nanotechnology.* 23(4), 045104 (2012).
36. Singh RP, Sharma G, Sonali, et al. Chitosan-folate decorated carbon nanotubes for site specific lung cancer delivery. *Mater. Sci. Eng. C.* 77, 446–458 (2017).
37. Dong X, Wei C, Liang J, Liu T, Kong D, Lv F. Thermosensitive hydrogel loaded with chitosan-carbon nanotubes for near infrared light triggered drug delivery. *Colloids Surf. B Biointerfaces.* 154, 253–262 (2017).
38. Seyfoori A, Sarfarazijami S, Seyyed Ebrahimi SA. pH-responsive carbon nanotube-based hybrid nanogels as the smart anticancer drug carrier. *Artif. Cells Nanomed. Biotechnol.* 47(1), 1437–1443 (2019).
39. Das M, Nariya P, Joshi A, et al. Carbon nanotube embedded cyclodextrin polymer derived injectable nanocarrier: a multiple faceted platform for stimulation of multi-drug resistance reversal. *Carbohydr. Polym.* 247, 116751 (2020).
40. Omurtag Ozgen PS, Atasoy S, Zengin Kurt B, Durmus Z, Yigit G, Dag A. Glycopolymer decorated multiwalled carbon nanotubes for dual targeted breast cancer therapy. *J. Mater. Chem. B.* 8(15), 3123–3137 (2020).
41. Thakur CK, Neupane R, Karthikeyan C, et al. Lysinated multiwalled carbon nanotubes with carbohydrate ligands as an effective nanocarrier for targeted doxorubicin delivery to breast cancer cells. *Molecules.* 27(21), 7461 (2022).
42. Dong Z, Wang Q, Huo M, et al. Mannose-modified multi-walled carbon nanotubes as a delivery nanovector optimizing the antigen presentation of dendritic cells. *ChemistryOpen.* 8(7), 915–921 (2019).
43. Feng H, Feng Y, Lin L, et al. Mannose receptor-mediated carbon nanotubes as an antigen delivery system to enhance immune response both in vitro and in vivo. *Int. J. Mol. Sci.* 23(8), 4239 (2022).
44. Gu L, Elkin T, Jiang X, et al. Single-walled carbon nanotubes displaying multivalent ligands for capturing pathogens. *Chem. Commun.* 7, 874–876 (2005).
45. Wang H, Gu L, Lin Y, et al. Unique aggregation of anthrax (Bacillus anthracis) spores by sugar-coated single-walled carbon nanotubes. *J. Am. Chem. Soc.* 128(41), 13364–13365 (2006).

46. Shrestha S, Shrestha BK, Ko SW, Kandel R, Park CH, Kim CS. Engineered cellular microenvironments from functionalized multiwalled carbon nanotubes integrating Zein/Chitosan @Polyurethane for bone cell regeneration. *Carbohydr. Polym.* 251, 117035 (2021).
47. Steel EM, Azar JY, Sundararaghavan HG. Electrospun hyaluronic acid-carbon nanotube nanofibers for neural engineering. *Materialia.* 9, 100581 (2020).
48. Xing J, Liu Z, Huang Y, et al. Lentinan-modified carbon nanotubes as an antigen delivery system modulate immune response in vitro and in vivo. *ACS Appl. Mater. Interfaces.* 8(30), 19276–19283 (2016).
49. Antonucci A, Kupis-Rozmysłowicz J, Boghossian AA. Noncovalent protein and peptide functionalization of single-walled carbon nanotubes for biodelivery and optical sensing applications. *ACS Appl. Mater. Interfaces.* 9(13), 11321–11331 (2017).
50. Comparetti EJ, Romagnoli GG, Gorgulho CM, Pedrosa V de A, Kaneno R. Anti-PSMA monoclonal antibody increases the toxicity of paclitaxel carried by carbon nanotubes. *Mater. Sci. Eng. C.* 116(July), 111254 (2020).
51. Namgung S, Kim T, Baik KY, Lee M, Nam JM, Hong S. Fibronectin–carbon-nanotube hybrid nanostructures for controlled cell growth. *Small.* 7(1), 56–61 (2011).
52. Vidal G, Delord B, Neri W, et al. The effect of surface energy, adsorbed RGD peptides and fibronectin on the attachment and spreading of cells on multiwalled carbon nanotube papers. *Carbon N Y.* 49(7), 2318–2333 (2011).
53. Pantarotto D, Partidos CD, Hoebeke J, et al. Immunization with peptide-functionalized carbon nanotubes enhances virus-specific neutralizing antibody responses. *Chem. Biol.* 10(10), 961–966 (2003).
54. Liu Z, Sun X, Nakayama-Ratchford N, Dai H. Supramolecular chemistry on water-soluble carbon nanotubes for drug loading and delivery. *ACS Nano.* 1(1), 50–56 (2007).
55. Zhao X, Tian K, Zhou T, Jia X, Li J, Liu P. PEGylated multi-walled carbon nanotubes as versatile vector for tumor-specific intracellular triggered release with enhanced anti-cancer efficiency: optimization of length and PEGylation degree. *Colloids Surf. B Biointerfaces.* 168, 43–49 (2018).
56. Liu Z, Chen K, Davis C, et al. Drug delivery with carbon nanotubes for in vivo cancer treatment. *Cancer Res.* 68(16), 6652–6660 (2008).
57. Sobhani Z, Dinarvand R, Atyabi F, Ghahremani M, Adeli M. Increased paclitaxel cytotoxicity against cancer cell lines using a novel functionalized carbon nanotube. *Int. J. Nanomed.* 6, 705 (2011).
58. Cao Y, Huang HY, Chen LQ, et al. Enhanced lysosomal escape of pH-responsive polyethylenimine-betaine functionalized carbon nanotube for the codelivery of survivin small interfering RNA and doxorubicin. *ACS Appl. Mater. Interfaces.* 11(10), 9763–9776 (2019).
59. Cheng Q, Blais MO, Harris G, Jabbarzadeh E. PLGA-carbon nanotube conjugates for intercellular delivery of caspase-3 into osteosarcoma cells. *PLoS One.* 8(12), e81947 (2013).
60. Vashist A, Kaushik A, Vashist A, et al. Advances in carbon nanotubes-hydrogel hybrids in nanomedicine for therapeutics. *Adv. Healthc. Mater.* 7(9), e1701213 (2018).
61. Wu J, Ren Y, Sun J, Feng L. Carbon nanotube-coated macroporous poly(N -isopropylacrylamide) hydrogel and its electrosensitivity. *ACS Appl. Mater. Interfaces.* 5(9), 3519–3523 (2013).
62. Cirillo G, Hampel S, Spizzirri UG, Parisi OI, Picci N, Iemma F. Carbon nanotubes hybrid hydrogels in drug delivery: a perspective review. *Biomed Res. Int.* 2014, 825017 (2014).
63. Servant A, Methven L, Williams RP, Kostarelos K. Electroresponsive polymer–carbon nanotube hydrogel hybrids for pulsatile drug delivery in vivo. *Adv. Healthc. Mater.* 2(6), 806–811 (2013).

64. Yu H, Zhao H, Huang C, Du Y. Mechanically and electrically enhanced CNT-collagen hydrogels as potential scaffolds for engineered cardiac constructs. *ACS Biomater. Sci. Eng.* 3(11), 3017–3021 (2017).
65. Singh RK, Patel KD, Kim JJ, et al. Multifunctional hybrid nanocarrier: magnetic CNTs ensheathed with mesoporous silica for drug delivery and imaging system. *ACS Appl. Mater. Interfaces.* 6(4), 2201–2208 (2014).
66. David ME, Ion RM, Grigorescu RM, et al. Hybrid materials based on multi-walled carbon nanotubes and nanoparticles with antimicrobial properties. *Nanomater.* 11(6), 1415 (2021).
67. Song J, Wang F, Yang X, et al. Gold nanoparticle coated carbon nanotube ring with enhanced Raman scattering and photothermal conversion property for theranostic applications. *J. Am. Chem. Soc.* 138(22), 7005–7015 (2016).
68. Hossein Panahi F, Peighambardoust SJ, Davaran S, Salehi R. Development and characterization of PLA-mPEG copolymer containing iron nanoparticle-coated carbon nanotubes for controlled delivery of Docetaxel. *Polymer (Guildf).* 117, 117–131 (2017).
69. Narain A, Asawa S, Chhabria V, Patil-Sen Y. Cell membrane coated nanoparticles: next-generation therapeutics. *Nanomedicine.* 12(21), 2677–2692 (2017).
70. Parodi A, Molinaro R, Sushnitha M, et al. Bio-inspired engineering of cell- and virus-like nanoparticles for drug delivery. *Biomaterials.* 147, 155–168 (2017).
71. Zhen X, Cheng P, Pu K. Recent advances in cell membrane-camouflaged nanoparticles for cancer phototherapy. *Small.* 15(1), e1804105 (2019).
72. Zeng Y, Li S, Zhang S, Wang L, Yuan H, Hu F. Cell membrane coated-nanoparticles for cancer immunotherapy. *Acta Pharm. Sin. B.* 12(8), 3233–3254 (2022).
73. Hu Q, Bu Y, Cao R, Zhang G, Xie X, Wang S. Stability designs of cell membrane cloaked magnetic carbon nanotubes for improved life span in screening drug leads. *Anal. Chem.* 91(20), 13062–13070 (2019).
74. Hu Q, Bu Y, Zhen X, et al. Magnetic carbon nanotubes camouflaged with cell membrane as a drug discovery platform for selective extraction of bioactive compounds from natural products. *Chem. Eng. J.* 364, 269–279 (2019).
75. Bregoli L, Movia D, Gavigan-Imedio JD, Lysaght J, Reynolds J, Prina-Mello A. Nanomedicine applied to translational oncology: a future perspective on cancer treatment. *Nanomed. Nanotechnol. Biol. Med.* 12(1), 81–103 (2016).

6 Carbon Nanotubes in Cancer Therapy

Ammu V.V.V. Ravi Kiran, Garikapati Kusuma Kumari, Praveen T. Krishnamurthy, and Renat R. Khaydrov

6.1 INTRODUCTION

Cancer initiates from mutations that lead to either "*loss-of-function*" in tumor-suppressor genes or "*gain-of-function*" in oncogenes or both [1]. Several therapeutic options including surgery, chemotherapy, radiotherapy, etc., have emerged to curb the uncontrolled proliferation of tumors but were confined due to off-target side effects, multidrug resistance, insolubility, stability issues, and patient noncompliance [2,3]. In recent years, nanotechnology has paved the way in many fields including medicine. Nanocarriers such as solid lipid nanoparticles (SLNs), carbon nanotubes (CNTs), and polymeric nanoparticles have been widely explored in many theragnostic fields (including in cancer) to potentiate therapeutic efficacy [4,5]. In 1993, Iijima and Ichihashi discovered CNTs as a by-product, which revolutionized many fields including cancer nanomedicine [6].

CNTs are tubular graphene sheets consisting of either single or multiple layers. Depending on the number of sheets/layers, CNTs are classified into single-walled (SWCNTs) or multi-walled CNTs (MWCNTs). CNTs possess many attractive features such as high payload capacity, ultrahigh aspect ratio, ease of surface functionalization, and so on, making them a promising nanocarrier for many anticancer drugs. CNTs have been widely used in different fields including cancer nanomedicine to encapsulate various drugs or biomolecules [6,7]. Synthesis of CNTs is achieved by three methods: chemical vapor deposition (CVD), laser ablation, and electric arc discharge [8–11]. Pristine CNTs (pCNTs) are reported to have detrimental issues due to their longer lengths, catalytic impurities, rapid aggregation, etc. [7,12]. Surface functionalization of CNTs (*f*-CNTs) with biocompatible polymers, functional moieties, or imaging agents. The inner cavity of CNTs could be used to entrap certain drugs that are sensitive or poorly water soluble, whereas the surface of CNTs could be surface engineered. This chapter focuses on the utilization of CNTs as drug/gene nanocarriers, especially in the field of cancer.

6.2 BASICS OF CNTS: MODIFICATIONS FOR MEDICAL USE

CNTs can be synthesized using three techniques: laser ablation, electric arc discharge, or CVD. CNTs are categorized into two types i.e., single-walled and multi-walled CNTs (SWCNTs). SWCNTs are rolled, single graphene sheet, whereas the MWCNTs

DOI: 10.1201/9781003386537-6

are rolled nesting of multiple sheets. SWCNTs are flexible and could help in imaging, whereas MWCNTs have high surface area, which could be more efficient in endohedral filling. CNTs possess innate lipophilicity and longer lengths, which are toxic to the physiological system. Proper functionalization of the surface CNTs (*f*-CNTs) and coating with suitable biomaterials could help in minimizing the toxic effects. The below are techniques that are used to functionalize the CNTs.

6.2.1 Exohedral Modifications

6.2.1.1 Covalent Modifications

Surface modification of covalent modification involves conversion of sp^2 to sp^3 configuration. Covalent modification involves direct sidewall functionalization or defect group formation. Reactions such as halogenation, cycloaddition of radicals, Prato reaction, and so on are usually preferred for covalent modifications. Prato reaction or cycloaddition of azomethine ylides is the most preferred reaction. Defect group surface functionalization is usually achieved by defect generation. Upon defect generation, various targeting groups of biomolecules are attached using amidation or esterification. Covalent functionalization involves single-, double- or triple-group functionalization. Single-group functionalization involves attachment of a single moiety on the surface, whereas multiple-group functionalization, i.e., double or triple, involves attachment of multiple moieties (Figure 6.1).

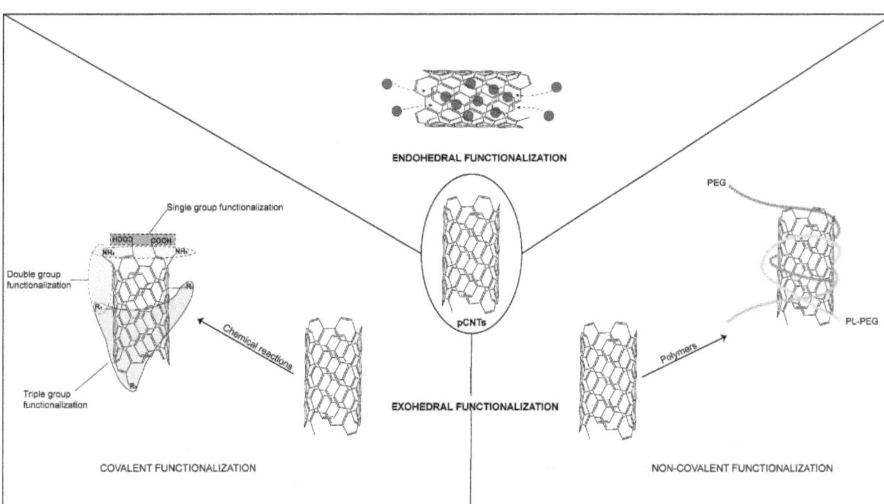

FIGURE 6.1 Different types of functionalizations of CNT: Exohedral functionalization involves both covalent and noncovalent modifications. (Reprinted (adapted) from Ref. [16]. Copyright © 2022, Springer Nature Switzerland AG.)

6.2.2 Non-Covalent Modifications

Non-covalent modification involves wrapping of biomolecules or adsorption phenomenon onto tubular surface (i.e., electrostatic, van der Waals interactions or π–π stacking). Contrastingly, non-covalent approach preserves the intrinsic properties such as optical or electrical properties. Polyethylene glycol and its derivatives are the most widely used functionalizing agents for wrapping of polymers. For instance, PEGylated CNTs demonstrated the longest circulation in the blood with less uptake reticuloendothelial system and excellent clearance [13,14]. Further, utilization of surfactants including Triton-X 100, sodium dodecyl sulfate (SDS), etc. aid in improving the dispersibility. Camouflaging CNTs with biopolymers is one of the widely accepted techniques that not only prevent toxicity issues but also improve delivery efficacy (Figure 6.1) [15].

6.3 ENDOHEDRAL MODIFICATION

Endohedral modification is majorly advantageous in loading of biologically sensitive anticancer agents inside the nanotubes. This helps in initial burst release followed by controlled release of agents at specific site (Figure 6.1) [16].

6.4 CARBON NANOTUBES IN CANCER THERAPY

Over decades, CNTs have been extensively used in cancer theragnostic including drug/gene therapy, imaging, and diagnostics due to their smaller size and high aspect ratio, enabling them to carry large cargo with suitable modifications [5,17–19].

6.5 CARBON NANOTUBES IN ANTICANCER DRUG DELIVERY

Delivery of anticancer drugs like paclitaxel, doxorubicin, etc. has been a major concern due to their aqueous insolubility, nonspecificity, off-target side effects, and poor pharmacokinetics [5,17,18,20–22]. The utilization of functionalized CNTs (*f*-CNTs) in delivering anticancer drugs is explained below.

6.5.1 Taxanes and Its Derivatives

Taxanes have been one of the best classes of anticancer agents, especially in treating many cancers including breast, colon, lung, etc. Although successful, taxanes are highly insoluble in aqueous systems and therefore require an additional component (like Cremophor or polyethoxylated castor oil), which is toxic to patients. In order to minimize the toxic effects and maximize the delivery, many researchers successfully delivered taxanes such as paclitaxel (PTX), docetaxel (DTX), taxoid-SB-T-1214, and so on to target using CNTs. For instance, Li and co-workers synthesized PEG-functionalized SWCNTs loaded with PTX, which not only minimized the side effects but also improved efficacy in lung cancer [23]. In another study, aspartic-acid-tethered docetaxel-loaded CNTs exhibited remarkable cytotoxicity, bioavailability, and reduced clearance in breast cancer cells [24]. SB-T-1214, a taxoid-loaded SWCNT that has been functionalized with biotin, delivers biotin specifically and increases cytotoxicity in leukemia and lung cancer models (Figure 6.2) [25]. Another study demonstrated that

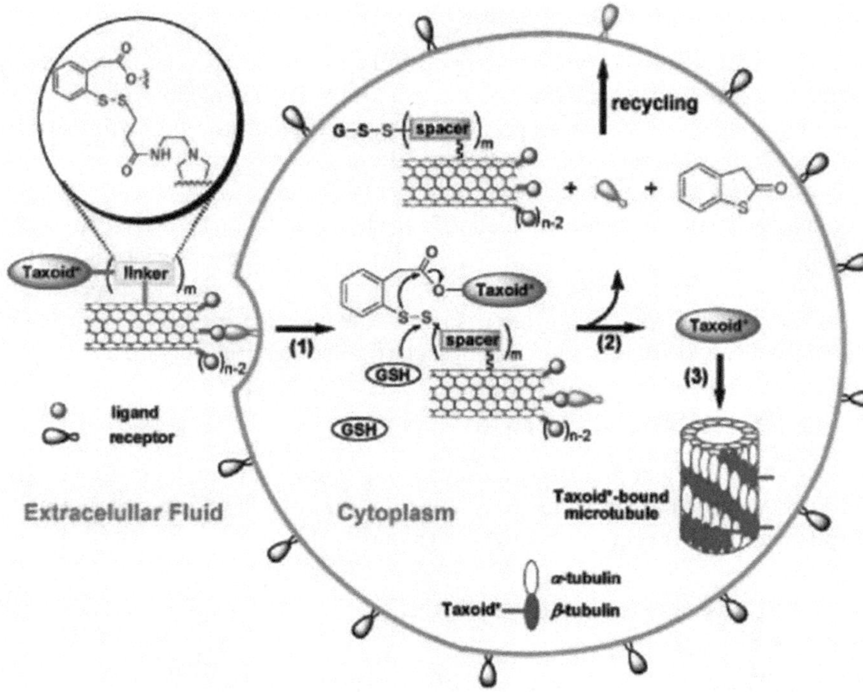

FIGURE 6.2 Schematic illustration of three key steps involved in the tumor-targeting drug delivery of biotin-linker-taxoid conjugate. (Reprinted (adapted) from Ref. [25]. Copyright © 2008, American Chemical Society.)

polycitric-acid-functionalized MWCNTs loaded with PTX had improved antitumor activity in *in vitro* lung and ovarian cancer models [26]. For site-specific delivery of docetaxel and coumarin-6 to lung cancer cells, chitosan-folate-coated MWCNTs were synthesized. The synthesized nanoformulation exhibited greater cellular uptake and 89-fold increase in anticancer activity than Docel™ [27].

6.5.2 Platinum-Based Anticancer Agents

Therapeutic efficacy of platinum-based anticancer drugs (cisplatin, carboplatin, etc.) exhibited greater off-target side effects such as nephrotoxicity, neurotoxicity, and myelosuppression, which forced clinicians to use lower doses. Employing the lower doses has not only decreased therapeutic outcomes but also promoted tumor drug resistance [28]. Many researches were performed in utilizing the CNTs for delivering the platinum-based anticancer agents [29–32]. The earliest study by Shanta Dhar and co-workers has demonstrated the novel phospholipid-polyethylene glycol (PL-PEG)-coated CNT-based delivery system for delivering platinum prodrug, i.e., Pt (IV). The prepared formulation exhibited 25-folds greater toxicity in testicular cancer cells (nTera-2) [32]. Further, in human choriocarcinoma (JAR) and nasopharyngeal carcinoma (KB) cell lines, SWCNT-Pt(IV)-FA demonstrated tumor-specific delivery via the folate receptor [33].

The utilization of endohedral cavities of CNTs for drug encapsulation was first demonstrated by Hampel and co-workers [34]. Endohedral-filled carboplatin in CNTs exhibited potent anticancer efficacy with controlled release in EJ28 bladder cancer cells [34]. Similarly, oxaliplatin (OX) was endohedrally loaded into PEG-MWCNTs (MWCNT-PEG-OX), which displayed sustained release profiles and greater anti-cancer efficacy [30]. Jian and colleagues created "CNT bottles" by injecting cisplatin endohedrally into CNT and coating them with functionalized gold nanoparticles from the external environment. The CNT bottles helped in increasing anticancer efficacy by lowering side effects [29].

6.5.3 Anthracycline Drugs

Among anthracyclines, doxorubicin (DOX) is the most conjugated drug due to its interaction with CNTs via π–π stacking and innate fluorescence. Liu and colleagues synthesized DOX-conjugated PEGlyated SWCNTs which showed reduced off-target side effects and higher retention in Ascites Sarcoma 180 [35]. For instance, DOX-loaded MWCNTs exhibited improved cytotoxicity in MCF-7 cell lines [36]. In another study, hyaluronate conjugated MWCNTs are conjugated with DOX via π–π stacking. Further testing in *in vitro* and *in vivo* cancer models such as HA-MWCNT-DOX showed greater tumor specificity, pH-triggered release, and increased cytotoxicity [37]. Triple functionalization of SWCNTs, i.e., with DOX, a monoclonal antibody and a fluorescent marker, was also designed for simultaneous detection and tumor killing (Figure 6.3) [38].

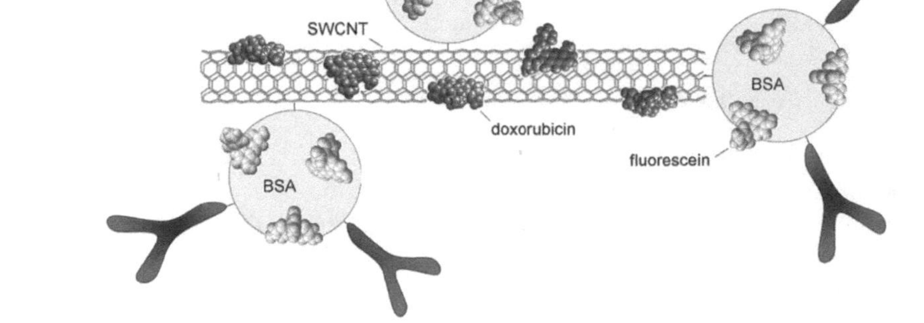

FIGURE 6.3 Schematic illustration of the doxorubicin–fluorescein–BSA–antibody-SWCNT complexes (Charcoal grey=doxorubicin, dark grey=fluorescein, light grey=BSA, Jet Black=antibodies). Insert: AFM image of doxorubicin–fluorescein–BSA–SWCNT complexes (without antibodies). (Reprinted (adapted) from Ref. [38]. Copyright © 2008, American Chemical Society.)

6.5.4 Other Anticancer Agents

Further from the above-discussed drugs, many research groups proved the successful delivery of other anticancer drugs such as raloxifene, tamoxifen curcumin, gemcitabine, and so on [39–44]. Gopinath and colleagues created alginate (ALG), and chitosan (CHI) functionalized SWCNTs loaded with curcumin (CUR) (CUR-CHI-ALG-SWCNT) which exhibited pH-dependent release and anticancer efficacy in A549 and NIH 3T3 cell lines [41]. Further, folic-acid-tethered CNTs loaded with raloxifene hydrochloride (RLX) (FA-RLX-CNT) were synthesized and tested in *in vitro* breast cancer models. Upon testing, FA-RLX-CNT demonstrated remarkable specificity, pH-dependent release, and enhanced cytotoxicity than RLX-CNT and naïve drugs [39]. In another study, MWCNT loaded with N-desmethyl tamoxifen and quercetin using tetraethylene glycol as a biodegradable linker (N-TAM-TEG-MWCNT-QT) showed enhanced cellular uptake and improved efficacy and biocompatibility [40]. Further, bromocriptine-conjugated functionalized MWCNTs (MWCNTs-BRCs Nf) were synthesized and tested its anticancer efficacy against lung cancer cells (i.e., A549 & QU-BD cells). The study reported that the prepared MWCNTs-BRCs-Nf exhibited a significant lethal effect on cancer cells, induced apoptosis, and specifically targeted delivery [17].

The studies above clearly demonstrate that CNTs have a larger potential for delivering numerous drug payloads to tumor tissues, with or without a targeting agent.

6.6 CARBON NANOTUBES IN ANTICANCER GENE DELIVERY

Unlike small drug molecules, biomolecules such as peptides, proteins, dsDNA, siRNA, miRNA, oligonucleotides, and aptamers lack stability (by plasma enzymes) and cannot readily permeate biological membranes. Many viral and non-viral vectors have been used to deliver these nucleic acids, but their ability to induce undesirable immunogenic effects is a challenge. f-CNTs have been used to deliver various nucleic acids to tumors due to their lack of immunogenicity. Covalent or non-covalent modification of CNTs for attachment to macromolecular conjugates has been instrumental in achieving intracellular delivery of biomolecules. However, endosome-disrupting agents are essential for releasing these biomolecules into endosomes.

6.6.1 Delivery of Plasmid DNA/dsDNA

f-CNT has been used to transfect dsDNA or plasmid DNA into tumor tissue. Functionalization such as ammonium ethylenediamine, PEI, PAMAM, and chitosan are the most preferred for delivery of plasmid DNA/dsDNA. Pantarotto and his colleagues demonstrated the earliest successful transfection or delivery of plasmid DNA using two different f-CNTs, such as ammonium- and lysine-functionalized. These results suggest that the charge density and surface area of CNTs are important factors affecting DNA–CNT complexes [45]. In another study, multi-functionalized SWCNTs (SWNTs-dsDNA-FAM-PL-PEG-FA) loaded with fluorescein-labeled (FAM) dsRNA and surface-modified folate-modified phospholipids for active targeting were designed. The developed formulation of SWNTs-dsDNA-FAM-PL-PEG-FA

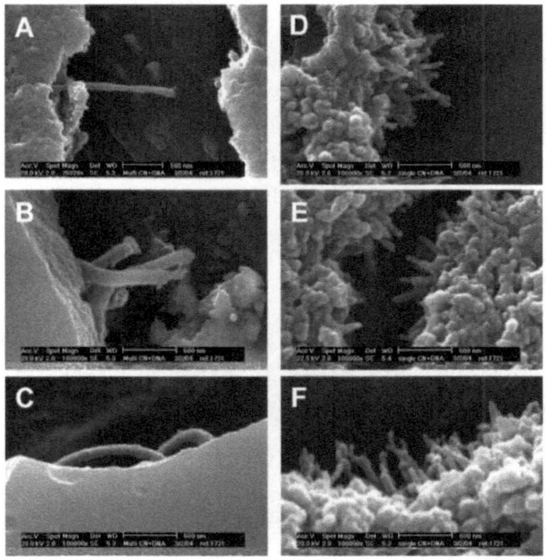

FIGURE 6.4 SEM images of CNT:DNA complexes formed at a 6:1 charge ratio: (a–c) MWNT-NH_3 +:DNA; (d–f) SWNT-NH_3+:DNA. (Reprinted (adapted) with permission from Ref. [46]. Copyright 2005 American Chemical Society.)

showed tumor-targeting properties and improved DNA transfection in HeLa cell lines (Figure 6.4) [46]. Karmakar et al. created ethylenediamine-functionalized SWCNTs (f-SWCNTs) carrying the p53 gene (f-SWCNTs-p53). When tested in MCF-7 breast cancer cells, f-SWCNT-p53 showed enhanced cellular uptake and induction of apoptosis [47].

6.6.2 Delivery of RNAi Components

In recent years, RNA interference has emerged as an excellent gene-silencing process using siRNA and miRNA at the post-transcriptional level. Despite its efficacy, RNAi is prone to stability problems (due to plasma endonucleases), immune-mediated disruption, and adverse effects. In recent years, many researchers have used CNT-based nano-delivery systems to deliver siRNA and miRNA [48]. Wang and colleagues first demonstrated the use of f-CNTs (cyclin A(2) siRNA-f-SWNT complexes) to deliver cyclin A(2) siRNA. The prepared formulation "Cyclin A (2) siRNA-f-SWNTs" successfully silenced the cyclin A2 gene in the leukemia cell line K562 [49]. In another study, cationic CNTs loaded with siRNA promoted apoptosis in a lung cancer model. Further, Al-Jamal and colleagues demonstrated the loading of two different siRNA, i.e., siTOX and siNEG siRNAs, into cationic dendron functionalized MWCNTs. The developed formulation triggered cellular apoptosis in the lung cancer model [36]. Successful delivery of miRNA using CNTs was first demonstrated by Haifend Dong and colleagues, using PEI-g-GMR-functionalized MWCNTs (LNA-m-MB/PEI-g-GNR). The study results suggest that PEI-g-GNR could be a promising nonviral vector for gene therapy [50]. Antonella and colleagues developed polymer-coated CNTs using two different polymers, i.e., polyethyleneimine (PEI) and polyamidoamine dendrimer

(PAMAM), to deliver miR-503, a miRNA that regulates the angiogenesis. Upon testing in endothelial cells, the polymer-coated CNT efficiently delivered miRNAs and regulated angiogenesis's target genes (Figure 6.5) [51]. Furthermore, the same group investigated the biological contributions of polymer-coated CNTs in delivering miRNAs to HEK-293 cells. The authors report that optimization of polyamine-coated CNTs is key to achieving efficient filler delivery [52].

6.6.3 Delivery of Oligonucleotides

Antisense therapy is one of the important strategies for treating genetic alterations in cancer. In this therapy, biologically active antisense oligonucleotides (asODNs) regulate gene expression at the transcriptional or translational level. Compared to siRNAs, ODNs have low target specificity [53,54]. Few studies have examined the use of CNTs as nano-vectors to deliver ODNs. Pan et al. prepared polyamidoamine dendrimer-modified MWCNTs (dMNTs) conjugated with FITC-tagged antisense c-myc ODN (asODN) (asODN-dMNT complex) to downregulate breast and breast c-myc genes and demonstrated inhibition of cell proliferation by liver cancer model [55]. Compared with other composites (CNT-NH2-asODNs; dendrimer-CNT-asODNs-), the CNT-based composite exhibited the greatest transfection and anti-tumor effects, demonstrating the potential of CNTs as gene delivery vectors for cancer therapy [55]. Jia et al. constructed functionalized MWCNTs loaded with asODNs and fluorescently labeled probe semiconductor quantum dots (CdTe) (MWCNT-PEI-asODNs-CdTe). The prepared nanocomposite, MWCNT-PEI-asODNs-CdTe, exhibited a His ODN delivery efficiency of approximately 95% and induced apoptosis in HeLa cells [56]. The author has compiled the data of Nano vector (CNT) mediated drug delivery in cancer therapy (Table 6.1) [57].

6.7 LIMITATION AND FUTURE PROSPECTS

Despite having many key advantages, pCNTs are reported to have toxic determinants which cause greater tissue damage, cellular oxidative stress, and DNA damage [58–61]. Longer CNTs trigger foreign immune response, thereby creating macrophage burden – "*frustrated phagocytosis.*" *Frustrated phagocytosis* in turn creates a massive oxidative burst (through MPO pathway) – increasing ROS levels and activating NF-κB pathway [62]. Catalytic residues, such as cobalt, nickel, and iron, which are either localized on the outer surface or in the interior cavities, are considered manufacturing impurities and are known to increase oxidative stress [63,64]. For example, MWCNT has been linked to major side effects such as hyperplasia, eye irritation, and Kupffer cell destruction, whilst SWCNT has been linked to acute inflammation, necrosis, and granuloma forms. The degree of CNT suspension will also determine their cellular fate and bioavailability. Freely suspended CNTs are reported to circulate and distribute to more regions and are easily cleared than aggregated CNTs. Surface morphological analysis of CNTs is essential to determine defects that might cause imperfect bonding and non-carbon element doping [57,60,65]. Alternate approaches like functionalization of CNTs with biopolymers, the right selection of CNT, reducing CNT length, optimizing their size, functionalizations, and suspension are notable for overcoming the limitations mentioned above [16].

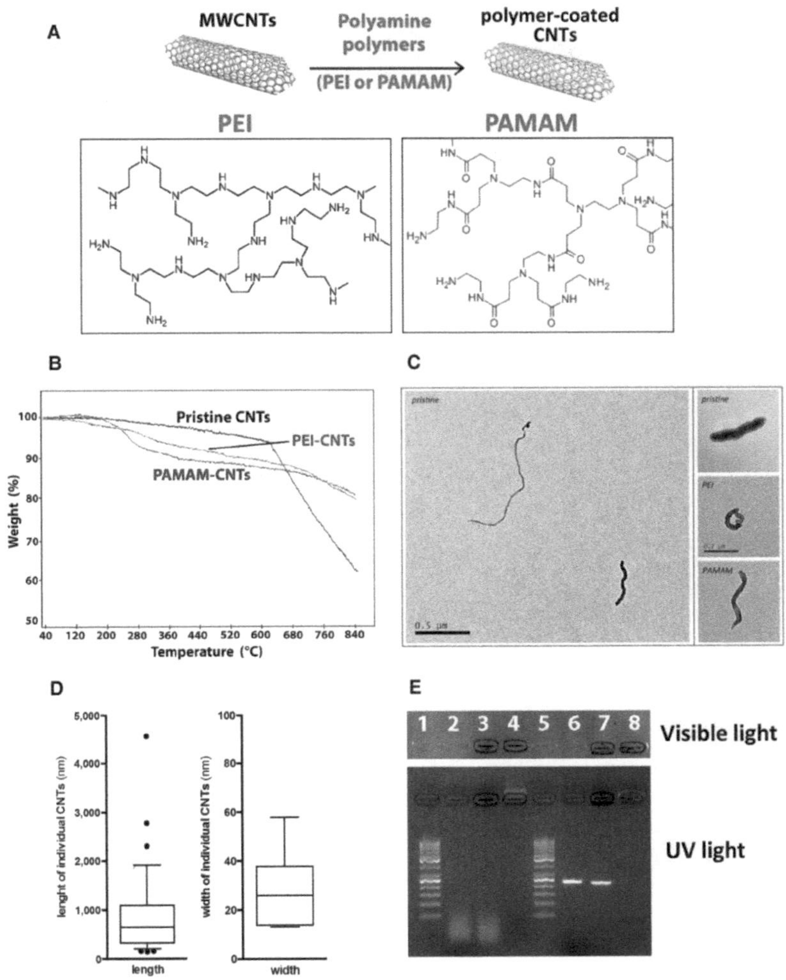

FIGURE 6.5 Polymer coating and dimensions of CNTs. (a) Schematic representation of polymer-coated (PEI and PAMAM) CNT preparation. Boxes report only a part of the PEI and PAMAM molecular structure. (b) Thermal analysis (TGA) performed on pristine CNTs, PEI-CNTs and PAMAM-CNTs. (c) Representative transmission electron microscopy (TEM) images of individual CNTs. Main panel showing inherent differences in CNT length and width, with small panels showing higher magnification of each type of CNT (scale bar = 200 nm). (d) Length and width of pristine CNTs derived from TEM images. Box and whisker plots with central bar indicating median value, box indicating interquartile range, and whiskers 10%–90% range (length: $N=33$ from $n=5$ subsamples; width: $N=6$ from high resolution images of $n=3$ subsamples). (e) Agarose gel (1%) stained with Hoechst 33258 showing that PEI-coated CNTs are able to bind DNA (5:1 w/w) (lanes 4 and 8) compared to uncoated/pristine CNTs (lanes 3 and 7). DNA ladder (50 bp) was loaded in lanes 1 and 5, fragmented DNA (<50 bp) in lane 2 and a well-characterized DNA fragment (~250 bp) in lane 6. (Reprinted (adapted) with permission from Ref. [51]. Creative Commons license Copyright 2016 Elsevier Ltd.)

TABLE 6.1
Examples of Drug or Vector CNT Based Delivery System for Delivery of Cancer Therapeutics

	Category	Functionalizations (*f*)	Type of Cancer and Preclinical Studies Done			Type of CNTs Preferred	Therapeutic Outcome
			Cancer Type	*In Vitro*	*In Vivo*		
			Anti-Cancer Agents				
1.	Platinum drug – Pt (IV), Prodrug	PL-PEG functionalized	Testicular cancer	Ntera-2		SWCNT	Increase cytotoxicity by 25 folds
2.	Cisplatin(CP)	Covalent amide linkage	Head and Neck cancer	HNCC		SWCNT	More effective than free CP and siRNA conjugate
3.	Carboplatin (CA)	Interior filling	Bladder cancer	EJ28		SWCNT & MWCNT	SWCNT conjugate showed controlled release like properties
4.	Oxaliplatin	PEG600 functionalized	Colorectal cancer	HT29		MWCNT	Delayed Cytotoxic activity
5.	Doxorubicin (DOX)	PEGylation functionalized		Ascites Sarcoma 180		SWCNT	Higher Retention of drug in tumor; Less effect on other tissues
6.		Non-functionalized	Breast cancer	MCF-7		MWCNT	Enhanced cytotoxicity
7.		P-gp functionalized	Leukemia	K562		SWCNT	Increased cellular uptake and cytotoxicity by 23 folds

(*Continued*)

TABLE 6.1 (*Continued*)
Examples of Drug or Vector CNT Based Delivery System for Delivery of Cancer Therapeutics

	Functionalizations (*f*)	Type of Cancer and Preclinical Studies Done			Type of CNTs Preferred	Therapeutic Outcome
Category	**Cancer Type**	***In Vitro***	***In Vivo***			
8. Paclitaxel (PTX)	PEGylated functionalized	Lung cancers	A549 & NCI-H460		SWCNT	Potentiated PTX activity
9.		Ovarian cancer	OVCAR3		SWCNT	Chemosensitizer. Enhanced cell death
10.	Polycitric acid functionalized	Lung cancer and ovary cancer	A549 and SKOV3		MWCNT	Enhanced potency
11.	PEGylated functionalized			Murine 4T1 model	SWCNT	Suppressed tumor growth and 10 folds higher uptake
12. SB-T-1214, taxoid	Biotin functionalization	Lung cancer and Leukemia	L1210FR, L1210 W138		SWCNT	More specifically targeted due to biotin
13. Gemcitabine (GEM)	Folic acid functionalized	Breast cancer	MCF-7 cell lines		MWCNT	More cytotoxic than free GEM
14.	PEGylated functionalized	Pancreatic cancer		BxPC-3 – BALB/c mice,		Decreasing metastatic lymph nodes
			Gene Therapy			
15. HIF-1α siRNA	Cationic functionalized			MiaPaCa-2 /HRE bearing mouse	SWCNT	Tumor growth suppression
16. Anti-EGFP siRNA	PEI and pyridinum functionalized	Lung cancer	H1299		MWCNT	Increased cytotoxicity
17. Cyclin A_2 siRNA	Amine (NH_3^+) functionalized	Leukemia	K562 cell lines		SWCNT	Decreased cell proliferation & induced apoptosis

(*Continued*)

TABLE 6.1 (*Continued*)
Examples of Drug or Vector CNT Based Delivery System for Delivery of Cancer Therapeutics

Category	Functionalizations (*f*)	Type of Cancer and Preclinical Studies Done			Type of CNTs Preferred	Therapeutic Outcome
		Cancer Type	*In Vitro*	*In Vivo*		
18. PLK 1 siRNA	Amine (NH_3^+) functionalized	Lung cancer	Calu6 cell lines	Lung cancer xenograft mice	MWCNT	Tumor eradication
19. siNEG siRNA	Amidation with putrescine, spermidine, and spermine	Lung cancer	A549 cell lines		SWCNT and MWCNT	Decrease in tumor proliferation
20. Bcl9l siRNA	Aptamer and piperazine-polyethylenimine derived functionalization	Breast cancer	MDA-MB-231 cell lines		SWCNT	Helped in achieving targeted silencing activity of the complex
21. GAPDH siRNA	Polyethylenimine (PEI) functionalized	Cervical cancer	HeLa cell lines		SWCNT	Induced cell death
22. siTOX and siNEG siRNAs	Cationic dendron functionalized	Cervical and lung cancer	HeLa and A549 cell lines		MWCNT	Minimized toxicity, effective delivery, and gene silencing capabilities
23. PLK-1 and siTOX siRNAs	Amino-functionalized or cationic liposomes	Breast, cervical, lung, melanoma, prostate, renal, kidney cancers	Calu6, A549, DU145, C-33A, MCF-7, HEK293, HeLa, NIH 3T3, B16F10 cell lines	Lung xenografted nude mice	MWCNT	Induced apoptosis, tumor suppression
24. PLK1 siRNA	Ammonium and guanidium dendron	Cervical cancer	HeLa cell lines		MWCNT	Showed significant silencing of PLK-1
25. hTERT siRNA with NIR phototherapy	Polyethylenimine functionalized	Prostate cancer	PC-3 cell lines	Tumor bearing mice	SWCNT	Showed severe apoptosis and strong tumor suppression

(*Continued*)

TABLE 6.1 (*Continued*)
Examples of Drug or Vector CNT Based Delivery System for Delivery of Cancer Therapeutics

Category	Functionalizations (*f*) Cancer Type	Type of Cancer and Preclinical Studies Done		Type of CNTs Preferred	Therapeutic Outcome
		In Vitro	***In Vivo***		
26. pDNA	Cationic glycopolymer functionalized	Cervical cancer	HeLa cell lines	SWCNT	The functionalization is biocompatible and efficient in transfection compared to lipofectamine 2000
27. p53 plasmids	Ethylenediamine functionalized	Breast cancer	MCF-7 cell lines	SWCNT	Induced apoptosis
28. pDNA	Ammonium functionalized	Cervical cancer	HeLa and CHO cell lines	SWCNT and MWCNT	First reported as gene cargo carrier
29. GFP gene	Polyamidoamine (PAMAM)	Cervical cancer	HeLa cell lines	MWCNT	Improved GFP gene transfection
30. dsDNA	Multi-functionalization	Epithelial cancer	Mouse ovarian surface epithelial cells	SWCNT	Targeted drug delivery
31. miRNA	PEI-g-GNR	Cervical cancer	HeLa cell lines	MWCNT	Provided a new nonviral vector for insitu detection
32. dsODN ~ NF-kB	Carbodiimide functionalization	Cervical cancer	HeLa cell lines	SWCNT	Successfully delivered the dsODN to target cells
33. Antisense *c-myc* ODN's	Polyamidoamine	Breast cancer and liver cancer	MCF-7, MDA-MB-435, HepG2 cell lines	MWCNT	Showed maximum gene efficiency
34. MUC-1 Apt	Carboxylate functionalization	Breast and lung cancer	MCF-7, Calu-6 cell lines	MWCNT	Internalization of aptamer was done successfully

Source: Reproduced with permission from Ref. [57]. Copyright 2020 Elsevier B.V.

6.8 CONCLUSION

Over the past few decades, CNTs have attracted numerous nano-oncologists and nano-pharmaceutical experts for their essential properties, including high aspect ratio, nano-needle-like cell entry, increased loading capacity, and ease of functionalization, which also contribute to multiple entities, including location-specific targeting. Attachment of different drug or organic moieties could be achieved through endohedral or exohedral functionalization. In addition, although they have good properties, extensive knowledge and a clear understanding of CNTs for pharmaco-toxicological aspects of clinical use are warranted before advising their daily use.

DECLARATIONS

Funding details: This work was supported by the Department of Science and Technology – Indo-Uzbek Joint Research Programme (INT/Uzbek/P-01), New Delhi.

Conflict of interest: None.

REFERENCES

1. Siegel, R. L., Miller, K. D. & Jemal, A. Cancer statistics, 2020. *CA. Cancer J. Clin.* **70**, 7–30 (2020).
2. DeVita, V. T. & Chu, E. A history of cancer chemotherapy. *Cancer Res.* **68**, 8643 (2008).
3. Miller, A. D. Human gene therapy comes of age. *Nature* **357**, 455–460 (1992).
4. Kumar, A., Mansour, H. M., Friedman, A. & Blough, E. R. *Nanomedicine in Drug Delivery.* CRC Press, Boca Raton, FL, 2013.
5. Mehra, N. K., Mishra, V. & Jain, N. K. A review of ligand tethered surface engineered carbon nanotubes. *Biomaterials* **35**, 1267–1283 (2014).
6. Iijima, S. & Ichihashi, T. Single-shell carbon nanotubes of 1-nm diameter. *Nature* **363**, 603–605 (1993).
7. Bianco, A., Kostarelos, K., Partidos, C. D. & Prato, M. Biomedical applications of functionalised carbon nanotubes. *Chem. Commun.* **5**, 571–577 (2005).
8. Scott, C. D., Arepalli, S., Nikolaev, P. & Smalley, R. E. Growth mechanisms for single-wall carbon nanotubes in a laser-ablation process. *Appl. Phys. A* **72**, 573–580 (2001).
9. Li, Y. et al. Purification of CVD synthesized single-wall carbon nanotubes by different acid oxidation treatments. *Nanotechnology* **15**, 1645–1649 (2004).
10. Su, M., Zheng, B. & Liu, J. A scalable CVD method for the synthesis of single-walled carbon nanotubes with high catalyst productivity. *Chem. Phys. Lett.* **322**, 321–326 (2000).
11. Dai, H. Carbon nanotubes: synthesis, integration, and properties. *ACC. Chem. Res.* **35**, 1035–1044 (2002).
12. Hirsch, A. & Vostrowsky, O. Functionalization of carbon nanotubes. In: Schlüter, A. D. (ed) *Functional Molecular Nanostructures*, Springer, Berlin Heidelberg, vol. 245, pp. 193–237, 2005.
13. Bilalis, P., Katsigiannopoulos, D., Avgeropoulos, A. & Sakellariou, G. Non-covalent functionalization of carbon nanotubes with polymers. *RSC Adv.* **4**, 2911–2934 (2014).
14. Prakash, S., Malhotra, M., Shao, W., Tomaro-Duchesneau, C. & Abbasi, S. Polymeric nanohybrids and functionalized carbon nanotubes as drug delivery carriers for cancer therapy. *Adv. Drug Deliv. Rev.* **63**, 1340–1351 (2011).
15. Moore, V. C. et al. Individually suspended single-walled carbon nanotubes in various surfactants. *Nano Lett.* **3**, 1379–1382 (2003).

16. Ravi Kiran, A. V., Kusuma Kumari, G., Krishnamurthy, P. T., Chintamaneni, P. K. & Pindiprolu, S. K. S. Carbon nanotubes in cancer therapy. In: Abraham, J., Thomas, S., Kalarikkal, N. (eds) *Handbook of Carbon Nanotubes*. Springer International Publishing, Cham, pp. 1739–1771, 2022.
17. Kamazani, F. M., Sotoodehnejad Nematalahi, F., Siadat, S. D., Pornour, M. & Sheikhpour, M. A success targeted nano delivery to lung cancer cells with multi-walled carbon nanotubes conjugated to bromocriptine. *Sci. Rep.* **11**, 24419 (2021).
18. Mehra, N. K. & Jain, N. K. Multifunctional hybrid-carbon nanotubes: new horizon in drug delivery and targeting. *J. Drug Target.* **24**, 294–308 (2016).
19. Singh, S., Mehra, N. K. & Jain, N. K. Development and characterization of the paclitaxel loaded riboflavin and thiamine conjugated carbon nanotubes for cancer treatment. *Pharm. Res.* **33**, 1769–1781 (2016).
20. Chary, P. S. et al. A review on emerging role of multifunctional carbon nanotubes as an armament in cancer therapy, imaging and biosensing. *J. Drug Deliv. Sci. Technol.* **85**, 104588 (2023).
21. Tang, L. et al. Insights on functionalized carbon nanotubes for cancer theranostics. *J. Nanobiotechnol.* **19**, 423 (2021).
22. Mehra, N. K., Verma, A. K., Mishra, P. R. & Jain, N. K. The cancer targeting potential of d-α-tocopheryl polyethylene glycol 1000 succinate tethered multi walled carbon nanotubes. *Biomaterials* **35**, 4573–4588 (2014).
23. Liu, Z. et al. Drug delivery with carbon nanotubes for in vivo cancer treatment. *Cancer Res.* **68**, 6652–6660 (2008).
24. Thotakura, N. et al. Aspartic acid tagged carbon nanotubols as a tool to deliver docetaxel to breast cancer cells: reduced hemotoxicity with improved cytotoxicity. *Toxicol. In Vitro* **59**, 126–134 (2019).
25. Chen, J. et al. Functionalized single-walled carbon nanotubes as rationally designed vehicles for tumor-targeted drug delivery. *J. Am. Chem. Soc.* **130**, 16778–16785 (2008).
26. Atyabi, F., Sobhani, Z., Dinarvand, R., & Ghahremani, R. Increased paclitaxel cytotoxicity against cancer cell lines using a novel functionalized carbon nanotube. *Int. J. Nanomedicine* **6**, 705–719 (2011). doi:10.2147/IJN.S17336.
27. Singh, R. P. et al. Chitosan-folate decorated carbon nanotubes for site specific lung cancer delivery. *Mater. Sci. Eng. C.* **77**, 446–458 (2017).
28. Wheate, N. J., Walker, S., Craig, G. E. & Oun, R. The status of platinum anticancer drugs in the clinic and in clinical trials. *Dalton Trans.* **39**, 8113 (2010).
29. Li, J. et al. Carbon nanotube bottles for incorporation, release and enhanced cytotoxic effect of cisplatin. *Carbon* N Y. **50**, 1625–1634 (2019). doi:10.1016/j.carbon.2011.11.043.
30. Wu, L. et al. PEGylated multi-walled carbon nanotubes for encapsulation and sustained release of oxaliplatin. *Pharm. Res.* **30**, 412–423 (2013).
31. Li, J. et al. In vivo biodistribution of platinum-based drugs encapsulated into multi-walled carbon nanotubes. *Nanomed. Nanotechnol. Biol. Med.* **10**, 1465–1475 (2014).
32. Feazell, R. P., Nakayama-Ratchford, N., Dai, H. & Lippard, S. J. Soluble single-walled carbon nanotubes as longboat delivery systems for platinum(IV) anticancer drug design. *J. Am. Chem. Soc.* **129**, 8438–8439 (2007).
33. Dhar, S., Liu, Z., Thomale, J., Dai, H. & Lippard, S. J. Targeted single-wall carbon nanotube-mediated Pt(IV) prodrug delivery using folate as a homing device. *J. Am. Chem. Soc.* **130**, 11467–11476 (2008).
34. Hampel, S. et al. Carbon nanotubes filled with a chemotherapeutic agent: a nanocarrier mediates inhibition of tumor cell growth. *Nanomedicine* **3**, 175–182 (2008).
35. Liu, H., Xu, H., Wang, Y., He, Z. & Li, S. Effect of intratumoral injection on the biodistribution and therapeutic potential of novel chemophor EL-modified single-walled nanotube loading doxorubicin. *Drug Dev. Ind. Pharm.* **38**, 1031–1038 (2012).

36. Al-Jamal, K. T. et al. Cellular uptake mechanisms of functionalised multi-walled carbon nanotubes by 3D electron tomography imaging. *Nanoscale* **3**, 2627 (2011).
37. Datir, S. R., Das, M., Singh, R. P. & Jain, S. Hyaluronate tethered, "smart" multiwalled carbon nanotubes for tumor-targeted delivery of doxorubicin. *Bioconjug. Chem.* **23**, 2201–2213 (2012).
38. Heister, E. et al. Triple functionalisation of single-walled carbon nanotubes with doxorubicin, a monoclonal antibody, and a fluorescent marker for targeted cancer therapy. *Carbon* **47**, 2152–2160 (2009).
39. Jawahar, N., De, A., Jubee, S. & Reddy, E. S. Folic acid-conjugated raloxifene hydrochloride carbon nanotube for targeting breast cancer cells. *Drug Dev. Res.* **81**, 305–314 (2020).
40. Kumar, M. et al. Desmethyl tamoxifen and quercetin-loaded multiwalled CNTs: a synergistic approach to overcome MDR in cancer cells. *Mater. Sci. Eng. C.* **89**, 274–282 (2018).
41. Singh, N., Sachdev, A. & Gopinath, P. Polysaccharide functionalized single walled carbon nanotubes as nanocarriers for delivery of curcumin in lung cancer cells. *J. Nanosci. Nanotechnol.* **18**, 1534–1541 (2018).
42. Razzazan, A., Atyabi, F., Kazemi, B. & Dinarvand, R. In vivo drug delivery of gemcitabine with PEGylated single-walled carbon nanotubes. *Mater. Sci. Eng. C.* **62**, 614–625 (2016).
43. Arsawang, U. et al. How do carbon nanotubes serve as carriers for gemcitabine transport in a drug delivery system? *J. Mol. Graph. Model.* **29**, 591–596 (2011).
44. Singh, R., Mehra, N. K., Jain, V. & Jain, N. K. Gemcitabine-loaded smart carbon nanotubes for effective targeting to cancer cells. *J. Drug Target.* **21**, 581–592 (2013).
45. Singh, R. et al. Binding and condensation of plasmid DNA onto functionalized carbon nanotubes: toward the construction of nanotube-based gene delivery vectors. *J. Am. Chem. Soc.* **127**, 4388–4396 (2005).
46. Yang, X. et al. Multi-functionalized single-walled carbon nanotubes as tumor cell targeting biological transporters. *J. Nanoparticle Res.* **10**, 815–822 (2008).
47. Karmakar, A. et al. Ethylenediamine functionalized-single-walled nanotube (f-SWNT)-assisted in vitro delivery of the oncogene suppressor p53 gene to breast cancer MCF-7 cells. *Int. J. Nanomedicine.* **6**, 1045–1055 (2011) doi: 10.2147/IJN.S17684.
48. Harvey, J. D. et al. A carbon nanotube reporter of microRNA hybridization events in vivo. *Nat. Biomed. Eng.* **1**, 0041 (2017).
49. Wang, X., Ren, J. & Qu, X. Targeted RNA interference of cyclin A2 mediated by functionalized single-walled carbon nanotubes induces proliferation arrest and apoptosis in chronic myelogenous leukemia K562 Cells. *ChemMedChem.* **3**, 940–945 (2008).
50. Dong, H., Ding, L., Yan, F., Ji, H. & Ju, H. The use of polyethylenimine-grafted graphene nanoribbon for cellular delivery of locked nucleic acid modified molecular beacon for recognition of microRNA. *Biomaterials* **32**, 3875–3882 (2011).
51. Masotti, A. et al. Regulation of angiogenesis through the efficient delivery of microRNAs into endothelial cells using polyamine-coated carbon nanotubes. *Nanomed. Nanotechnol. Biol. Med.* **12**, 1511–1522 (2016).
52. Celluzzi, A. et al. Biophysical and biological contributions of polyamine-coated carbon nanotubes and bidimensional buckypapers in the delivery of miRNAs to human cells. *Int. J. Nanomedicine.* **13**, 1–18 (2017).
53. Uhlmann, E. & Peyman, A. Antisense oligonucleotides: a new therapeutic principle. *Chem. Rev.* **90**, 543–584 (1990).
54. Lima, W. F., Vickers, T. A., Nichols, J., Li, C. & Crooke, S. T. Defining the factors that contribute to on-target specificity of antisense oligonucleotides. *PLoS One.* **9**, e101752–e101752 (2014).

55. Pan, B. et al. Synthesis and characterization of polyamidoamine dendrimer-coated multi-walled carbon nanotubes and their application in gene delivery systems. *Nanotechnology* **20**, 125101 (2009).
56. Jia, N. et al. Intracellular delivery of quantum dots tagged antisense oligodeoxynucleotides by functionalized multiwalled carbon nanotubes. *Nano Lett.* **7**, 2976–2980 (2007).
57. Ravi Kiran, A. V. V. V., Kusuma Kumari, G. & Krishnamurthy, P. T. Carbon nanotubes in drug delivery: focus on anticancer therapies. *J. Drug Deliv. Sci. Technol.* **59**, 101892 (2020).
58. Beg, S. et al. Advancement in carbon nanotubes: basics, biomedical applications and toxicity. *J. Pharm. Pharmacol.* **63**, 141–163 (2011).
59. Shvedova, A. et al. Exposure to carbon nanotube material: assessment of nanotube cytotoxicity using human keratinocyte cells. *J. Toxicol. Environ. Health A.* **66**, 1909–1926 (2003).
60. Pulskamp, K., Diabaté, S. & Krug, H. F. Carbon nanotubes show no sign of acute toxicity but induce intracellular reactive oxygen species in dependence on contaminants. *Toxicol. Lett.* **168**, 58–74 (2007).
61. Boyles, M. S. P. et al. Multi-walled carbon nanotube induced frustrated phagocytosis, cytotoxicity and pro-inflammatory conditions in macrophages are length dependent and greater than that of asbestos. *Toxicol. In Vitro.* **29**, 1513–1528 (2015).
62. Kagan, V. E. et al. Carbon nanotubes degraded by neutrophil myeloperoxidase induce less pulmonary inflammation. *Nat. Nanotechnol.* **5**, 354 (2010).
63. Bethune, D. S. et al. Cobalt-catalysed growth of carbon nanotubes with single-atomic-layer walls. *Nature* **363**, 605–607 (1993).
64. Sharifi, S. et al. Toxicity of nanomaterials. *Chem. Soc. Rev.* **41**, 2323–2343 (2012).
65. Charlier, J.-C. Defects in carbon nanotubes. *ACC Chem. Res.* **35**, 1063–1069 (2002).

7 Carbon Nanotubes in Diagnostic and Imaging Application

K. M. Gijith Mohan and Keerti Jain

ABBREVIATIONS

AFM	atomic force microscopy
CA	contrasting agent
CNT	carbon nanotube
CT	computed tomography
DWCNT	double-walled carbon nanotube
EGF	epidermal growth factor
GO	glucose oxidase
HA	hyaluronic acid
MRI	magnetic resonance Imaging
MWCNT	multi-walled carbon nanotube
NEMS	nanoelectromechanical systems
NPs	nanoparticles
PA	photoacoustic
PEG	polyethylene glycol
PET	positron emission tomography
RGD	arginylglycylaspartic acid
RS	Raman spectroscopy
SERS	surface-enhanced Raman scattering
SPECT	single-photon emission computed tomography
SPION	superparamagnetic iron oxide nanoparticle
SWCNT	single-walled carbon nanotube
TAT	thermo acoustic tomography

7.1 INTRODUCTION

Carbon nanotubes (CNTs) belong to the class of nanomaterials, composed of carbon atoms forming tube-like structures whose diameter falls in the nanometer scale usually less than 100 nm [1]. CNTs are made exclusively of sp^2 hybridized carbon atoms organized hexagonally in a sequence of condensed benzene rings at an interatomic spacing of 1.4 Å. This forms graphene, a two-dimensional sheet-like material that is capable of being rolled up into a tubular shape giving rise to CNTs

 DOI: 10.1201/9781003386537-7

[2–4]. CNTs come under the family of fullerenes, which are allotropes of carbon [4,5]. This tubular structure, which can act as a cage for many active molecules, imparts properties to act as a diagnostic and imaging agent [6]. The graphene sheets are rolled in specific angles. Together the angle and the radius of the tube formed are the major factors that influence the properties of CNTs. The major applications of CNTs are contributed by their unique properties like specificity of functionalization and biocompatibility, and these entities find a range of applications in the biomedical field [7,8].

Advancements in the field of nanotechnology laid the pathway for the development of so many applications for CNTs including imaging, diagnosis, targeted drug delivery, biosensing, bioelectronics, etc. [9]. Graphene-derived or oligonucleotide noncomplex is used in *in vivo* detection of biomolecules, detection of DNA, bioimaging, protein identification and characterization, identification of pathogens, etc. [10]. Nanomedicine is one of the most important emerging research fields, which is successful in prevention, imaging, diagnosis, and treatment of many diseases where most of the conventional methods have limited access [11–13]. Nanomedicines may also facilitate the targeted delivery of bioactive molecules including drugs, genes, and imaging agents to the target site [8,14,15]. Also, this technology is applied in the development of nanodevices and nanorobots, which are capable of early diagnosis of many diseases like cancer and Alzheimer's disease and can identify pathogens in case of infectious diseases [7,16,17]. There are several characteristic properties possessed by the CNTs like high elasticity, high thermal conductivity, low density, chemical inertness, small size, large surface area, chemical stability, biocompatibility, and flexibility, which make them play important roles in diagnosis and bioimaging [8,18]. Functionalizing their surface by tagging with pharmacologically active particles, their properties can be tuned to produce biomedically relevant molecules possessing various applications [1,19].

7.2 BIOMEDICAL APPLICATIONS OF CNTs

Successful treatment for life-threatening diseases begins with early detection and diagnosis followed by the right choice of medications and treatments. CNTs can fulfill this requirement as they possess several characteristic attributes that make them suitable for many biomedical applications. Compared to conventional medications, CNTs possess following qualities:

- light weight
- high biocompatibility
- chemical inertness
- big surface area
- high strength (~37 GPa)
- thermal conductivity (~3,500 W/m/K) [20].
- fast electron transfer kinetics
- semi and metallic conductive properties
- high elastic modulus compared to all other materials [21].

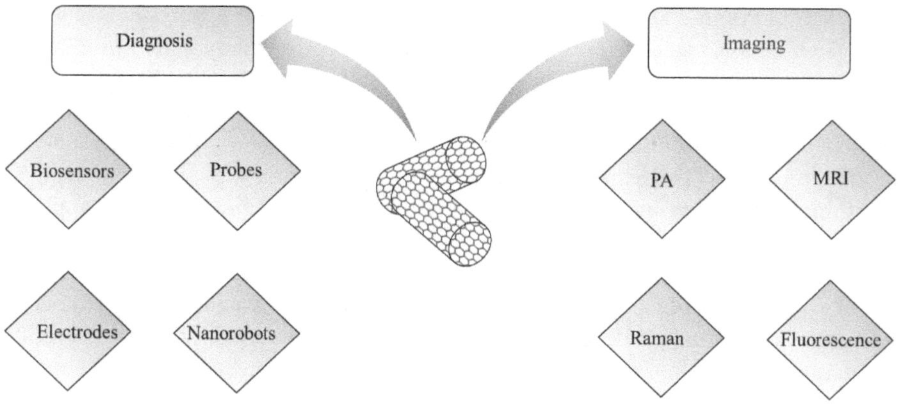

FIGURE 7.1 Diagnostic and imaging applications of CNTs.

CNTs are effectively used as diagnostic as well as imaging agents by virtue of these properties, which provide them with a high sensitivity and specificity. Functionalized CNTs have a high binding affinity towards the particular proteins expressed during the early stages of many diseases like cancer and Alzheimer's disease providing an option for early detection [22] (Figure 7.1).

The diagnostic applications of CNTs make use of nanofluidic systems, implantable nanodevices such as nanosensors and nanorobots, and biosensor diagnostics and analysis. Single-walled carbon nanotubes (SWCNTs) strongly absorb the electromagnetic radiations, in the regions of ultraviolet (UV) to near-infrared (NIR), which lays the foundation for the photoacoustic (PA) imaging [23–25]. CNTs possessing coatings and the time-lapse visualization of S-SWCNTs (semiconducting SWCNTs) and M-SWCNTs (metallic SWCNTs) in live cells and animals can be used for imaging techniques [26]. In the following sections, we have discussed the diagnostic, imaging, and biosensing applications of CNTs.

7.3 CNTs AND DIAGNOSIS

Diagnostic applications of the CNTs are mainly attributed to their characteristic electronic and optical properties. By virtue of electronic properties, CNTs can be incorporated as a key element in electrochemical sensors and many types of label-free CNT-based biosensors [27,28]. Their surfaces can be attached with many imaging probes, thus making them capable of being used in many diagnostic applications. They can also be used as contrasting agents (CA) in many imaging techniques [8]. Common diagnostic techniques like X-ray, X-ray-computed tomography scan (CT-scan), magnetic resonance imaging (MRI), single-photon emission computed tomography (SPECT), and positron emission tomography (PET) are used in combination to get high-resolution images which could help us to have a sensitive and accurate diagnosis of the cancers at the early stages [29]. CNT-based nanoplatforms have a dual advantage as it can be used for imaging and therapeutic activities,

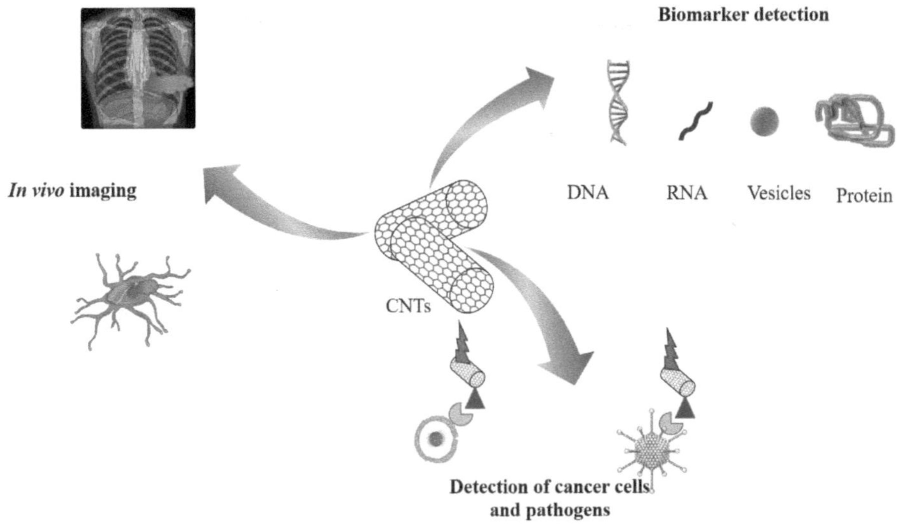

FIGURE 7.2 CNT-based diagnosing techniques. (Prepared using images from Servier Medical Art. https://creativecommons.org/licenses/by/4.0/.)

making them suitable in the field of biomedicine as a 'theranostic' modality [30]. CNTs coupled with glucose oxidase can be used as biosensors for monitoring blood sugar levels in diabetic patients and a more efficient treatment can be offered with higher accuracy. Other examples of CNT-based biosensors include dehydrogenase biosensors or peroxidase, and catalase biosensors are applied in drug therapeutics and monitoring [31]. Bioengineering is applied to transform CNTs into various forms which find a variety of uses in the medical field (Figure 7.2).

7.3.1 Biosensors

Biosensors are specialized agents used for the detection and analysis of biomarkers. Analysis of biomarkers like bioreceptors, which include antibodies, enzymes, and aptamers, provides exact information about the biofunction and the pathological status, which can be used in deciding the best treatment regimen in clinical aspects [6,32]. CNTs are advantageous in biosensing applications as they can increase sensitivity and can narrow down the detection limits as they have small size and a large surface/volume ratio [33]. CNTs are found to be useful as biological sensors for the detection and monitoring of bacterial infectious agents, carcinoma, diabetes, Alzheimer's disease, etc. [1]. CNT-based biosensors have fast electron-transfer properties due to their characteristic structural features and hence can be applied for diagnosis with high sensitivity. CNTs can be used as probes in the construction of nanolithography, nanoelectrodes, high-resolution structural imaging, nanosensors, etc. [29,34]. Several types of biosensors have been developed that are utilized in the analysis of biomolecules which help in the assessment of many disease conditions (Figure 7.3).

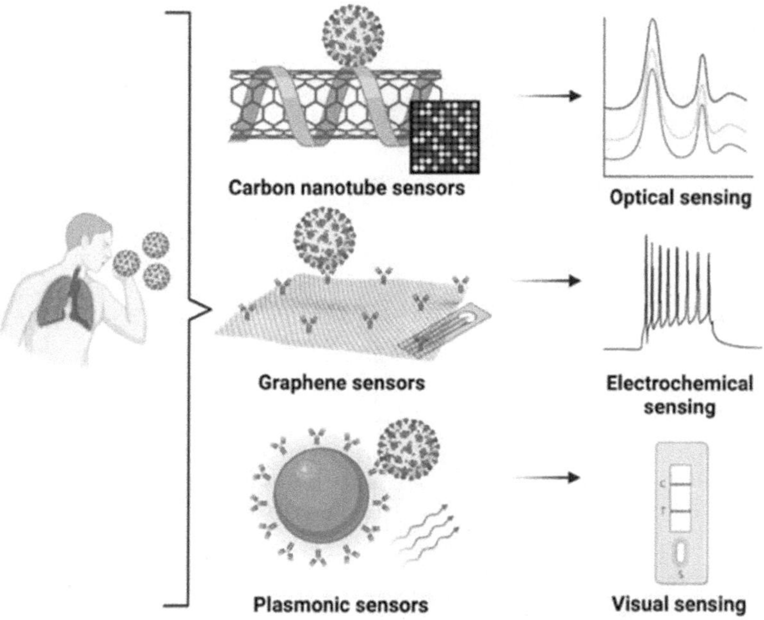

FIGURE 7.3 Schematic presentation of CNT-, graphene-, and plasmonic-based biosensors. (Reproduced with permission from Ref. [35].)

7.3.1.1 Enzyme Biosensors

The first developed CNT-based glucose sensor used the conjugation of the enzyme glucose oxidase (GO) to CNTs, along with a Teflon binder. Apart from glucose biosensors, CNTs are also used in the detection of fructose, galactose, neurotransmitters, amino acids, etc. CNT-based screen-printed amperometric biosensors can be effectively utilized for the rapid assay of organophosphorus (OP) compounds. Nickel-based metal organic framework (Ni-MOF) and CNT composite (Ni-MOF/CNTs) are promising candidates in glucose sensing applications [36].

7.3.1.2 DNA Biosensors

DNA Biosensors are used to test the genetics behind the diseases. Transducer electrodes are modified using CNTs conjugated with single-stranded (ss-) DNA probe sequences. These CNT-based electrodes can act as electrochemical DNA biosensors as they can generate transduction signals by forming a duplex. Many molecules are used as labels for the enhancement of the detection sensitivity, like enzymes, fluorophore, magnetic beads, or radioactive compounds [29]. These approaches offer a promising future for the treatment of many genetic disorders.

7.3.1.3 Cancer Biosensors

DNA, RNA, extracellular vesicles, or proteins that are present in the bodily fluids can serve as the specific biomarkers for cancer and aid in the early diagnosis and treatment [29]. A biosensing system can effectively identify the specific cancer biomarker antigens (CBAs). Thus, they can identify specifically the prostate-specific antigen

(PSA) which will be helpful in the early detection and appropriate treatment. Volatile organic components (VOC) can be detected in respiratory samples, which can serve as a reliable marker for the detection of lung cancer. An array of CNT sensors can be used to detect the increase in the rate of exhalation of these components in persons affected by lung cancer. Usually, SWCNTs, coated with non-polymeric organic substances, are used to detect VOC changes as lung cancer biomarkers [37,38]. A similar application is SWCNTs-Indocyanine Green (ICG) conjugated with cyclic ArgGly-Asp (RGD) peptides to target alpha (v) beta (3) integrins, which are utilized as a PA contrast agent in the cancer diagnosis [33].

7.3.2 Probes

Probes are used to study and obtain information about a remote region or cavity. CNT-based detection devices are highly suitable materials to be used as probes. One of the important applications of CNTs is the inclusion of atomic force microscopy (AFM) probes in detecting biomarkers and cells involved in cancer. The image obtained from AFM depends on the shape of the probe tip and the surface morphology of the sample of interest. Nanoprobes made of CNTs can be used to have high-resolution images owing to their cylindrical shape with small diameter so that morphological changes associated with disease conditions can be identified accurately. Also, this makes them advantageous for imaging of narrow and deep cavities where conventional methods are not effective in obtaining the details of the pathological state [24].

7.3.3 Implantable Nanosensors and Nanorobots

Implantable nanosensors and nanorobots can be used for the assessment of health, especially by monitoring the pulse, temperature, glucose, etc. Implantable nanosensors are used to monitor heart activity that can give an early indication of cardiac dysfunction and even these sensors can be combined with a defibrillator to regulate heartbeats, which is a lifesaving application [24]. A nanorobot is intended to detect cancer cells at a very early stage autonomously and further can be used to deliver the therapeutic agent at the target or diseased site specifically to have a more efficient treatment with less side effects [39]. They can be used to differentiate between healthy and malignant human tissues during direct *in vivo* surgery with the least amount of invasiveness and maximum level of precision, resulting in the least degree of damage. They can also be used to detect pathogens causing infectious diseases where they can selectively ingest and destroy them. Nanoelectromechanical systems (NEMS), which form the basis of micro and nanorobots, are derived from the CNTs, which provide high strength and chemical inertness and are not affected by the host's immune system [7].

7.3.4 CNT-Based Electrodes

CNTs have a superior electron conducting capacity and they can increase the electron transfer when they are used as electrode. It is evident that CNTs are better electrode material as compared to the conventional electrodes. CNTs present at the surface of

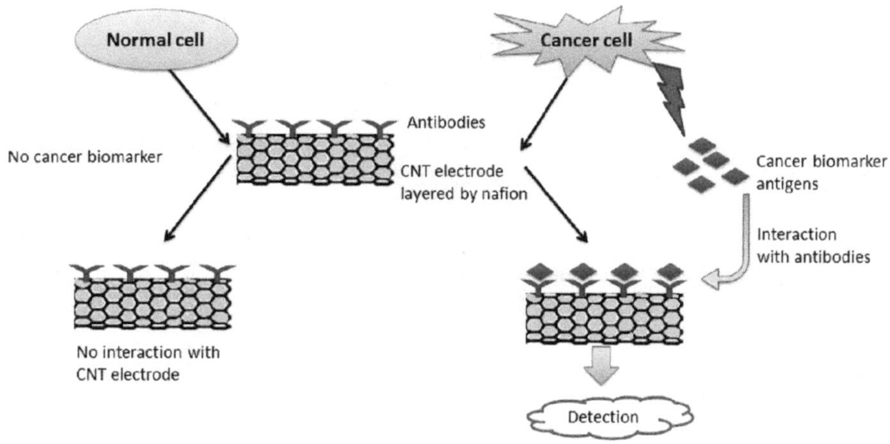

FIGURE 7.4 Detection of cancer cells using CNT-based electrode. (Reproduced with permission from Ref. [29].)

the electrodes may promote and improve the sensitivity of electrochemical reactions. Such CNT electrodes are used for the analysis of urine samples to detect uric acid with high degree of selectivity and sensitivity. Similarly, antibodies can also be chemically linked to the CNTs, which makes them a sensitive probe for the complementary counterpart, cancer antigens, in the given specimen for efficient detection [40] (Figure 7.4).

7.4 CNTs AND IMAGING

CNT-based imaging techniques are now evolving owing to their many advantages over the prevailing conventional techniques. SWCNTs and multi-walled carbon nanotubes (MWCNTs) are extensively used in biomedical imaging as nanoprobes with extremely large scattering cross sections [41]. Due to the intrinsic optical properties and modifiability, SWCNTs are used as optical probes for bioimaging [42,43]. The quasi one-dimensional (1D) semiconducting SWCNTs possess fluorescent properties, which make them suitable for both *in vitro* and *in vivo* imaging applications [44]. These structures have a narrow band gap of about 1 eV and produce fluorescence emission in the NIR regions [45]. As CNTs possess excellent imaging capabilities, they are combined in various imaging modalities.

7.4.1 Photoacoustic Imaging

PA imaging is a bioimaging technique to visualize biological systems based on the ability to absorb irradiated light. PA imaging is done by measuring the ultrasounds generated by tissues upon interaction with the CNTs. When the administered CNTs are exposed to NIR stimulation, it leads to the emission of heat, and this further causes the surrounding tissues to expand generating the ultrasounds [27]. It has got advantages like high spatial resolution, visualization of deeper tissues, and high contrast and safety [46]. PA imaging is found to be superior to conventional imaging techniques, in that it detects sound instead of light and as a result, much improved tissue penetration and enhanced spatial resolutions can be obtained [45]. A proper CA should be there to obtain the image with natural contrast, as it is helpful in the early detection of many diseases including cancers, which influence the treatment and recovery. SWCNTs can be modified into target specific imaging agents by conjugating with RGD (cyclic Arg-Gly-Asp). When administered intravenously, it showed eight times higher PA signal in the tumor bearing mice as compared to the unconjugated SWCNTs [47]. Similarly, it gave a twofold image amplification in the thermo acoustic tomography (TAT) [48].

Specific contrast agents can be conjugated with SWCNTs to visualize specific tissues, e.g., antibodies specific toward the tumor cells can be used for the functionalization of the SWCNTs and these can specifically target and identify an early stage tumor by using PA molecular imaging [49]. PA imaging technique can be used to elucidate structural and functional information in case of many cancers. They can also be utilized in many biological processes like gene overexpression, monitoring vascular damage, imaging of cerebrovascular activities, etc. Usually, NIR illumination in the range of 700–1,000 nm is used as these wavelengths can penetrate several centimeters into the tissues. The absorption properties of SWCNTs depend more upon their rigid structure rather than on molecular orbital electronic transitions [50]. PA imaging of SWCNTs can be applied for the non-invasive cancer imaging and monitoring of nanotherapeutics [47].

7.4.2 Fluorescence Imaging

Fluorescence imaging is a technique used to get images of tissues for medical diagnostic purposes. The live cells bearing the label-free CNTs can be visualized by this technique at high speed. Fluorescent CNTs are effectively used in the imaging of lymphatic or cardiovascular systems or stem or progenitor cells [19,51,52]. One of the major drawbacks with this technique is the limitation of the penetration depth. This can be overcome by using a fluorescence probe that has emission/excitation wavelengths in the NIR spectrum, in which biological tissues are transparent [53,54]. SWCNTs covalently linked to visible-wavelength fluorophores gives a real-time visualization of circulation through various internal organs like kidneys and lungs [55–57]. This property can be applied as SWCNTs-based fluorescence imaging to be used as an intraoperative surgical tool, which helps in the identification and resection of microscopic tumors. This makes the treatment more efficient and resulted in an increased post-operative survival. Apart from the oncological applications, high fluorescent video rate imaging and PCA analysis are done for the vessel imaging *in vivo*. Multicolor fluorescence imaging can be done using SWCNTs possessing purified single chirality and these exhibits separated excitation and emission wavelengths [45].

7.4.3 Raman Imaging

SWCNTs can be used as Raman probes as they have strong resonance Raman scattering with extremely large scattering cross section and can be used as sensing and imaging probes [45]. Raman spectroscopy (RS) has laid an important basis for the study and characterization of CNTs and has been utilized for tracking the CNTs in cancer-targeted delivery [26,45,56]. RS is based on the inelastic scattering of light and is used for the characterization of advanced materials and devices [57]. SWCNTs used as a Raman tag produce peaks that are simple, narrow, and intense, with a full width at half-maximum less than 2 nm, and they are easy to differentiate from the autofluorescent background. The Raman signals of SWCNTs are robust, and there is no quenching or bleaching; they are capable of long-term tracking and imaging. The isotope composition within SWCNTs can be changed to induce a modulation, which can be utilized in the multiplexed Raman imaging to generate multicolored images. Improved sensitivity and reduced imaging time can be obtained by combining the resonance Raman signals of SWCNTs and surface-enhanced Raman scattering (SERS) technique. Live cell tracking using SWCNTs, wrapped with DNA oligonucleotides as markers, with RS without using any fluorescent labels is one of the advantages [45]. Targeting ligand conjugated SWCNTs can be used for the *in vivo* Raman imaging of tumors [58].

7.4.4 Cathode for Imaging X-Ray Radiation in Computed Tomography

Vertically arranged SWCNT bundles can be used as the electron emitter (cathode) through field emission [59]. They are also used for the focusing of the X-rays with minimum energy loss and producing images with high resolution. CNT-based equipment is easy to operate and can be used to have repetitive imaging essential for the development of 3D images [60]. CNTs can be used in the construction of microfocus X-ray tube that is utilized in the development of compact systems for non-invasive imaging in biomedical research [7]. It is estimated that cold cathode X-ray tube can be a promising integral part of future X-ray sources, leading to the development of portable and mobile X-ray medical systems, and advanced X-ray CT systems will have huge applications in the healthcare sector [61,62]. CNT cathodes can stably perform a tomosynthesis imaging and are used in cases of breast cancer [63,64].

7.4.5 Magnetic Resonance Imaging

Magnetic resonance imaging (MRI) is a scanning technique which utilizes strong magnetic fields and radio waves to obtain detailed images of the internal organs of the body. It is a widely accepted tomography technique in diagnosis due to the non-invasive character and the high spatial resolution. It is possible to have soft tissue imaging using this technique. Gadolinium complexes and superparamagnetic iron oxides are used as CA to highlight the organs and the morphological changes caused due to disease state. As these CAs are found to interfere with so many other biological functions, the best alternative found was the MWCNTs. These were found to have properties which match with an ideal CA like morphology, electronic and magnetic properties, photoluminescence and functionalization to impart additional properties, etc. [65]. To use CNTs as CA in MRI scanning, a paramagnetic or superparamagnetic iron oxide nanoparticles (SPIONs) must be conjugated to or encapsulated within the CNT platform [66]. CNTs can also be used for stem cell labelling and tracking [45].

7.4.6 Nuclear Imaging

As the SWCNTs have other specialized surface properties, they can be tagged with external labels such as radio isotopes, which makes the CNTs to be utilized as imaging probes. There are so many advantages for the radio-labeled SWCNTs over the conventional modalities. There is no limit for the depth of tissue penetration so that we can obtain a much-detailed image of the internal organs. Regardless of how deep is the evaluation, this method provides images with high sensitivity and resolution [45,67]. CNTs and specific antibodies can be conjugated with ^{111}In and are used to target CD20 epitope of human Burkitt lymphoma cells. These conjugates bind specifically to those cells whose surface is overexpressed with such cancer antigens [68,69]. Many other radio labels are also used such as ^{131}I, ^{64}Cu, and ^{86}Y [43]. The main techniques that use CNT-based agents are single-photon emission computed tomography (SPECT) and positron emission tomography (PET). SWCNTs can be filled by suitable radioactive material, and after administration, tissue localization can be easily determined [66].

7.5 CONCLUSION

CNTs since their invention have been found to have several applications in the biomedical field, especially in the past decades, which include theranostics, bioimaging, biosensing, nanorobotics, fluorescent tags, optical tags, and many more. Their extreme small size, shape, electronic configuration, structure, and helicity provide the physicochemical, thermal, optical, fluorescent, electrical, and mechanical properties of CNTs, which make them efficiently used as diagnostic and imaging agents. Functionalization is one of the characteristic features of the CNTs where we can conjugate the surface of the CNTs with specific ligands so that special biological functionalities can be imparted. Due to its small size and high surface area to volume ratio, the penetration and interaction capabilities are much higher compared to the normal molecules. Even though there are a few hurdles like the stability and solubility, some of the toxicological aspects of the CNTs are still yet to be unveiled, and they have a huge scope in the biomedical field. CNTs will gain momentum in their applications, and it can overrule the conventional diagnostic and imaging methods.

ACKNOWLEDGMENTS

We acknowledge the Department of Pharmaceuticals (DOP), Ministry of Chemicals and Fertilizers, Government of India, for providing facilities for writing this book chapter. The authors are also grateful to Servier Medical Arts for providing figure templates (Unported License: https://creativecommons.org/licenses/by/4.0/). NIPER – Raebareli communication number for this chapter is 500.

REFERENCES

1. Anzar N, Hasan R, Tyagi M, Yadav N, Narang J. Carbon nanotube – a review on synthesis, properties and plethora of applications in the field of biomedical science. *Sensors International*. 2020;1(December 2019):100003.
2. Hirlekar R, Yamagar M, Garse H, Vij M. Carbon nanotubes and its applications: a review. *AJPCR*. 2009;2(December 4):17–27.

3. Murjani BO, Kadu PS, Bansod M, Vaidya SS, Yadav MD. Carbon nanotubes in biomedical applications: current status, promises, and challenges. *Carbon Letters.* 2022;32(5):1207–26.
4. Chavan R, Desai U, Mhatre P, Chinchole R. A review: carbon nanotubes. *International Journal of Pharmaceutical Sciences Review and Research.* 2012;13(1):124–34.
5. He Y, Hu C, Li Z, Wu C, Zeng Y, Peng C. Bio multifunctional carbon nanomaterials for diagnostic applications in infectious diseases and tumors. *Materials Today Bio.* 2022;14(March):100231.
6. Polizu S, Savadogo O, Poulin P, Yahia LH. Applications of carbon nanotubes-based biomaterials in biomedical nanotechnology. *Journal of Nanoscience and Nanotechnology.* 2006;6(7):1883–904.
7. Hussein LS. A general review on carbon nanotubes: advantages, disadvantages, biomedical applications and analytical techniques. *AMJ.* 2022;62(08):3593–603.
8. Jain K. Nanohybrids of dendrimers and carbon nanotubes: a benefaction or forfeit in drug delivery? *Nanoscience &Nanotechnology-Asia.* 2018;9(1):21–9.
9. Kaur S, Mehra NK, Jain K, Jain NK. Development and evaluation of targeting ligand-anchored CNTs as prospective targeted drug delivery system. *Artificial Cells, Nanomedicine, and Biotechnology.* 2017;45(2):242–50.
10. Kumar N, Chamoli P, Misra M, Manoj MK, Sharma A. Advanced metal and carbon nanostructures for medical, drug delivery and bio-imaging applications. *Nanoscale.* 2022;14:3987–4017.
11. Jain K. Dendrimers: smart nanoengineered polymers for bioinspired applications in drug delivery. In: Jana S, Maiti S, Jana S (eds) *Biopolymer-Based Composites.* Woodhead Publishing, Cambridge, 2017, pp. 169–220.
12. Tandel H, Bhatt P, Jain K, Shahiwala A, Misra A. Chapter 1: *In-Vitro and In-Vivo* Tools in Emerging Drug Delivery Scenario: Challenges and Updates. In International Book *In-Vitro and In-Vivo Tools in Drug Delivery Research for Optimum Clinical Outcomes.* CRC Press, USA, 2018; 1-24. ISBN: 9781351368728, https://doi.org/10.1201/b22448-1
13. Bajwa N, Mehra NK, Jain K, Jain NK. Targeted anticancer drug delivery through anthracycline antibiotic bearing functionalized Quantum Dots. *Artificial Cells, Nanomedicine and Biotechnology.* 2016;44(7):1774-82. https://doi.org/10.3109/21691401.2015.1102740.
14. Ahmad J, Rizwanullah Md, Suthar T, Albarqi HA, Ahmad MZ, Vuddanda PR, et al. Receptor-targeted surface-engineered nanomaterials for breast cancer imaging and theranostic applications. *Critical Reviews in Therapeutic Drug Carrier Systems.* 2022;39(6):1–44.
15. Patel P, Kumar K, Jain VK, Popli H, Yadav AK, Jain K. Nanotheranostics for diagnosis and treatment of breast cancer. *Current Pharmaceutical Design.* 2023;29(10):732–47.
16. Suthar T, Jain K. Nutraceuticals against neurodegeneration: understanding the mechanistic pathways. *Nutraceuticals for Aging and Anti-Aging.* 2021;14:133–68.
17. Suthar T, Jain K. Nanostructure-Based Molecules as Diagnostic and Theranostic Tools in Alzheimer's Disease. In: Sharma A, Modi GP (eds) *Natural Product-based Synthetic Drug Molecules in Alzheimer's Disease: Therapeutic & Theranostic Agents.* Springer, Singapore, 2023, pp. 133–53.
18. Mehra NK, Jain K, Jain NK. Multifunctional carbon nanotubes in cancer therapy and imaging.In: Grumezescu AM (eds) *Nanobiomaterials in Medical Imaging.* William Andrew Publishing, Norwich, NY, 2016, pp. 421–53.
19. Sharma P, Kumar Mehra N, Jain K, Jain NK. Biomedical applications of carbon nanotubes: a critical review. *Current Drug Delivery.* 2016;13(6):796–817.
20. Rao R, Pint CL, Islam AE, Weatherup RS, Hofmann S, Meshot ER, Wu F, Zhou C, Dee N, Amama PB, Carpena-Nuñez J. Carbon nanotubes and related nanomaterials: critical advances and challenges for synthesis toward mainstream commercial applications. *ACS Nano.* 2018;12(12):11756–84.

21. Alshehri R, Ilyas AM, Hasan A, Arnaout A, Ahmed F, Memic A. Carbon nanotubes in biomedical applications: factors, mechanisms, and remedies of toxicity: miniperspective. *Journal of Medicinal Chemistry*. 2016;59(18):8149–67.
22. Sharma S, Mehra NK, Jain K, Jain NK. Effect of functionalization on drug delivery potential of carbon nanotubes. *Artificial Cells, Nanomedicine, and Biotechnology*. 2016;44(8):1851–60.
23. Lamberti M, Pedata P, Sannolo N, Porto S, De Rosa A, Caraglia M. Carbon nanotubes: properties, biomedical applications, advantages and risks in patients and occupationally-exposed workers. *International Journal of Immunopathology and Pharmacology*. 2015;28(1):4–13.
24. Almeida MR, Nunes JC, Cristóvão RO, Faria JL, Tavares AP, Silva CG, Freire MG. Carbon nanotubes for biomedical applications. *Nanotechnology for Biomedical Applications*. 2022;32:285–331.
25. Mehra NK, Jain K, Jain NK. Pharmaceutical and biomedical applications of surface engineered carbon nanotubes. *Drug Discovery Today*. 2015;20(6):750–9.
26. Tong L, Liu Y, Dolash BD, Jung Y, Slipchenko MN, Bergstrom DE, Cheng JX. Label-free imaging of semiconducting and metallic carbon nanotubes in cells and mice using transient absorption microscopy. *Nature Nanotechnology*. 2012;7(1):56–61.
27. Simon J, Flahaut E, Golzio M. Overview of carbon nanotubes for biomedical applications. *Materials*. 2019;12(4):624.
28. Sharma P, Jain K, Jain NK, Mehra NK. Ex vivo and in vivo performance of anti-cancer drug loaded carbon nanotubes. *Journal of Drug Delivery Science and Technology*. 2017;1(41):134–43.
29. Kumar S, Rani R, Dilbaghi N, Tankeshwar K, Kim KH. Carbon nanotubes: a novel material for multifaceted applications in human healthcare. *Chemical Society Reviews*. 2017;46(1):158–96.
30. Saliev T. The advances in biomedical applications of carbon nanotubes. *C*. 2019;5(2):29.
31. He H, Pham-Huy LA, Dramou P, Xiao D, Zuo P, Pham-Huy C. Carbon nanotubes: applications in pharmacy and medicine. *BioMed Research International*. 2013;2013:1–12.
32. Yang W, Thordarson P, Gooding JJ, Ringer SP, Braet F. Carbon nanotubes for biological and biomedical applications. *Nanotechnology*. 2007;18(41):412001.
33. Tîlmaciu CM, Morris MC. Carbon nanotube biosensors. Frontiers in Chemistry. 2015;3:59.
34. Ojha B, Jain VK, Mehra NK, Jain K. Nanotechnology: introduction and basic concepts. *Dendrimers in Nanomedicine*. 2021;22:1–17.
35. Bardhan NM, Jansen P, Belcher AM. Graphene, carbon nanotube and plasmonic nanosensors for detection of viral pathogens: opportunities for rapid testing in pandemics like COVID-19. *Frontiers in Nanotechnology*. 2021;3:733126.
36. Zhang X, Xu Y, Ye B. An efficient electrochemical glucose sensor based on porous nickel-based metal organic framework/carbon nanotubes composite (Ni-MOF/CNTs). *Journal of Alloys and Compounds*. 2018;767:651–6.
37. Sheikhpour M, Naghinejad M, Kasaeian A, Lohrasbi A, Shahraeini SS, Zomorodbakhsh S. The applications of carbon nanotubes in the diagnosis and treatment of lung cancer: a critical review. *International Journal of Nanomedicine*. 2020;15:7063–78.
38. Peng G, Trock E, Haick H. Detecting simulated patterns of lung cancer biomarkers by random network of single-walled carbon nanotubes coated with nonpolymeric organic materials. *Nano Letters*. 2008;8(11):3631–5.
39. Yamaan Saadeh BS, Dinesh Vyas MD. Nanorobotic applications in medicine: current proposals and designs. *American Journal of Robotic Surgery*. 2015;1(1):4–11.
40. He P, Dai L, Cambridge R, October UK, December A. Aligned carbon nanotube – DNA electrochemical sensors. *Chemical Communications*. 2004;3:1–2.

41. Deshmukh MA, Jeon JY, Ha TJ. Carbon nanotubes: an effective platform for biomedical electronics. *Biosensors & Bioelectronics.* 2020;150(September):111919.
42. Liu Z, Tabakman S, Welsher K, Dai H. Carbon nanotubes in biology and medicine: in vitro and in vivo detection, imaging and drug delivery. *Nano Research.* 2009;2:85–120.
43. Tang L, Xiao Q, Mei Y, He S, Zhang Z, Wang R, Wang W. Insights on functionalized carbon nanotubes for cancer theranostics. *Journal of Nanobiotechnology.* 2021;19:1–28.
44. Mohammadi H, Hafezi M, Hesaraki S, Sepantafar MM. Preparation and characterization of Sr-Ti-hardystonite (Sr-Ti-HT) nanocomposite for bone repair application. *Nanomedicine Journal.* 2015;2(3):203–10.
45. Gong H, Peng R, Liu Z. Carbon nanotubes for biomedical imaging: the recent advances. *Advanced Drug Delivery Reviews.* 2013;65(15):1951–63.
46. Siregar S, Nagoka R, Ishikawa K, Saijo Y. Carbon nanotubes as potential candidate for photoacoustic imaging contrast agent. *Proceedings of Meetings on Acoustics Acoustical Society of America.* 2018;32(1):020018.
47. De La Zerda A, Zavaleta C, Keren S, Vaithilingam S, Bodapati S, Liu Z, Levi J, Smith BR, Ma TJ, Oralkan O, Cheng Z. Carbon nanotubes as photoacoustic molecular imaging agents in living mice. *Nature Nanotechnology.* 2008;3(9):557–62.
48. Chen Z, Zhang A, Wang X, Zhu J, Fan Y, Yu H, et al. The advances of carbon nanotubes in cancer diagnostics and therapeutics. *Journal of Nanomaterials.* 2017;2017:1–13.
49. Liu Z, Cai W, He L, Nakayama N, Chen K, Sun X, Chen X, Dai H. In vivo biodistribution and highly efficient tumour targeting of carbon nanotubes in mice. *Nature Nanotechnology.* 2007;2(1):47–52.
50. Xiang L, Yuan Y, Xing D, Ou Z, Yang S, Zhou F. Photoacoustic molecular imaging with antibody-functionalized single-walled carbon nanotubes for early diagnosis of tumor. *Journal of Biomedical Optics.* 2009;14(2):021008.
51. Ashiba H, Iizumi Y, Okazaki T, Wang X, Fujimaki M. Carbon nanotubes as fluorescent labels for surface plasmon resonance-assisted fluoroimmunoassay. *Sensors.* 2017;17(11):2569.
52. Mao B, Calatayud DG, Mirabello V, Kuganathan N, Ge H, Jacobs RM, Shepherd AM, Ribeiro Martins JA, Bernardino De La Serna J, Hodges BJ, Botchway SW. Fluorescence-lifetime imaging and super-resolution microscopies shed light on the directed-and self-assembly of functional porphyrins onto carbon nanotubes and flat surfaces. *Chemistry–A European Journal.* 2017;23(41):9772–89.
53. Hwang Y, Park SH, Lee JW. Applications of functionalized carbon nanotubes for the therapy and diagnosis of cancer. *Polymers.* 2017;9(1):13.
54. Zhou P, Wang J, Du X, Huang T, Nallathamby PD, Yang L, Zou W, Zhou Y, Jault JM, Chen S, Ding F. Nanoparticles in biomedicine-focus on imaging applications. *Engineered Science.* 2018;5(2):1–20.
55. Cherukuri P, Bachilo SM, Litovsky SH, Weisman RB. Near-infrared fluorescence microscopy of single-walled carbon nanotubes in phagocytic cells. *Journal of the American Chemical Society.* 2004;126:15638–9.
56. Dresselhaus MS, Jorio A, Hofmann M, Dresselhaus G, Saito R. Perspectives on carbon nanotubes and graphene Raman spectroscopy. *Nano Letters.* 2010;10(3):751–8.
57. Welsher K, Sherlock SP, Dai H. Deep-tissue anatomical imaging of mice using carbon nanotube fluorophores in the second near-infrared window. *Proceedings of the National Academy of Sciences.* 2011;108(22):8943–8.
58. Zavaleta C, De La Zerda A, Liu Z, Keren S, Cheng Z, Schipper M, Chen X, Dai H, Gambhir SS. Noninvasive Raman spectroscopy in living mice for evaluation of tumor targeting with carbon nanotubes. *Nano Letters.* 2008;8(9):2800–5.
59. Lee YZ, Burk L, Wang KH, Cao G, Lu J, Zhou O. Carbon nanotube based X-ray sources: applications in pre-clinical and medical imaging. *Nuclear Instruments and Methods in Physics Research Section A: Accelerators, Spectrometers, Detectors and Associated Equipment.* 2011;648:S281–3.

60. Puett C, Inscoe C, Hartman A, Calliste J, Franceschi DK, Lu J, Zhou O, Lee YZ. An update on carbon nanotube-enabled X-ray sources for biomedical imaging. *Wiley Interdisciplinary Reviews: Nanomedicine and Nanobiotechnology*. 2018;10(1):e1475.
61. Han JS, Lee SH, Go H, Kim SJ, Noh JH, Lee CJ. High-performance cold cathode X-ray tubes using a carbon nanotube field electron emitter. *ACS Nano*. 2022;16(7):10231–41.
62. Parmee RJ, Collins CM, Milne WI, Cole MT. X-ray generation using carbon nanotubes. *Nano Convergence*. 2015;2:1–27.
63. Li, X, Liu, M, Zhou, J, Wu, X, Wu, Q, Ren, L, Zheng, J, Chen, Z. (2019). High definition X-ray source based on carbon nanotube emitter for micro CT application. *International Symposium on Digital Industrial Radiology and Computed Tomography*, (p. 338). Germany
64. Gidcumb E, Gao B, Shan J, Inscoe C, Lu J, Zhou O. Carbon nanotube electron field emitters for x-ray imaging of human breast cancer. *Nanotechnology*. 2014;25(24):245704.
65. Kuźnik N, Tomczyk MM. Multiwalled carbon nanotube hybrids as MRI contrast agents. *Beilstein Journal of Nanotechnology*. 2016;7(1):1086–103.
66. Hernández-Rivera M, Zaibaq NG, Wilson LJ. Toward carbon nanotube-based imaging agents for the clinic. *Biomaterials*. 2016;101:229–40.
67. Wang H, Wang J, Deng X, Sun H, Shi Z, Gu Z, Liu Y, Zhaoc Y. Biodistribution of carbon single-wall carbon nanotubes in mice. *Journal of Nanoscience and Nanotechnology*. 2004;4(8):1019–24.
68. Alsewidi WM, Alfahdawi IH, Alsewidi WM. Carbon nanotubes : properties and medical applications. *Auctores Publishing*. 2021;4(03):1–7.
69. McDevitt MR, Chattopadhyay D, Kappel BJ, Jaggi JS, Schiffman SR, Antczak C, Njardarson JT, Brentjens R, Scheinberg DA. Tumor targeting with antibody-functionalized, radiolabeled carbon nanotubes. *Journal of Nuclear Medicine*. 2007;48(7):1180–9.

8 Cellular Trafficking Mechanism of Carbon Nanotubes in Drug Delivery

Sri Pada Datta Chigurupati, Rohit Bhawale, Nalla Usha Kumari, and Neelesh Kumar Mehra

ABBREVIATIONS

CME	clathrin-mediated endocytosis
CNTs	carbon nanotubes
DRM	detergent-resistant membrane
DWCNTs	double-walled carbon nanotubes
EFTEM	energy-filtered transmission electron microscopy
ER	endoplasmic reticulum
f-CNTs	functionalized carbon nanotubes
f-MWCNTs	functionalized multi-walled carbon nanotubes
FA-SWNTs	folate-conjugated single-walled carbon nanotubes
Mf-CNTs	multifunctional carbon nanotubes
MWCNTs	multi-walled carbon nanotubes
polymer-g-MWNTs	polymer grafted multi-walled carbon nanotubes
SWCNTs	single-walled carbon nanotubes

8.1 INTRODUCTION

Carbon nanotubes (CNTs) are the allotropes of carbon where the sheets of graphene are embedded into each other. Carbon nanotubes have gained vast importance over the years due to their widespread application in the field of drug delivery. Its high surface area, tunable properties, and capacity to deliver multiple types of cargo make it an excellent choice for drug delivery. Carbon nanotubes are utilized for targeting by conjugating with certain ligands, allowing them to target specific regions of the body. Modified carbon nanotubes have better therapeutic properties and no immunogenic effect and are biocompatible. Based on the number of walls present in CNTs, there are three types of CNTs, namely, single-walled carbon nanotubes (SWCNTs), double-walled carbon nanotubes (DWCNTs), and multiple-walled carbon nanotubes (MWCNTs) [1] different carbon bonds can form structures with entirely distinct

 DOI: 10.1201/9781003386537-8

characteristics. A layered structure with strong in-plane limits and weak out-of-plane bonding of the van der Waals kind is created by the sp^2 hybridization of carbon [2]. Carbon nanotubes can be modified in two ways, namely, covalent functionalization and noncovalent functionalization. In noncovalent functionalization, the aromatic groups of the drug moieties will form a π–π stacking interaction with the pristine CNTs and get adsorbed onto the surface. Occasionally, a charge developed over the surface of CNTs after chemical treatment will help get the drug adsorbed onto the surface of the CNT. Noncovalent functionalization is superior as it doesn't disturb the electronic framework of the CNTs. Covalent functionalization is a promising strategy to derivatize and modify CNTs. Carbon nanotubes can be modified by oxidizing using strong acids, which leads to the generation of free carboxylic groups, making them dispersible in aqueous media; after oxidizing the CNTs, the addition reactions can be made by employing specific reagents such as thionyl chloride and carbodiimide [3]. The emerging trend is the employment of the multifunctional carbon nanotubes (mf-CNTs) for theragnostic purposes and for treating multiple diseases, such as breast cancer and colon cancer, as they will be specifically taken up by cancer cells due to the presence of the ligand. Another significant application of CNTs is photothermal therapy and photodynamic therapy, where functionalized carbon nanotubes (f-CNTs) respond to the irradiated infrared radiation and generate heat or free radicals to kill the cancerous cells. Cytotoxic medications like doxorubicin and prodrugs like paclitaxel and cisplatin can be transported across the cellular membrane by CNTs acting as biological carriers through endocytosis. Drugs may be delivered to (specific) cancer cells via CNTs, maybe producing better effects than medications without a carrier. Drug molecules may be concentrated in tumors, and systemic toxicity may be decreased with targeted drug delivery. Furthermore, by adjusting the pH, it is possible to control the release of active moiety from CNTs, potentially increasing their therapeutic efficacy. CNT-drug complexes may be released readily because many solid tumors have an acidic microenvironment [4].

Cells are the basic structural and functional unit of any living organism. The cells in nature are of two types depending on the presence or absence of a nuclear envelope, i.e., they are prokaryotes and eukaryotes. In prokaryotes, the plasma membrane is involved in cellular functions, whereas eukaryotes have specifically allotted organelles to carry out cellular functions [5]. The cells communicate with other cells or send signals to the in-situ organelles by sending the ligands via the cargo or carrier molecules to the target site. The cargoes transported are mainly by the two pathways, namely, the exocytic pathway and endocytic pathway. In the exocytic pathway, proteins synthesized from the cytoplasm are translocated into the endoplasmic reticulum (ER), whereas the proteins and membrane are internalized from the cell environment by the aid of endosomes and finally taken to the lysosomes. The movement of proteins, ligands, and biomolecules through the cell organelles and cellular compartments is called cell trafficking. The aforementioned processes such as the exocytic and endocytic processes together constitute cellular trafficking [6].

There are two major pathways in cellular trafficking, they are exocytic and endocytic pathways. Proteins produced in the cytoplasm are translocated into the ER via the exocytic pathway. All secreted proteins and resident proteins for every compartment linked by vesicular transport are synthesized in the rough ER. The majority of the lipids in cells are also synthesized in the ER. Membrane vesicles transport material to the Golgi apparatus from the ER. Cargo obtained from the ER passes through the medial and trans-cisternae of the Golgi after entering through the cis-cisterna. Proteins that are meant to be displayed on the plasma membrane or secreted are bundled into secretory vesicles in the trans-Golgi, which then fuse with the plasma membrane. Proteins and membranes are ingested from the cell environment via a set of early and late endosomes and transported to the lysosome via the endocytic route. One important location for the breakdown of cellular and internalized proteins is the lysosome. Therefore, cellular proteins can enter lysosomes through the endocytic pathway from the PM or the cytoplasm through the routes of autophagy and cytoplasm-to-vacuole targeting [8].

8.2 CELLULAR INTERNALIZATION OF MULTIFUNCTIONAL CNTs

Carbon nanotubes are used for drug delivery, but the mechanism followed by them is quite fascinating as multiple pathways are involved. The pathways involved in the CNT internalization are depicted in Figure 8.1. Depending on the energy requirement, there

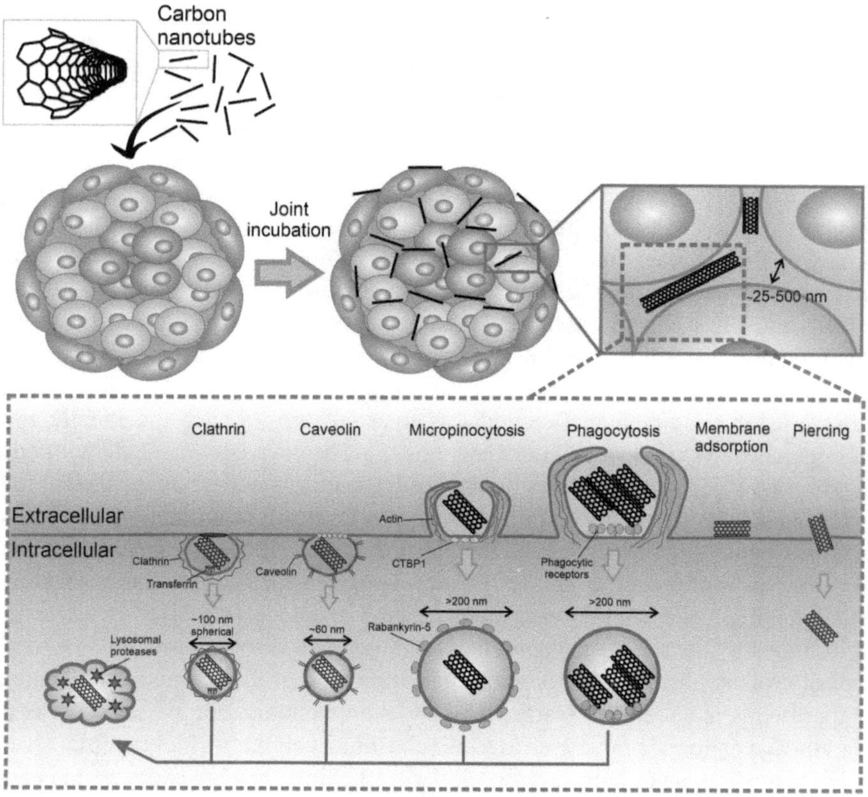

FIGURE 8.1 Overview of the primary mechanisms of uptake of CNTs into cellular spheroid and cell [7].

are mainly two types of pathways for the uptake of CNTs, namely, energy-dependent endocytosis and energy-independent endocytosis. Energy-independent endocytosis includes passive diffusion, whereas energy-dependent endocytosis includes phagocytosis, pinocytosis, and receptor-mediated endocytosis, such as clathrin-dependent endocytosis, caveolae-dependent endocytosis, or clathrin-/caveolae-independent endocytosis [9].

8.2.1 Energy-Independent Endocytosis

Energy-independent endocytosis or passive diffusion of the CNTs across the cell membrane is one of the cellular trafficking pathways for transporting them across the cell. The passive diffusion procedure comprises two important steps, namely, CNTs get lodged over the surface of the lipid bilayer membrane accompanied by the transmembrane configuration. Passive diffusion is unaffected by the endocytosis inhibitors and is independent of the concentration of nanotubes present in the vicinity of the cell [9,10].

8.2.2 Energy-Dependent Endocytosis

Energy-dependent endocytosis of nanoparticles is a critical process in nanoparticle-cell interactions, particularly in designing nanoparticles for applications such as intracellular sensing and drug delivery. This process involves two stages: anchoring and internalization. The anchoring stage refers to the attachment of nanoparticles to the plasma membrane after random collisions during Brownian motion in the culture medium. The internalization stage relies on the deformation of the cell membrane and is an energy-dependent process. The process involves the utilization of energy for the endocytosis process to take place [10]. Phagocytosis is an endocytic process where phagocytes such as macrophages, monocytes, and neutrophils take foreign moieties from the extracellular environment [11].

8.2.2.1 Internalization through Phagocytosis

Phagocytosis is of different types based on the type of receptor involved in the process. Fc receptors initiate the reaching or zippering phagocytosis, and complement receptors initiate the sinking phagocytosis. The tethering/tickling model explains how the cell decides to either remain connected to a ligand or engulf it. Complement receptors, $\alpha5\beta1$ integrin receptors, and F_C receptors are involved in opsonic phagocytosis; $\alpha V5\beta$ receptor, dectin receptor, macrophage receptor, and scavenger receptor are involved in nonopsonic phagocytosis; and toll-like receptors are involved in triggered phagocytosis [12]. Phagocytosis depends upon the length of the CNTs; shorter and tangled ones get easily engulfed by the phagocytes, whereas the long and rigid CNTs are not phagocytosed completely, which may cause harmful effects due to accumulation [13].

8.2.2.2 Internalization through Pinocytosis

Pinocytosis is the consumption of extracellular solute particles and fluid. The term pinocytosis is derived from the Greek word pino, which means to drink, and cyto, which means cell. This means that pinocytosis can also be termed as cellular drinking. Pinocytosis is an energy-consuming process where ATP is utilized by cells. Also, a form of an endocytic process as the external substances are being brought into the cell. Receptors are not involved in this process; hence, it is nonspecific. There are five important sub-steps involved in the pinocytosis process. They are the attraction of positively charged molecules to the negatively charged cell membrane, molecular binding, shape change, invagination, and degradation of vesicular contents by lysosome [14].

8.2.2.3 Internalization through Endocytosis

Clathrin-mediated endocytosis (CME) reigns supreme as the most understood pathway for bringing essential nutrients like iron and cholesterol into mammalian cells. While the older term "receptor-mediated endocytosis" is less accurate, CME showcases surprising adaptability. It leverages specific interactions between receptors and cargo molecules, but other pathways (like FEME) employ similar mechanisms. The process begins with activated receptors on the cell surface recruiting adaptor proteins, as depicted in Figure 8.2. This sets off a tightly regulated cascade of interactions between proteins and lipids, particularly with a molecule called phosphatidylinositol 4,5-bisphosphate (PtdIns P2) (4,5). Within minutes, a distinctive clathrin-coated pit forms. This intricate system acts as a filter, ensuring that only designated cargo enters the cell. The pit swiftly transforms into a clathrin-coated vesicle, which pinches off from the membrane with the help of dynamin, a large protein with a key role in cellular mechanics. The size of these vesicles, around 100 nm in diameter, limits the dimensions of particles that can be internalized through this pathway. Interestingly, larger particles can exploit a clever trick. They hijack CME components and utilize actin-based mechanisms to gain entry into the cell.

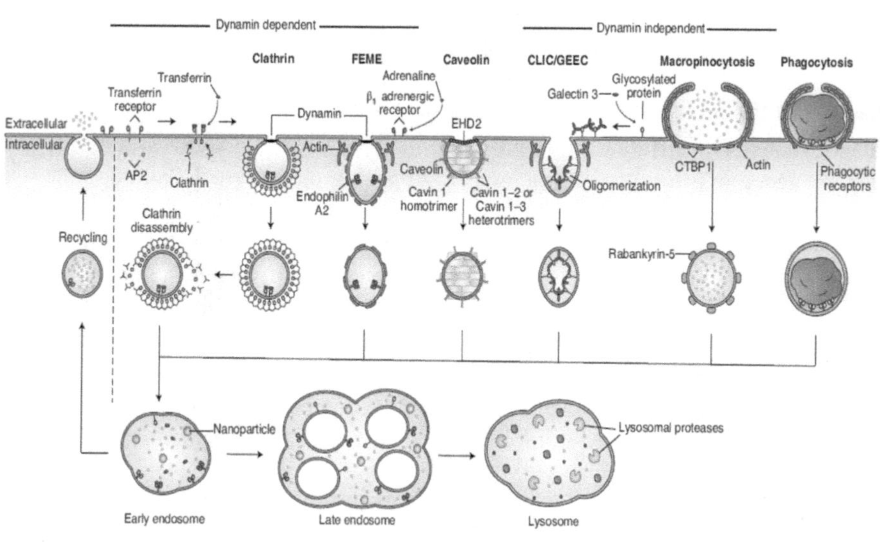

FIGURE 8.2 Various pathways are involved in clathrin-mediated endocytosis [15].

The alternative method comes at a cost – the uptake process slows down compared to standard CME. The unique structure of the clathrin-coated pit, the existence of tools to manipulate CME, and the presence of cargo completely dependent on this pathway for uptake all contribute to its status as a well-studied and essential cellular process [15]. Caveolae-mediated endocytosis involves the caveolae, that was discovered in 1953 and are tiny flask-shaped indentations, just 50–100nm wide, dotting the surface of many eukaryotic cells [16]. These specialized membrane regions, enriched with cholesterol and specific lipids, are essentially a type of lipid raft. Caveolin-1, a key structural protein, acts like a scaffold, anchoring caveolae to the cell membrane. Caveolae plays a vital role in various cellular processes, including signal transduction pathways for facilitating communication between cells, Fat (lipid) level regulation and maintains proper balance of lipids within the cell [17]. Also, transports the materials into the cell thereby bringing in necessary molecules through endocytosis. Interestingly, caveolae dysfunction has been linked to the development of some diseases. Caveolae exhibit a variety of shapes, including elongated flasks and curved invaginations, suggesting a possible role in endocytosis alongside other pathways. However, despite this morphological evidence, there's ongoing debate about whether caveolae can detach from the cell surface during endocytosis. Studies with GFP-tagged caveolin-1 show a surprisingly slow exchange between the plasma membrane and internal caveolin pools, implying a potential lack of mobility for caveolae [18]. Additionally, research suggests that caveolin-1 overexpression might stabilize caveolae on the cell surface, as it inhibits the uptake of a caveolae-localized receptor. Interestingly, recent evidence highlights a different aspect: interaction with specific ligands appears to trigger the rapid internalization of caveolae, indicating a potential role in ligand-specific endocytosis [19].

8.2.2.4 Internalization through Multiple Pathways

Drug molecules internalized through clathrin-independent pathways were once thought to rely solely on lipid rafts for this process, as depicted in Figure 8.3. These specialized membrane microdomains, enriched in cholesterol and sphingolipids, were hypothesized to act as sorting and signaling platforms due to their dependence on these specific lipids. However, the concept of lipid rafts in living cells is more intricate than previously envisioned. While artificial membranes exhibit a clear link between cholesterol levels and phase separation into distinct liquid-ordered and liquid-disordered domains, replicating this structure in living cells remains challenging. Detergent-resistant membrane (DRM) fractions, initially considered a reliable proxy for lipid rafts, might not perfectly capture their actual composition in living cells. This highlights the need for novel approaches to explore how lipids organize within living cell membranes [20].

While DRMs offer some valuable insights, they likely represent an imperfect model for these assemblies. The true picture of lipid rafts in endocytosis is expected to be more nuanced, involving a dynamic interplay between pre-existing small lipid assemblies and larger, induced structures. Future research should focus on developing new methods that can provide a more accurate representation of these dynamic lipid assemblies in living cells. Also, under clathrin-independent endocytosis, several pathways are involved, as shown in Table 8.1 [21].

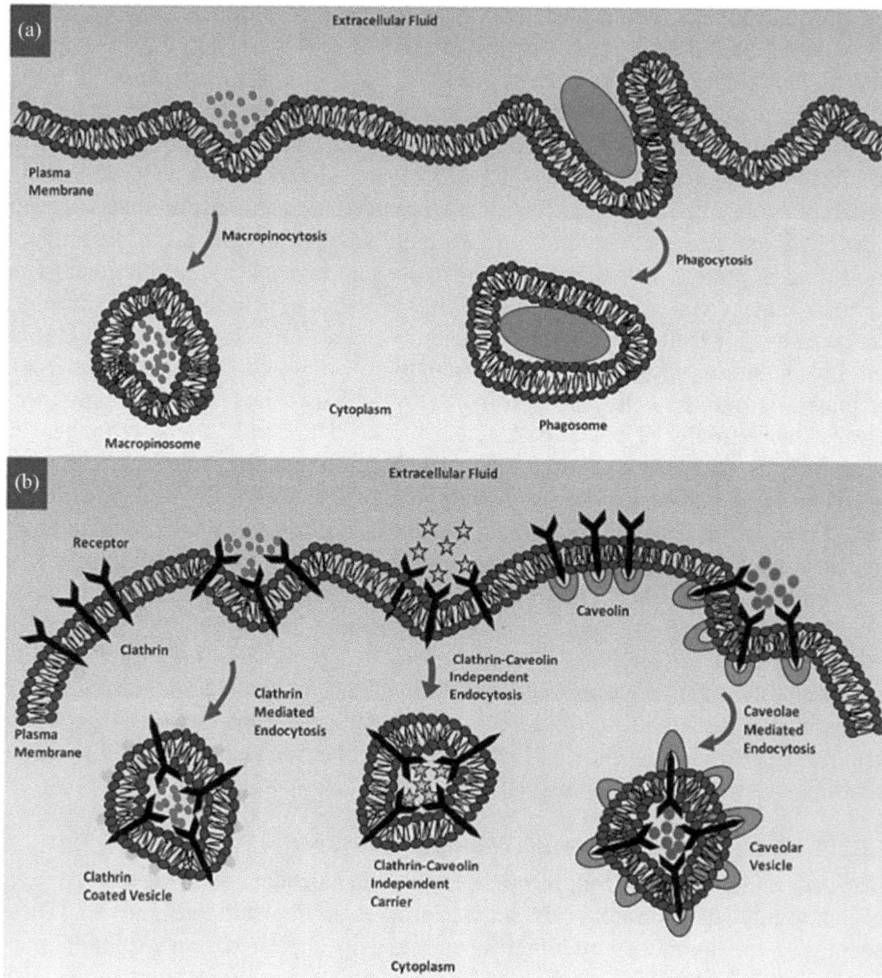

FIGURE 8.3 Diagrammatic representation of clathrin-independent pathway [11].

8.3 FACTORS AFFECTING CNT UPTAKE

Carbon nanotube uptake into the cell is affected by several factors such as the length of the CNT, the functional group attached to the CNT, the environment in the vicinity of the cell, a surface charge on CNT, degree of agglomeration and aggregation, and the type of cell used to study the uptake of the CNT as shown in Figure 8.4. CNT dimensions are undeniably a crucial factor influencing cellular uptake [22].

8.3.1 Size

As research suggests, shorter CNTs are more efficiently internalized by cells. Active endocytosis, particularly clathrin-mediated, serves as the primary mechanism for this efficient uptake of shorter CNTs. Interestingly, shortening CNTs also increases

TABLE 8.1
Types of Clathrin-Independent Pathways

Characteristics	ARF6 Regulated	Caveolar	CDC 42 Regulated	RhoA-Regulated
Cargo	MHC1, β1 integrin, E-cadherin, CPE, GPI-AP, IL-2Rα (Tac)	Modified albumins GSL analogs SV40; EV1 some integrins (for example, β1) CtxB Ab-clustered GPI-AP	GPI-AP Fluid-phase markers, CtxB, VacA toxin, Aerolysin	IL-2R-β, IL-2R-αβ, γ_c-cytokine receptor, IgE receptor
Identity of earliest accessor compartment	Early tubular recycling compartment	Caveosome or early endosome	GPI-AP enriched early endosomal compartments (GEECs)	Early endosome
Primary carriers	Unknown	50–80 nm diameter caveolin-coated vesicles and "grapes-like" aggregates	Clathrin and dynamin independent carriers (CLICs)	Uncoated vesicles
Structural components and machinery involved with the pathway	ARF6, Flotillin-1, RAB11, RAB22, PI3K	CAV1, PV1, Src, PKC-α, Dynamin, RAB5	CDC42, Flotillin-1, RAB5, PI3K, EEA1	RhoA, Dynamin

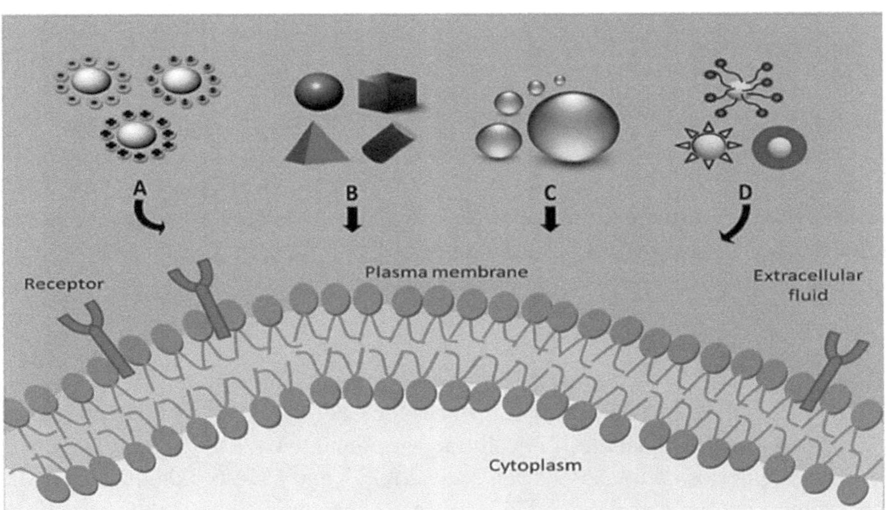

FIGURE 8.4 Physicochemical factors that affect cellular uptake of NP. (a) Surface charge, (b) shape, (c) size, and (d) surface chemistry [22].

the possibility of passive internalization [23]. Zhang et al.'s study exemplifies this trend. Their investigation compared the uptake of eight different CNT types (including SWCNTs and MWCNTs) by RAW264.7 macrophages, with particle sizes ranging from 30 to 400 nm. Their findings revealed that larger dynamic particle size led to increased cellular uptake by macrophages, but with a concerning rise in cytotoxicity. This suggests a potential link between increasing CNT size and higher cytotoxicity [24].

8.3.2 Degree of Agglomeration and Aggregation

Agglomeration may enhance uptake. A study performed by Song et al. observed that larger agglomerates of O-MWCNT (200–700 nm long, 10–20 nm diameter) were taken up more readily by HeLa cells. This increased uptake was attributed to the larger surface area of agglomerates facilitating interaction with the cell membrane and promoting endocytosis. Interestingly, at lower concentrations (less than 150 µg/mL), this enhanced uptake didn't lead to significant cytotoxicity. This suggests exciting possibilities for utilizing controlled CNT agglomeration to improve drug or gene delivery with minimal side effects [25]. Aggregation also influences uptake: Kuroda et al. reported increased uptake of aggregated CNT in RAW264 cells. In this work, considerable levels of aggregated CNT were taken up by RAW264 cells, which may have prevented cell division and produced inflammatory cytokines, therefore suppressing cell proliferation. On the other hand, RAW264 cells showed some uptake of FT9110 that was widely distributed in FBS but in a smaller quantity than that of aggregated CNT or CNHs. Since FT9110 did not build up in the cells significantly, decreased cell viability was not seen. Further research is needed to understand the specific characteristics of the aggregation and its impact on cytotoxicity [26].

8.3.3 Surface Charge

The surface charge affects the uptake of CNTs. Considering the electrostatic interactions and the dispersity of CNT, interaction varies. In the case of cationic CNT, more cellular uptake was observed, whereas in the case of anionic CNT, there was more interaction with the serum proteins; due to protein corona formation, there was a much lower amount of uptake into the cell. The protein corona complex formed might hinder the cell identification process of CNT [27].

8.3.4 Type of Cells

The type of cells used also greatly influences the CNT uptake. Zare et al. conducted a study and found out that the J774 cell line was found to have the highest amount of CNT uptake, followed by A549 cells. Next, MCF-7 and Calu-6 didn't show significant cellular uptake differences. J774 cells are phagocytic cells that tend to absorb the larger particles and CNT agglomerates [22]. Many researchers have performed research on elucidating the mechanism of CNTs cellular uptake, some of which are being discussed here (Table 8.2).

TABLE 8.2
Summary of Various CNT Uptake Mechanisms

Types of CNTs	Surface Modification	Length (*L*)/Diameter (*D*)	Mechanism	Cell Types
SWCNTs	Pluronic- F108: CNT	*L*: >1 μm	Phagocytosis	Macrophage cells
SWCNTs MWCNTs	–	30–400 nm	Phagocytosis	Murine macrophage RAW264.7 cells
MWCNTs	C6-CNTPC	*L*: 453.5 ± 6.2	Phagocytosis	A549 cells
SWCNTs	–	*L*:195–630 nm	Macropinocytosis	Murine macrophage cells
MWCNTs	p(4VP)-MWCNTs	*L*: >300 nm	Macropinocytosis	TT1 cells
SWCNTs	AO–SWCNTs	*L*: <500 nm/*D*: 1–2 nmL	Clathrin-mediated endocytosis	Hela cells
MWCNTs	–	*L*:10 μm/*D*: 60 nm	Clathrin- and caveolae-mediated endocytosis	BEAS-2B cells, HBEpCs
SWCNTs	SWCNT-DNA, SWCNT- proteins	*L*: 0.05–0.2 μm/<1 μm	Clathrin-dependent endocytosis	HeLa cell
MWCNTs	Oxidized-MWCNTs	*D*: ~200 nm	Clathrin-dependent endocytosis	Caco-2 cell
CNTs	PEG-CNTABT737 nanodrug	*L*: 250–400 nm	Clathrin-mediated endocytosis	A549 nonsmall cell lung cancer cells
SWCNTs	FA-SWCNTs	*L*: 100–200 nm/*D*: 1–3 nm	Folate receptor-mediated endocytosis	HepG2 cells
SWCNTs	FA-SWCNTs	*L*:50–100 nm/*D*: 1–3 nm	Folate receptor-mediated endocytosis	HepG2 cells
MWCNTs	–	*L*: 3–10 Lm/*D*: 50–300 nm	Lectin receptor mediated endocytic pathway	Human neonatal epidermal keratinocytes (HEK)
SWCNTs	FA-SWCNTs	*L*: 50–200 nm/D: 1–3 nm	Passive diffusion for 50 nm or less SWCNTs Clathrin-mediated endocytosis for 100–200 nm SWCNTs	HepG2 cells
MWCNTs	FITC-MWCNTs	*D*: 20–30 nm	Passive diffusion	Plant cells

8.4 CELLULAR TRAFFICKING IN VARIOUS CNTs

8.4.1 Single-Walled Carbon Nanotubes

Single-walled carbon nanotubes comprise a single sheet of graphene folded into a seamless cylinder with a diameter of 1 nm and a length of 50 nm. SWCNTs are altered in several ways to make them fit to deliver cargo to the site of interest. According to a study, uptake was presumed to be a spontaneous process as there is a cation-pi interaction between the CNT basic groups and the amino acid residue of the membrane proteins [28]. Kostarelos et al. had conducted a study in which they made several CNTs with different functional groups of ylides over them. The interaction between the f-CNTs and various kinds of cells was observed. The cellular uptake was the same for all kinds of CNTs with modification, and it was proven that the uptake is independent of modification and cell type used for the study [29]. Yaron et al. had evaluated the entry mechanism of the SWCNTs, where the SWCNTs were labeled with the fluorescent endosomes and evaluated with the help of confocal Raman spectroscopy. A temperature-dependent endocytosis assay had proven that the cellular uptake of the SWCNTs is an energy-dependent process. However, the electrochemical impedance spectroscopy and Langmuir monolayer film measurements showed that the Pluronic SWCNTs had been associated with the membrane but did not have enough energy to penetrate through the membrane. The possible mechanism by which the SWCNTs are taken into the cell is by altering the tension in the outer layer of the membrane; later endocytosis gets stimulated, and the SWCNTs are taken into the cytoplasm [30]. Kang et al. conducted a study on the intracellular uptake and cellular trafficking of the folate-conjugated single-walled carbon nanotubes (FA-SWCNTs). SWCNT was functionalized with chitosan noncovalently, and later, folic acid was linked. HepG2 cells were grown and then incubated with these FA-SWCNTs. There was fluorescence in the cytoplasm, disproving the fact of transport through the nuclear envelope. Uptake was achieved by the folate receptor-mediated pathway, which was proven by less fluorescence due to pretreated folic acid and no transport of the FA-SWCNTS. Lysosomes were filled with FA-SWCNTs and discharged into the cytoplasm, as observed in TEM images. The uptake of FA-SWCNTs was observed to be concentration-dependent [31]. Feazell et al. developed platinum (IV) complexes that are conjugated with soluble SWCNTs to deliver the platinum into the cancerous cells, wherein the uptake mechanism has also been investigated by co-tethering a fluorescein-based fluorophore (6-carboxy-2′,7′-dichlorofluorescein -3′,6′-diacetatesuc-cinimidyl ester). After one and a half hours, the SWCNT-Pt (IV)/FI conjugates had been observed in the vesicles, as shown in Figure 8.5. This proves that the SWCNTs are being taken up via endocytosis. Atomic absorption spectra analysis had shown that the nucleus comprises 36 ng of platinum and cytosol with 23 ng of platinum concentration in cells that are being pretreated with 1 μM of platinum. The concentration of platinum was found 6–8 times higher compared to that of the untethered complex and also twice that of the concentration of platinum found in cis-[Pt-(NH3)2C12]-treated cells [32].

Antonelli et al. conducted a study to improve the uptake of the SWCNTs by functionalizing them. Small micelle-like constructs were given to human macrophages

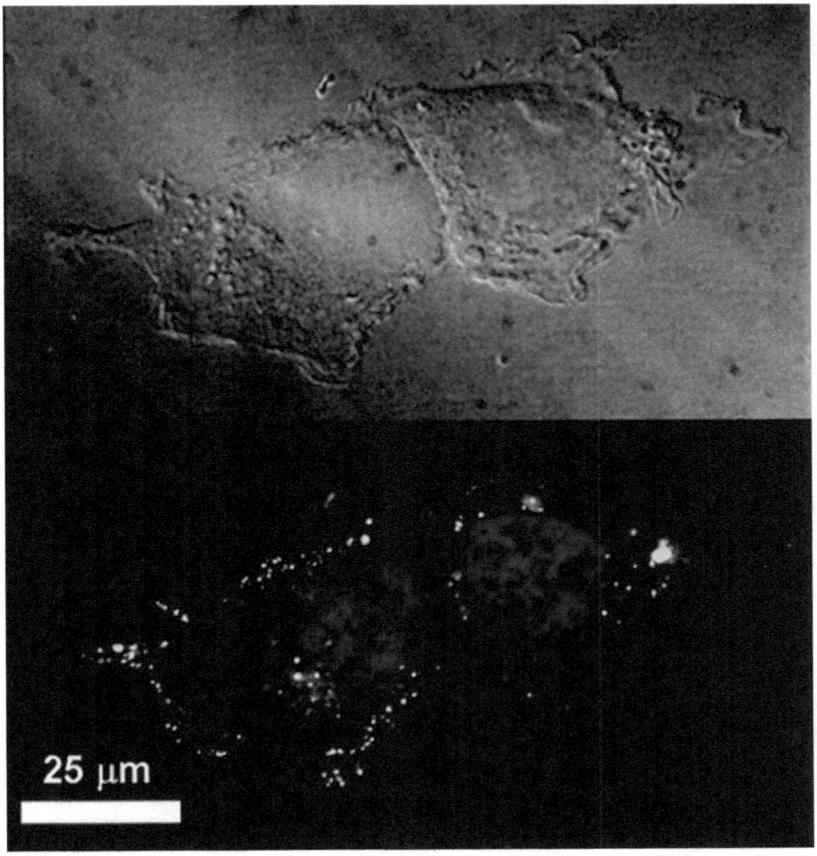

FIGURE 8.5 TEM image of CNTs internalized into the cells, black color portion indicating the CNTs present in the cells [32].

to evaluate the cellular uptake and confirm the pathway involved. After incubating it for over four hours with micelle-like structures, the images depict that there is no co-localization involved with the lysosome dye, and FITC-SWCNTs–$COOC_{18}H_{37}$ constructs can pass through the cell membrane without the formation of endocytic vacuoles. For particles greater than 400 nm, endocytosis was the mechanism of uptake, whereas for particles of size less than 400 nm, a passive diffusion mechanism was involved [33]. Effect of the length of CNTs over the exocytosis and the type of endocytosis process involved in internalization into the murine macrophage cells was observed in a study conducted by Xue Jing et al. Endocytosis inhibitors such as methyl-β-cyclodextrin and chlorpromazine were used to find the pathway for uptake into RAW 264.7 cells. The uptake rate declined upon increasing the concentration of SWCNT, indicating the saturable nature of the process. Macropinocytosis, caveolae-mediated endocytosis, and CME are the priority sequence of uptake pathways followed by the RAW264.7 cells. Fascinatingly, CME was a pathway for the uptake of shorter SWCNTs [34]. Becker et al. did a study to observe the effect of the

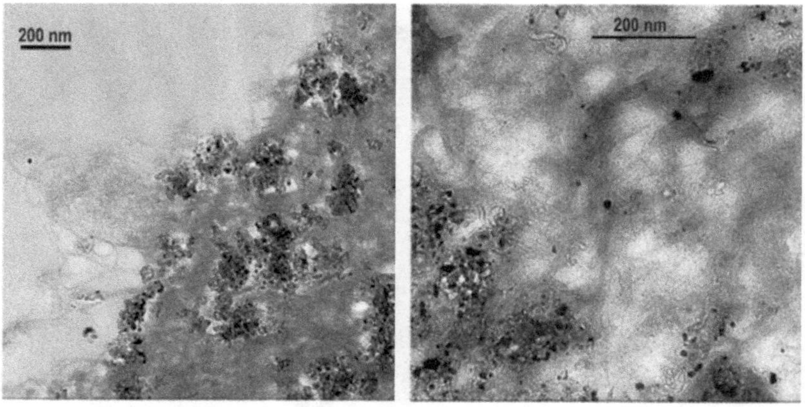

FIGURE 8.6 TEM images of cells with internalized CNTs [35].

length of DNA-wrapped CNTs over cellular uptake in IMR 90 cells. Wst-1 assay was performed with the SWCNTs of length 335 ± 27 nm and 253 ± 26 nm. These longer fractions did not show any impact on the viability of cells. The nanotubes less than 335 ± 27 nm had an effect which was evidenced by the decreased metabolic activity, and within the given incubation time, cells were found dead in staining as shown in Figure 8.6. The supernatant was filled with CNTs of length greater than 189 ± 17 nm. The CNTs had shown length-dependent uptake, and nanotubes of length less than 189 ± 17 nm showed higher uptake and toxicity [35].

Kong et al. designed smart CNTs, PEI-Chol modified SWCNTs (PCS), with laser control to deliver genes via a nondigestive pathway. The mechanism involved in the cellular uptake was assessed with the help of endocytosis inhibitors such as chlorpromazine (CPZ) and Filipin. The cellular internalization of PEI_{25K}/DNA into HeLa cells and HEK293 cells was inhibited by the pretreatment of CPZ and Filipin into the cells, which was further confirmed by observing the fluorescence signal of Cy3 and FITC in complexes; there was prominent intensity of FITC and Cy3 fluorescence signals from the CPZ-treated HeLa and HEK293 cells. As CPZ is the inhibitor of CME, we can conclude that caveolae-mediated endocytosis is the only pathway for cellular uptake of PEI_{25K}/DNA. The PCS was able to deliver the DNA through the caveolae-mediated endocytosis pathway, which is safer when compared to that of PEI/DNA, which gets internalized through the CME where extensive damage of the cargo DNA occurs [36]. Das et al. performed a study to find out the effect of surface chemistry on the trafficking and the therapeutic performance of drug-loaded CNTs. The influence of endocytosis-inhibiting conditions was assessed using sodium azide-pretreated A549 cells and incubated with pegylated and f-CNTs for three hours. Quantitative data has proven the fact that there is a nominal effect on cellular uptake of p and f-CNTs even after treating cells with sodium azide to block the energy-dependent pathway. The pegylated and amino group-modified CNTs had been uptaken by passive diffusion, whereas the rest of them had been taken up by the endocytosis pathway. This proves the nanoneedle mechanism involved in the incorporation of CNTs [37]. Cai et al. had done a study to improve gene delivery

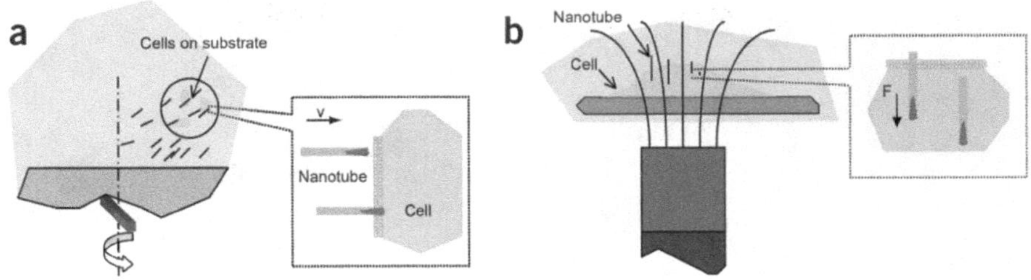

FIGURE 8.7 The mechanism involved in the spearing of CNTs. (a) In the first step, a rotating magnetic field drives nanotubes (short black lines) to spear the cells (white) on a substrate. The boxed inset is a close-up of one cell with the nanotube penetrating the membrane. The plasma membrane is illustrated as an assembly of. (b) In the second step, a static field persistently pulls nanotubes into the cells. F, magnetic force; v, the velocity of nanotubes [38].

into the cells by utilizing the magnetic spearing concept with aid of DNA utilizing nickel-containing carbon nanotubes (CNTs) as a vector. The experiments conducted showed that the magnetic field did not show any fluorescence, indicating that it did not affect cellular uptake, and no endocytosis or phagocytosis was involved. As illustrated in Figure 8.7, the mechanical spearing was solely mediated by the DNA uptake [38].

Kam et al. conducted a study on the internalization of protein conjugates of SWCNTs into the cells. The fluorescent protein attached to the SWCNT protein conjugate was found in the endosome, indicating endocytosis as a method for uptake into cells. In microscopy, it was observed that the compound penetrated the nucleus via the cytoplasm [39]. Ankit Gangarade et al. developed an injectable carbon nanotube-impregnated silk-based multifunctional hydrogel for localized targeting and anti-cancer drug delivery. The uptake mechanism was evaluated using the KB and A549 cell lines. These cell lines were selected based on the criteria of the presence or absence of folate receptors. Here, doxorubicin-loaded folic acid functionalized single-walled carbon nanotube (SWCNT-FA/DOX) is a formulation developed for delivering the drug to cancerous cells. The specific endocytosis inhibitors have been used to evaluate the exact pathway of cellular uptake of these SWCNTs. The FA-positive KB cell only showed the uptake of FITC labeled SWCNT-FA. Whereas uptake was absent in the cells when they had been treated with the folic acid, which indicates that the cellular uptake proceeds via the receptor-mediated endocytosis pathway. The SWCNT had determined its uptake pathway, a shorter one via the energy-independent diffusion and a longer one through the energy-dependent endocytosis [40]. Meng et al. developed a SWCNTs-based drug delivery system to target the cancer cells using doxorubicin. The acridine orange functionalized single-walled carbon nanotubes (AO-SWCNTs) uptake was observed by utilizing various endocytosis pathway inhibitors such as chlorpromazine, cytochalasin, and nystatin, as depicted in Figure 8.8. The uptake was only affected in the presence of the chlorpromazine, suggesting that the uptake was mediated by CME, as shown below in the image [41].

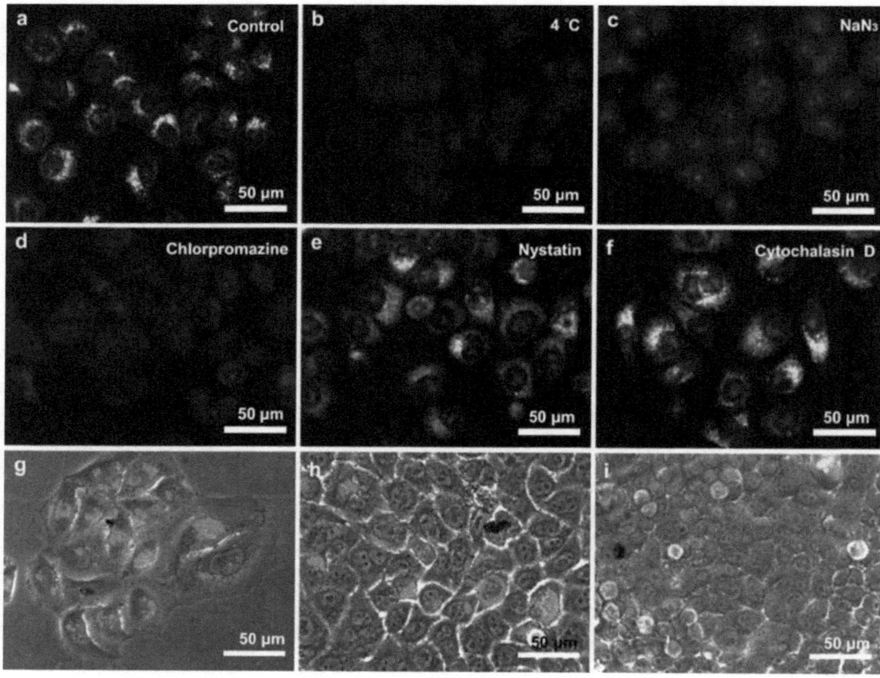

FIGURE 8.8 (a–f) Fluorescence images of HeLa cells incubated with cut SWCNTs modified with an acridine gray fluorescent tag, termed AO-SWCNTs, under various endocytosis-inhibiting conditions. Before being incubated with AO-SWCNTs, cells were pretreated for one hour under various endocytosis-inhibiting conditions (10 mg/mL, SWCNT-concentration) for another 30 minutes (g–i). Long-term observation of AO-SWCNTs inside HeLa cells. (g) Merged image between bright field image and fluorescence image of HeLa cells incubated with AO-SWCNTs (20 mg/mL) for one hour and (h) culture as continued for seven days after the nanotube containing media was replaced with fresh media (i) 12 days [41].

8.4.2 Double-Walled Carbon Nanotubes

Double-walled carbon nanotubes are coaxial nanostructures composed of exactly two SWCNTs, one nuzzled in another. Based upon the scientific observations, some found that DWCNT had been uptaken by cells via the nanoneedle mechanism, whereas the DNA-wrapped DWCNT had been taken up by exocytosis, and others have been taken up via the endocytosis process [42]. Neves et al. had done a trial to understand the cellular localization, trafficking, and accumulation of DWCNTs in prostate cancer cells (PC3 cells). From the Raman spectroscopy and fluorescence data, it was observed there is an intracellular localization, and the endocytic pathway was involved along with the capture by clathrin-coated vesicles. Specific antigens have helped to track the stages of CNT complexes involved in cellular trafficking, as depicted in the below image. CNT was found in the Rab 11 positive vesicles commending an explicit route for the exocytosis shown in Figure 8.9 [43].

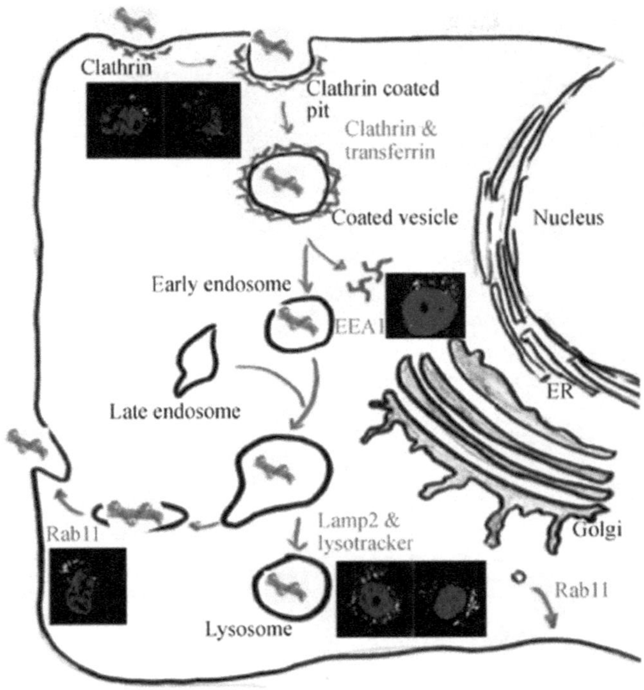

FIGURE 8.9 Cellular internalization mechanism of CNTs [43].

8.4.3 Multi-Walled Carbon Nanotubes

Multi-walled carbon nanotubes are composite structures that constitute the nanotubes embedded in each other. The process of uptake of MWCNT cannot be generalized as many processes had been observed for specific uptake of the MWCNT for the respective formulation developed. Here are some of the remarks regarding the uptake of the MWCNT. Jamal et al. investigated the mechanisms of cellular uptake of functionalized multi-walled carbon nanotubes (f-MWCNTs) in the A549 cells with the aid of 3D electron tomography imaging. They observed the piercing capability of the f-MWCNTs through the cell membrane with the assistance of energy-filtered transmission electron microscopy (EFTEM) maps of osmicated A549 cells as observed in Figure 8.10. It was observed that the cells were localized intracellularly after 24 hours of treatment with the MWCNTs. The uptake was energy-dependent endocytosis, translocation of tubes directly into the cytoplasm, and captured by lipid-rich vesicular components such as lysosomes and endosomes [44].

Antonino et al. observed the intracellular trafficking and the therapeutic outcome of the MWCNTs modified with cyclodextrins and polyethyleneimine. To elucidate the cellular uptake mechanism, the Vero and Hep-2 cells were incubated with MWCNT-CD-PEI-Rhod and observed by fluorescence microscopy. Vero cells were incubated with endocytosis inhibitors sucrose and filipin were employed as CME

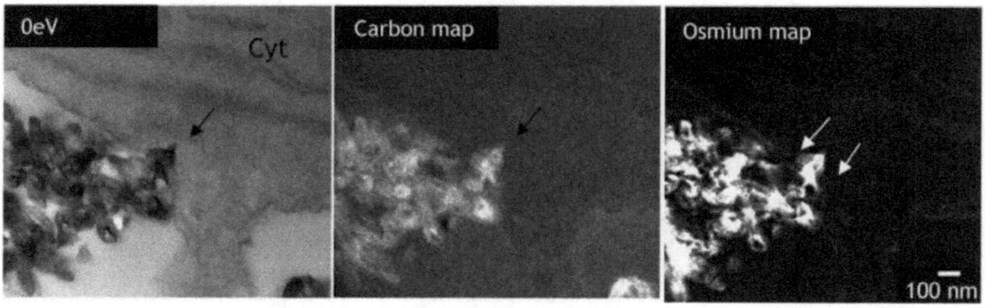

FIGURE 8.10 EFTEM maps of osmicated A549 cells [44].

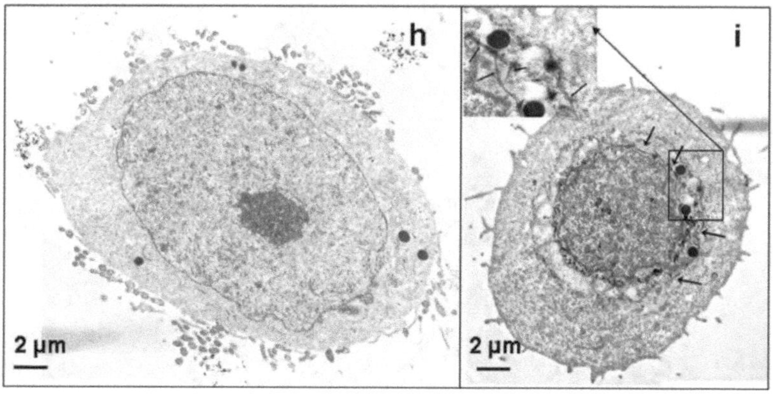

FIGURE 8.11 Internalization of CNTs into cells treated with bioconjugate [47].

inhibitors and caveolae-mediated endocytosis inhibitors, respectively, to sort out the mechanism of cellular uptake. MWCNT-CD-PEI-Rhod uptake was affected by sucrose presence and it was left unaffected even in the presence of filipin. This indicates that CD/PEI f-MWCNTs uptake crosses the cell membrane by the clathrin-dependent pathway [45]. Kamazani et al. synthesized bromocriptine-conjugated MWCNTs to deliver a drug to lung cancer cells. The treated cells were observed for the expression of proteins such as Bax and Bcl-2. It was observed that Bax protein levels had decreased in lung cancer cell lines compared to that of MRC 5 cells, due to expression of DRD2 on the MRC5 cell surface, which causes less interaction and uptake of the drug into the cell. Due to the great diameter of MWCNT, they escape the endosomes and piercing lysosome, enter into the cytoplasm, and interact with the nucleus [46]. Boucetta et al. conducted a study to observe the uptake of surface-modified multi-walled carbon nanotubes and their impact on the cells. The modified lactate dehydrogenase assay was performed to evaluate the toxicity of these modified MWCNTs. The adsorption phenomenon between MWCNTs and cells is accountable for uptake. Chemically modified MWCNTs had more uptake than the Pluronic F-127-modified MWCNTs. The Pluronic F-127-modified MWCNTs had shown more toxic effects in the modified LDH assay. The below Figure 8.11 depicts the higher uptake of chemically modified MWCNTs on the left side and Pluronic

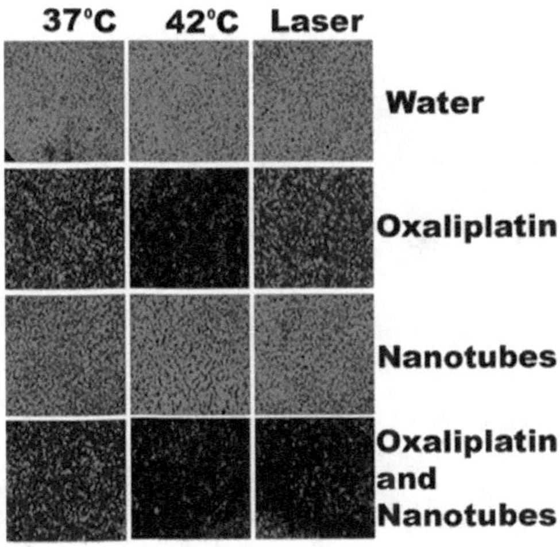

FIGURE 8.12 RKO cell live population, compared when exposed to IR radiation [49].

F-127-modified MWCNTs on the right side [47]. Bhirde et al. developed CNTs that target the EGFR receptor and kill the cancerous cells. The results suggested that the uptake was done by receptor-mediated endocytosis. Uptake was limited in the cells without EGFR expression due to knockdown by the siRNA, indicating the importance of EGFR-EGF interaction. Only the SQE cells showed dark circular cylindrical structures, whereas the cells treated with SWCNT-Qdot only [48].

Nicole et al. developed a rapid photothermal intracellular drug delivering MWCNTs. The cellular uptake was measured using RKO and HCT 116 cell lines. Whenever IR radiation hits, there is an increase in the molecular frequency generating heat in the surrounding environment that further enhances the drug localization. Oxaliplatin uptake was enhanced after achieving rapid hyperthermia. Also, a noticeable thing was a reduction of the time for MWCNTs uptake into the cell; viability of cells was reduced to seconds. Platinum accumulation was higher in the cells at 42°C, as shown in the below Figure 8.12 [49].

Nunes et al. had conducted research on hybrid polymer grafted multi-walled carbon nanotubes (polymer-g-MWCNTs) for in vitro gene delivery. Cationic polymer-g-MWCNTs to complex and transfer the plasmid DNA (pcmv-βgal) without causing cytotoxicity. The fluorescein-labeled DNA uptake was observed with the help of confocal microscopy; here, the polymer-g-MWCNTs showed better uptake compared to that of the plain MWCNTs as shown in Figure 8.13. This suggests that there might be an orientation by the polymer-g-MWCNTs towards the endocytosis pathway that will aid in increasing the trafficking to the nuclear region. The difference in the localization of polymer-g-CNT:pDNA and the absence of co-localization of vectors into endosomes suggests that either energy-independent translocation or phagocytosis (receptor-mediated endocytosis) had been the main mechanism for uptake into cells. Also, we can conclude that CME was not involved in the uptake process [50].

Al-Jamal et al. observed the cellular uptake mechanisms of f-MWCNTs by 3D electron tomography imaging. Cells capable of endocytosis and cells that are not

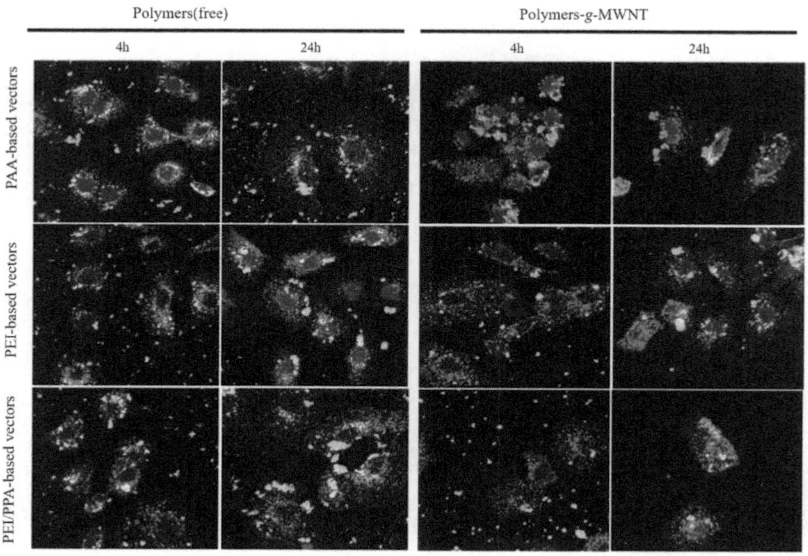

FIGURE 8.13 Intracellular trafficking of polymer-g-MWCNT:pDNA or polymer-pDNA complexes following 4 and 24 hours of A549 cell incubation. pDNA uptake was monitored using fluorescently labeled pDNA (black), endosomes stained with anti-EEA1 antibodies labeled with Cy3 (gray), and nuclei were counterstained with TO-PRO 3 [50].

capable of endocytosis are used to determine the pathway for the uptake of the f-MWCNTs. Several types of mechanisms, such as direct membrane translocation, membrane wrapping, and transportation as clusters via vesicles, were observed. EFTEM maps had shown that modified MWCNTs were able to traverse the membrane by piercing through and extending into the cytoplasm, i.e., the process termed endosomal escape [51]. Wang et al. conducted a study to assess the pathway of the MWCNTs with recombinant protein toxin, ricin A chain tagged with the enhanced black fluorescent protein (MWCNT-EGFP-RTA). MWCNT-EGFP-RTA was tracked with the help of fluorescence generated, and several tests were performed to trace out the mechanism involved. When treated with nystatin, chlorpromazine, and cytochalasin D, it was observed that the uptake was inhibited in the HeLa cells that were treated with chlorpromazine, the rest of them being unaffected in the uptake process. This suggests that the MWCNT-EGFP-RTA was following the CME to reach inside the cell, also, the MWCNTs are found accumulated in the ER of the cells [52].

8.5 SUMMARY

The study of carbon nanotubes in drug delivery systems demonstrates their enormous potential owing to their distinct structural and functional features. These graphene-based allotropes have a large surface area, customizable characteristics, and the potential to transport a variety of therapeutic cargoes, making them excellent candidates for targeted and controlled drug administration. The internalization processes of CNTs, which include both energy-dependent and energy-independent channels, highlight the intricacy of their cellular transport. CME, caveolae-mediated endocytosis, and macropinocytosis are important mechanisms for CNT uptake and intracellular behavior. The effectiveness of CNT absorption is affected by several aspects, such as size, level of agglomeration and aggregation, surface charge, and the specific kind of cells used. Smaller CNTs are taken up into cells more effectively, with the methods of uptake depending on their size and surface changes, which may be either passive or active. The level of clustering may improve absorption, but it can also raise the risk of cell toxicity, requiring a careful equilibrium between clumping and compatibility with living organisms. The presence of a surface charge affects the interaction with cellular membranes and the creation of protein corona, which in turn affects the efficiency of absorption. Different cell types have varied degrees of CNT absorption, with phagocytic cells being the most efficient. Understanding these parameters is critical for optimizing CNT-based drug delivery systems that enable targeted distribution while minimizing harmful effects. Future research should concentrate on improving the functionalization and delivery methods of CNTs to increase their specificity and safety. By solving these obstacles, CNTs may be efficiently used in clinical applications, revolutionizing the area of targeted drug administration and opening up new treatment options for a variety of disorders.

REFERENCES

1. A. Bianco, K. Kostarelos, and M. Prato, "Applications of carbon nanotubes in drug delivery," *Curr. Opin. Chem. Biol.*, vol. 9, no. 6, pp. 674–679, Dec. 2005, doi: 10.1016/J.CBPA.2005.10.005.
2. A. Eatemadi et al., "Carbon nanotubes: properties, synthesis, purification, and medical applications," *Nanoscale Res. Lett.*, vol. 9, no. 1, p. 393, 2014, doi: 10.1186/1556-276X-9-393.
3. W. Zhang, Z. Zhang, and Y. Zhang, "The application of carbon nanotubes in target drug delivery systems for cancer therapies," *Nanoscale Res. Lett.*, vol. 6, no. 1, p. 555, 2011, doi: 10.1186/1556-276X-6-555.
4. C. Zhang, L. Wu, M. de Perrot, and X. Zhao, "Carbon nanotubes: a summary of beneficial and dangerous aspects of an increasingly popular group of nanomaterials," *Front. Oncol.*, vol. 11, p. 693814, Jul. 2021, doi: 10.3389/fonc.2021.693814.
5. "What Is A Cell? – Definition, Structure, Types, Functions." https://byjus.com/biology/cells/ (accessed Jun. 14, 2024).
6. A. A. Tokarev, A. Alfonso, and N. Segev, "Overview of Intracellular Compartments and Trafficking Pathways," Springer, New York, 2013, Accessed: Dec. 07, 2023. Available: https://www.ncbi.nlm.nih.gov/books/NBK7286/
7. R. A. Anisimov, D. A. Gorin, and A. A. Abalymov, "3D cell spheroids as a tool for evaluating the effectiveness of carbon nanotubes as a drug delivery and photothermal therapy agents," *C-J. Carbon Res.*, vol. 8, no. 4, p. 56, Dec. 2022, doi: 10.3390/C8040056.
8. A. A. Tokarev, A. Alfonso, and N. Segev, "Overview of intracellular compartments and trafficking pathways," *Traffick. Insid. Cells.*, vol. 1, pp. 3–14, 2009, doi: 10.1007/978-0-387-93877-6_1.
9. P. S. Chary et al., "A review on emerging role of multifunctional carbon nanotubes as an armament in cancer therapy, imaging and biosensing," *J. Drug Deliv. Sci. Technol.*, vol. 85, p. 104588, Aug. 2023, doi: 10.1016/J.JDDST.2023.104588.

10. E. Vives, "Cellular utake of the Tat peptide: an endocytosis mechanism following ionic interactions," *J. Mol. Recognit.*, vol. 16, no. 5, pp. 265–271, Sep. 2003, doi: 10.1002/JMR.636.
11. P. Foroozandeh and A. A. Aziz, "Insight into cellular uptake and intracellular trafficking of nanoparticles," *Nanoscale Res. Lett.*, vol. 13, no. 1, pp. 1–12, Oct. 2018, doi: 10.1186/S11671-018-2728-6.
12. D. M. Underhill and H. S. Goodridge, "Information processing during phagocytosis," *Nat. Rev. Immunol.*, vol. 12, no. 7, pp. 492–502, Jun. 2012, doi: 10.1038/nri3244.
13. J. Simon, E. Flahaut, and M. Golzio, "Overview of carbon nanotubes for biomedical applications," *Materials (Basel).*, vol. 12, no. 4, Feb. 2019, doi: 10.3390/MA12040624.
14. L. Guo, "PinocytosisWhat Is It, How It Occurs, and More," 2022. Available: https://www.osmosis.org/answers/pinocytosis
15. J. J. Rennick, A. P. R. Johnston, and R. G. Parton, "Key principles and methods for studying the endocytosis of biological and nanoparticle therapeutics," *Nat. Nanotechnol.*, vol. 16, no. 3, pp. 266–276, Mar. 2021, doi: 10.1038/s41565-021-00858-8.
16. G. E. Palade, "Fine structure of blood capillaries," *J. Appl. Phys.*, vol. 24, pp. 1424–1448, 1953, Accessed: Jun. 14, 2024. Available: https://cir.nii.ac.jp/crid/1573668924125256320
17. J. B. Helms and C. Zurzolo, "Lipids as targeting signals: lipid rafts and intracellular trafficking," *Traffic*, vol. 5, no. 4, pp. 247–254, Apr. 2004, doi: 10.1111/J.1600-0854.2004.0181.X.
18. B. Van Deurs, K. Roepstorff, A. M. Hommelgaard, and K. Sandvig, "Caveolae: anchored, multifunctional platforms in the lipid ocean," *Trends cell Biol.*, vol. 13, no. 2, pp. 92–100, 2003. Accessed: Jun. 14, 2024. Available: https://www.cell.com/trends/cell-biology/fulltext/S0962-8924(02)00039-9?large_figure=true&code=cell-site
19. A. L. Kiss and E. Botos, "Endocytosis via caveolae: alternative pathway with distinct cellular compartments to avoid lysosomal degradation?" *J. Cell. Mol. Med.*, vol. 13, no. 7, p. 1228, Jul. 2009, doi: 10.1111/J.1582-4934.2009.00754.X.
20. A. I. Magee and I. Parmryd, "Detergent-resistant membranes and the protein composition of lipid rafts," *Genome Biol.*, vol. 4, no. 11, p. 234, 2003, doi: 10.1186/GB-2003-4-11-234.
21. S. Mayor and R. E. Pagano, "Pathways of clathrin-independent endocytosis," *Nat. Rev. Mol. Cell Biol.*, vol. 8, no. 8, pp. 603–612, Aug. 2007, doi: 10.1038/nrm2216.
22. H. Zare et al., "Carbon nanotubes: smart drug/gene delivery carriers," *Int. J. Nanomedicine*, vol. 16, pp. 1681–1706, 2021, doi: 10.2147/IJN.S299448.
23. K. Fujita et al., "Size effects of single-walled carbon nanotubes on in vivo and in vitro pulmonary toxicity," *Inhal. Toxicol.*, vol. 27, no. 4, p. 207, Mar. 2015, doi: 10.3109/08958378.2015.1026620.
24. M. Zhang et al., "Size-dependent cell uptake of carbon nanotubes by macrophages: a comparative and quantitative study," *Carbon N. Y.*, vol. 127, pp. 93–101, Feb. 2018, doi: 10.1016/J.CARBON.2017.10.085.
25. Z. M. Song, L. Wang, N. Chen, A. Cao, Y. Liu, and H. Wang, "Biological effects of agglomerated multi-walled carbon nanotubes," *Colloids Surf. B Biointerfaces*, vol. 142, pp. 65–73, Jun. 2016, doi: 10.1016/J.COLSURFB.2016.02.032.
26. C. Kuroda et al., "Different aggregation and shape characteristics of carbon materials affect biological responses in RAW264 cells," *Int. J. Nanomedicine*, vol. 13, p. 6079, 2018, doi: 10.2147/IJN.S172493.
27. W. Jiang et al., "Effects of charge and surface defects of multi-walled carbon nanotubes on the disruption of model cell membranes," *Sci. Total Environ.*, vol. 574, pp. 771–780, Jan. 2017, doi: 10.1016/J.SCITOTENV.2016.09.150.
28. C. Caoduro et al., "Carbon nanotubes as gene carriers: focus on internalization pathways related to functionalization and properties," *Acta Biomater.*, vol. 49, pp. 36–44, Feb. 2017, doi: 10.1016/J.ACTBIO.2016.11.013.

29. K. Kostarelos et al., "Cellular uptake of functionalized carbon nanotubes is independent of functional group and cell type," *Nat. Nanotechnol.*, vol. 2, no. 2, pp. 108–113, 2007, doi: 10.1038/NNANO.2006.209.
30. P. N. Yaron, B. D. Holt, P. A. Short, M. Lösche, M. F. Islam, and K. N. Dahl, "Single wall carbon nanotubes enter cells by endocytosis and not membrane penetration," *J. Nanobiotechnol.*, vol. 9, no. 1, pp. 1–15, Sep. 2011, doi: 10.1186/1477-3155-9-45.
31. B. Kang, D. C. Yu, S. Q. Chang, D. Chen, Y. D. Dai, and Y. Ding, "Intracellular uptake, trafficking and subcellular distribution of folate conjugated singlewalled carbon nanotubes within living cells," *Nanotechnology*, vol. 19, no. 37, p. 375103, Aug. 2008, doi: 10.1088/0957-4484/19/37/375103.
32. R. P. Feazell, N. Nakayama-Ratchford, H. Dai, and S. J. Lippard, "Soluble single-walled carbon nanotubes as longboat delivery systems for platinum(IV) anticancer drug design," *J. Am. Chem. Soc.*, vol. 129, no. 27, pp. 8438–8439, Jul. 2007, doi: 10.1021/ja073231f.
33. A. Antonelli et al., "Improved cellular uptake of functionalized single-walled carbon nanotubes," *Nanotechnology*, vol. 21, no. 42, p. 425101, Sep. 2010, doi: 10.1088/0957-4484/21/42/425101.
34. X. Cui, B. Wan, Y. Yang, X. Ren, and L. H. Guo, "Length effects on the dynamic process of cellular uptake and exocytosis of single-walled carbon nanotubes in murine macrophage cells," *Sci. Rep.*, vol. 7, no. 1, pp. 1–13, May 2017, doi: 10.1038/s41598-017-01746-9.
35. M. L. Becker et al., "Length-dependent uptake of DNA-wrapped single-walled carbon nanotubes," *Adv. Mater.*, vol. 19, no. 7, pp. 939–945, Apr. 2007, doi: 10.1002/ADMA.200602667.
36. F. Kong et al., "Smart carbon nanotubes with laser-controlled behavior in gene delivery and therapy through a non-digestive trafficking pathway," *Small*, vol. 12, no. 48, pp. 6753–6766, Dec. 2016, doi: 10.1002/SMLL.201601092.
37. M. Das, R. P. Singh, S. R. Datir, and S. Jain, "Surface chemistry dependent 'switch' regulates the trafficking and therapeutic performance of drug-loaded carbon nanotubes," *Bioconjug. Chem.*, vol. 24, no. 4, pp. 626–639, Apr. 2013, doi: 10.1021/bc300598z.
38. D. Cai et al., "Highly efficient molecular delivery into mammalian cells using carbon nanotube spearing," *Nat. Methods*, vol. 2, no. 6, pp. 449–454, May 2005, doi: 10.1038/nmeth761.
39. N. W. S. Kam, T. C. Jessop, P. A. Wender, and H. Dai, "Nanotube molecular transporters: internalization of carbon nanotube-protein conjugates into mammalian cells," *J. Am. Chem. Soc.*, vol. 126, no. 22, pp. 6850–6851, Jun. 2004, doi: 10.1021/ja0486059.
40. A. Gangrade and B. B. Mandal, "Injectable carbon nanotube impregnated silk based multifunctional hydrogel for localized targeted and on-demand anticancer drug delivery," *ACS Biomater. Sci. Eng.*, vol. 5, no. 5, pp. 2365–2381, May 2019, doi: 10.1021/acsbiomaterials.9b00416.
41. L. Meng, X. Zhang, Q. Lu, Z. Fei, and P. J. Dyson, "Single walled carbon nanotubes as drug delivery vehicles: targeting doxorubicin to tumors," *Biomaterials*, vol. 33, no. 6, pp. 1689–1698, Feb. 2012, doi: 10.1016/j.biomaterials.2011.11.004.
42. V. Neves et al., "Uptake and release of double-walled carbon nanotubes by mammalian cells," *Adv. Funct. Mater.*, vol. 20, no. 19, pp. 3272–3279, Oct. 2010, doi: 10.1002/ADFM.201000994.
43. V. Neves et al., "Cellular localization, accumulation and trafficking of double-walled carbon nanotubes in human prostate cancer cells," *Nano Res.*, vol. 5, no. 4, pp. 223–234, Feb. 2012, doi: 10.1007/s12274-012-0202-9.

44. K. T. Al-Jamal et al., "Cellular uptake mechanisms of functionalised multi-walled carbon nanotubes by 3D electron tomography imaging," *Nanoscale*, vol. 3, no. 6, pp. 2627–2635, Jun. 2011, doi: 10.1039/C1NR10080G.
45. A. Mazzaglia et al., "Intracellular trafficking and therapeutic outcome of multiwalled carbon nanotubes modified with cyclodextrins and polyethylenimine," *Colloids Surf. B Biointerfaces*, vol. 163, pp. 55–63, Mar. 2018, doi: 10.1016/J.COLSURFB.2017.12.028.
46. F. M. Kamazani, F. S. Nematalahi, S. D. Siadat, M. Pornour, and M. Sheikhpour, "A success targeted nano delivery to lung cancer cells with multi-walled carbon nanotubes conjugated to bromocriptine," *Sci. Rep.*, vol. 11, no. 1, pp. 1–15, Dec. 2021, doi: 10.1038/s41598-021-03031-2.
47. H. Ali-Boucetta et al., "Cellular uptake and cytotoxic impact of chemically functionalized and polymer-coated carbon nanotubes," *Small*, vol. 7, no. 22, pp. 3230–3238, Nov. 2011, doi: 10.1002/SMLL.201101004.
48. A. A. Bhirde et al., "Targeted killing of cancer cells in vivo and in vitro with EGF-directed carbon nanotube-based drug delivery," *ACS Nano*, vol. 3, no. 2, pp. 307–316, Feb. 2009, doi: 10.1021/nn800551s.
49. N. H. Levi-Polyachenko, E. J. Merkel, B. T. Jones, D. L. Carroll, and J. H. Stewart IV, "Rapid photothermal intracellular drug delivery using multiwalled carbon nanotubes," *Mol. Pharm.*, vol. 6, no. 4, pp. 1092–1099, Aug. 2009, doi: 10.1021/MP800250E.
50. A. Nunes et al., "Hybrid polymer-grafted multiwalled carbon nanotubes for in vitro gene delivery," *Small*, vol. 6, no. 20, pp. 2281–2291, Oct. 2010, doi: 10.1002/SMLL.201000864.
51. K. T. Al-Jamal et al., "Cellular uptake mechanisms of functionalised multi-walled carbon nanotubes by 3D electron tomography imaging," *Nanoscale*, vol. 3, no. 6, pp. 2627–2635, Jun. 2011, doi: 10.1039/C1NR10080G.
52. M. Wang, S. Yu, C. Wang, and J. Kong, "Tracking the endocytic pathway of recombinant protein toxin delivered by multiwalled carbon nanotubes," *ACS Nano*, vol. 4, no. 11, pp. 6483–6490, Nov. 2010, doi: 10.1021/nn101445y.

9 Pharmacokinetic and Toxicological Aspects of Carbon Nanotubes

Sonali Batra, Sumit Sharma, Navjot Kanwar, Abhinav Kanwal, Gulsheen, Lalit Kumar, and Sachin

9.1 INTRODUCTION

A form of nanomaterial known as carbon nanotubes (CNTs) comprises the carbon allotrope known as buckminsterfullerenes, or fullerenes (C_{60}). Fullerenes contain sp^2 hybridized carbon atoms, which are organized in a cylindrical fashion. CNTs are the tubular form of fullerenes and have a broad range of applications from electronics to biotechnology. Their diameter can be measured on a nanometer scale; that is, they are approximately ten thousand times thinner than a human hair but are one of the toughest and stiffest materials yet discovered. Their discovery was revolutionary as these are a hundred times robust as compared to steel at the molecular level and weigh only one-sixth the weight of steel, besides having a high aspect ratio. The discovery of CNTs initiated the way for a completely new branch of material science—nanoscience. The presence of a tubular fullerene was suggested by Rick Smalley in 1990 during a workshop on carbon-carbon composites. He ideated that a C_{60} molecule could be stretched to form a bucky tube. At a fullerene workshop in Philadelphia in August 1991, Dresselhaus continued the conversation by giving an oral presentation regarding the suggested symmetry for CNTs with fullerene hemisphere caps on each end. The experimental evidence for the existence of CNTs was given by Lijima in 1991 while he was researching the synthesis of fullerenes using an electric arc discharge approach. Lijima used high-resolution transmission electron microscopy (HRTEM) to analyze these and found that they were multi-walled CNTs (MWNTs). These had at least two graphitic layers with an internal diameter of 4 nm in general. After two years, Lijima and Ichihashi of NEC Corporation, Japan, as well as Bethune and associates of IBM Almaden Research Center, California, synthesized single-walled CNTs (SWNTs) concurrently and independently. These were created using the same method as MWNTs, but by mixing certain metal particles, such as cobalt, with carbon electrodes. The synthesized SWNTs had a very small diameter (~1 nm) and were looped and coiled instead of straight, which gave them a considerably different appearance from MWNTs [1].

DOI: 10.1201/9781003386537-9

While Ijima receives official recognition for discovering CNTs, Bacon at Union Carbide in Parma, Ohio, is thought to have seen CNTs 30 years earlier. Bacon, while pursuing his research on carbon arcs to observe the characteristics of carbon fibers in 1956, presumably observed CNTs in his samples. He described the discovery of carbon nanowhiskers having a scroll-like structure during SEM analysis of his material in his paper published in the year 1960. The scientific community did not acknowledge the significance of CNTs until the discovery of C60 and theoretical investigations into potential fullerene configurations, even though CNTs have been identified for over 40 years. Since this ground-breaking achievement, the study of CNTs has advanced into a pre-eminent field of nanotechnology that is rapidly increasing [2].

9.2 STRUCTURE AND PROPERTIES

SWNTs and MWNTs are the two basic kinds of CNTs. Despite their striking similarities, SWNTs and MWNTs have quite diverse physical characteristics because of how they are structurally different. The single layer of graphene that makes up the nanotube wall is the key characteristic that sets SWNTs apart. In other words, if we think about SWNTs as hollow cylinders, they are formed from a smoothed-up, tubular shell of graphene sheets made of carbon atoms that have hexagonal rings similar to those seen in benzene. They are frequently referred to as graphene nanotubes (GNTs) for this reason. Consequently, half-fullerene molecules form the caps at the ends of the tubular shell, which is mostly composed of hexagonal rings of carbon atoms organized in a sheet. While the curvature of the end caps results from topological (pentagonal rings) imperfections in the normally hexagonal underlying lattice structure, the natural curvature of the side walls is created by the sheet rolling into the tubular structure. An end-to-end tube seal is facilitated by the positive (convex) curvature that a pentagonal ring defect imparts to the surface. MWNTs, unlike SWNTs, are made up of numerous layers of graphene that have been flawlessly wrapped up into a tube shape. The ends are left open or capped with half-fullerenes that resemble SWNTs. Due to these variations between SWNTs and MWNTs, their characteristics and subsequent effects on materials are noticeably different because of the two-dimensional organization of carbon atoms in the sheet of graphene, which permits significant out-of-plane deformation while maintaining exceptional strength against any in-plane distortion or fracture. Single- and multi-wall nanotubes have remarkable elasto-mechanical properties. The structural and material properties of nanotubes suggest that they could be used to create the next generation of composite materials that are incredibly light but both robust and highly elastic.

Nanotubes have a high aspect ratio and advantageous mechanical and electrical properties. Thus, two commercial uses of nanotubes that are starting to take shape are field-emission displays and scanning probe microscopic tips for metrology. Nanotubes have been suggested as a light weight, high surface area packing material for the storage of gas and hydrocarbon-based fuel storage devices, liquid or gas filtration systems, nanoscale drug delivery containers, and casting structures for nanowires and nanocapsules due to their hollow, tubular, and caged molecule nature. Carbon is the sole element that can be found in a wide range of shapes and forms with different physical and chemical properties when it comes to structural and functional materials. Therefore, the best materials to use as molecular-level building blocks in the creation and design of nanoscale systems are those based on carbon. For example, two common types of carbon are diamond and layered graphite, which can also be found in the following shapes: nano-conical, helical spring, rolled-up tubes, planar sheets, and rectangular hollow boxes. Covering any carbon-based nanoscale device with biological lipid layers or protein molecules would also allow one to participate in the quickly developing field of bio-nanotechnology [3].

9.2.1 Electrical Properties

CNTs have remarkable conductivity, and their electronic properties depend upon the following various factors: (i) tubular structure, (ii) chirality, (iii) different diameters, and (iv) crystallinity. Since each carbon atom in the lattice structure is covalently bound to three neighboring carbon atoms via sp^2 hybridization, the fourth valence electron, which is free, is responsible for the electrical property of CNTs. The conductivity of CNTs is attributed to the delocalized free electrons. Therefore, CNTs can be either conductive or semi-conductive depending on the sort of chirality. CNTs' electrical properties make them useful for switching applications in transistors and other advanced electronic devices, including biosensors [4].

9.2.2 Mechanical Properties

CNTs are regarded as the toughest material in nature, especially in axial direction, and have Young's modulus in the range 270–950 GPa. These also have high tensile strength between the range 11 and 63 GPa. This is explained by the fact that graphite contains carbon-carbon bonds. However, CNTs are found to be soft in radial direction, a characteristic known as radial elasticity. This property has significant importance, mainly in the formation of nanocomposites. The mechanical property of CNTs is directly proportional to their diameter [4].

9.2.3 Thermal Properties

The performance and stability of nano and micro devices are greatly influenced by the thermal conduction property of CNTs. For this reason, controlling thermal properties has become crucial, particularly for nano devices At room temperature, multi-walled carbon nanotubes (MWNTs) exhibited a thermal conductivity of approximately 3,000 W/m·K, surpassing that of graphite. This value was also nearly two orders of magnitude greater than the thermal conductivity observed in bulk MWNT samples. Moreover, individual single-walled carbon nanotubes (SWNTs) demonstrated thermal conductivity values around 200 W/m·K higher than individual MWNTs. An inverse relationship—with an exponent of 0.69—was identified between vacancy concentration and thermal conductivity [4].

9.3 TOXICOLOGICAL ASPECTS OF CNTs

CNTs have gained significant attraction due to their special properties, but concerns have been expressed about their potential of causing cardiovascular toxicity. Comprehending the cardiovascular toxicity features of CNTs is necessary for evaluating the hazards linked to their application and guaranteeing the safety of human health and the environment. The heart and blood vessels, collectively known as the cardiovascular system, are essential for preserving general health. Research on the effects of CNTs exposure on the cardiovascular system has shown several possible toxicological implications. The toxicological effects of CNTs on the cardiovascular system are complicated and rely on multiple variables, such as the kind of CNTs, their length, surface changes, dosage, and exposure routes. To completely understand the reasons behind CNTs cardiovascular toxicity and create suitable safety recommendations for their use in a wide range of applications, more research is required.

9.3.1 Toxicological Effects of CNTs on Heart Rate and Blood Pressure

The effects of CNTs on heart rate and blood pressure have been the subject of numerous investigations. In a study investigating the effects of commonly used nanomaterials on heart rate and blood pressure in conscious rats, intravenous administration of mesoporous silicon nanoparticles, Eudragit® nanoparticles, and single-walled carbon nanotubes (SWCNTs) resulted in decreased blood pressure and heart rate. The effects were independent of the nanomaterials size and zeta potential. The physicochemical characteristics of the NMs were proposed by the authors as potential contributors to various modes of action. More specifically, it was discovered that Eudragit® NPs may modify the constrictor and dilator responses of blood arteries, whereas SWCNTs may influence endothelium-dependent vasodilation. Chen et al. studied the effects of different kinds of MWCNTs on the heart rates and blood pressure of rats in a distinct investigation. It was shown that in rats with spontaneous hypertension, but not in normal rats, intratracheal instillation of MWCNTs at a dose of 600 μg/(kg.day) for two days in a row resulted in a reduction in heart rate and blood pressure. The effects were observed 30 days after the exposure. The study also documented enhanced indicators of vascular effects and pathological alterations in the animal arteries, along with the transfer of MWCNTs from the lungs to systemic organs. These studies collectively indicated that exposure to high levels of CNTs could lead to alterations in blood pressure and heart rate in both normal and diseased animals [5]. Recent findings demonstrated that low doses (20 ng or 100 ng) of intracerebroventricularly administered micronucleated multi-walled carbon nanotubes (MWCNTs) in rats led to reductions in blood pressure and heart rate, suppression of sympathetic nervous activity, and decreased brain activity. As the authors proposed, the mechanism involved the acetylation and nuclear translocation of NF-κB as well as promoting the activity of *n*NOS, thus leading to decreased sympathetic nerve activity and hypotension [6]. On the other hand, there have been reports suggesting that the exposure of CNT may increase blood pressure. Zheng et al. investigated the effects of inhalational exposure to MWCNTs (1.4 μm) on heart rate and blood pressure in rats at a concentration of 5 mg/m^3 for five hours. They discovered that during the exposure period, exposure to MWCNTs raised both the diastolic and systolic blood pressure. However, these values reverted to normal levels one and seven days after the exposure. The authors suggested that pulmonary exposure to MWCNTs increased parasympathetic nervous activity [7]. In a study performed on American workers, a positive association between resting heart rate and levels of inhalable or respirable elemental carbon was observed. Additionally, the study discovered no correlation between exposure to CNTs and the majority of clinical deficits in the vascular or pulmonary systems. These findings suggest that the CNT industry has taken steps to minimize CNT exposure to workers in their workplaces [8]. Numerous factors, including the physicochemical properties of CNTs, animal models employed, and the exposure routes, could be responsible for the contradictory outcomes shown across different research. The specific mechanisms through which CNTs may impact blood pressure and heart rate are still not fully understood. It is suggested that CNTs could potentially influence these cardiovascular parameters through multiple mechanisms, such as induction of systemic inflammation, dysfunction of the nitric oxide synthase (NOS) system, and alterations in nerve activity [9,10]. Moreover, although previous research has shown that CNT exposure has no significant effect on blood pressure or cause major clinical impairments in subjects, it is essential to carefully monitor the vascular systems of human beings under controlled conditions for the potential adverse effects. In order to protect those who are exposed to CNTs and to have a thorough understanding of the cardiovascular effects of CNT exposure, more research is required.

9.3.2 Toxicological Effects of CNTs (CT) on Vascular Dysfunction

The proper functioning of blood vessels, which includes their ability to dilate and constrict, is crucial for the circulatory system. Exposure to particles can disrupt this vasomotor function [11]. Research into the effects of single-walled carbon nanotubes (SWCNTs) on vascular function in rats has revealed endothelial dysfunction, evidenced by diminished vasorelaxation and heightened vasoconstriction. The findings suggest that oxidative stress from reactive oxygen species (ROS) production, along with inflammation, plays a significant role in the development of this vascular impairment. Vasorelaxation was seen by Vlasova et al. after mesenteric artery segments were incubated with 0.2 mg/mL SWCNTs. Nevertheless, there was no discernible alteration in vasorelaxation or vasoconstriction in isolated thoracic aorta rings in rats given intravenous injections of 5 mg SWCNTs. The vasorelaxation induced by SWCNTs was found to be dependent on nitric oxide (NO), as the effects were completely abolished in the presence of NO synthase (NOS) inhibitors [12]. In isolated rat aortic rings exposed to SWCNTs (0.1 to 10 µg/mL), vasorelaxation was also observed. However, this response was found to be independent of the endothelium and remained unaffected by nitric oxide synthase (NOS) inhibitors. The authors proposed that the observed effects may result from a direct interaction between SWCNTs and vascular smooth muscle cells [13]. On the other hand, Vesterdal et al. found that intratracheal instillation of SWCNTs (0.5 mg/kg, twice) in atherosclerotic $ApoE^{-/-}$ mice led to lung inflammation but did not alter vasorelaxation in aorta segments, indicating no effect on endothelial function [14].

The inconsistent findings among these studies could be attributed to differences in exposure routes or the physicochemical properties of SWCNTs. For instance, to produce stable suspensions, Frame et al. coated the SWCNTs surface with the biocompatible polymer PEG-DSPE. They found that aggregated SWCNTs induced acute vasorelaxation, while non-aggregated SWCNTs caused acute vasoconstriction, but both types led to long-lasting endothelial dysfunction [15]. These outcomes suggest that the aggregation status of SWCNTs may play a role in determining their effects on vasomotor dysfunction. In addition, *in vitro* evaluation of CNTs on vascular endothelial function in mice by Wang and its co-workers suggested oxidative stress, inflammation, and apoptosis in endothelial cells, leading to endothelial dysfunction [16]. A recent meta-analysis of studies examining the cardiovascular effects of engineered nanomaterials, including carbon nanotubes (CNTs), found consistent evidence linking CNT exposure to impaired endothelial function and elevated markers of vascular dysfunction [17].

9.3.3 Toxicological Effects of CNTs on Atherosclerosis Plaque

Atherosclerosis is a complex disease involving the buildup of plaque within arterial walls. The possible toxicological impact of CNTs on the development of atherosclerotic plaques has been investigated. These studies demonstrate the possible negative consequences of exposure to SWCNTs and MWCNTs on the development of atherosclerotic plaque, including elevated levels of inflammation, oxidative stress, plaque susceptibility, and plaque formation. It is noteworthy that additional investigation is still required to completely comprehend the processes and consequences of CNT-induced impacts on atherosclerosis. SWCNTs influenced the progression of atherosclerotic plaque in ApoE-deficient mice, with repeated intrapharyngeal instillation leading to accelerated plaque development. Interestingly, this effect was not associated with alterations in lipid profiles or overt signs of inflammation, but rather with mitochondrial DNA damage. Furthermore, in normal mice, SWCNT exposure triggered heme oxygenase-1 activation, mitochondrial DNA damage, and oxidative protein damage, indicating a key role for oxidative stress in the observed effects [18]. In a similar vein, Suzuki et al. discovered more plaque development in atherosclerotic ApoE-deficient mice following repeated pharyngeal aspiration of both SWCNTs and double-walled CNTs. The mechanism behind this observation involved the promotion of monocyte adhesion onto endothelial cells by inducing the expression of adhesion molecules and dysfunction of endothelial progenitor cells, impeding their migration from the bone marrow to sites of endothelial damage [19]. Regarding MWCNTs, Xu et al. discovered that IV administration in a rat model accelerated plaque progression and aortic calcification. MWCNTs, both *in vivo* and *in vitro,* caused endothelial cell injury, which may have contributed to the faster growth of the plaque, even while serum lipid levels stayed stable. Catheter sclerotic ApoE-deficient mice were shown by Cao et al. to exhibit enhanced plaque progression with intratracheal instillation of MWCNTs. The lungs and livers exhibited up-regulated inflammatory genes, suggesting both local and systemic inflammatory responses, despite the absence of significant changes in blood cytokine levels or lipid peroxidation. The authors suggested that direct contact with MWCNTs could promote key events associated with the early development of atherosclerosis, such as inducing adhesion molecules, monocyte adhesion onto endothelial cells, and foam cell formation [20]. Han et al. demonstrated that pharyngeal aspiration of MWCNTs to ApoE-deficient mice only caused pulmonary injury and inflammation without affecting the progression of atherosclerotic plaques or total plasma cholesterol concentrations [21]. The effects of MWCNTs on plaque progression are still inconclusive and require further investigation [21]. Indeed, recent studies have indicated a potential association between CNTs exposure and cardiovascular biomarkers in humans, providing further support to the findings from animal studies. It was found that the biomarker of endothelial activation, serum intercellular adhesion molecule-1 (ICAM-1), was positively correlated with workplace exposure to MWCNTs. However, they did not discover any meaningful correlations between MWCNTs and other cardiovascular indicators. Similarly, [22] conducted a study on CNT workers and observed a consistent association between CNT exposure by inhalation and cardiovascular biomarkers in blood samples. Additionally, they discovered relationships in sputum and blood samples between CNT exposure levels and oxidative stress and inflammatory indicators. These epidemiological studies offer proof that alterations in cardiovascular indicators, including endothelial activation, oxidative stress, and inflammation, may be linked to occupational exposure to CNTs. These findings support the notion that pulmonary exposure to CNTs could potentially impair cardiovascular function. However, further research is necessary to better understand the mechanisms and long-term implications of CNT-induced effects on the cardiovascular system in humans.

9.3.4 Immunological Responses to CNTs

Immunological responses to CNTs have been investigated thoroughly due to their potential applications in different areas, including electronics, materials science, and biomedicine. The physical and chemical characteristics of the nanotubes, the route, the dose, and the length of exposure all affect the intricate interactions that CNTs have with the immune system. However, the potential interactions between CNTs and the immune system have raised concerns regarding their safety and biocompatibility. Here is a detailed overview of the immunological response to CNTs:

9.3.4.1 Innate Immune Response

The initial line of protection against foreign objects is the innate immune system. Immune cells like neutrophils and macrophages are stimulated by CNTs, which results in the release of chemokines and pro-inflammatory cytokines like TNF-α and IL-1β. This response is influenced by the physical characteristics (size, length, surface area) and surface functionalization of CNTs [23]. The activation of toll-like receptors (TLRs) plays a significant role in initiating the immune response to CNTs. TLR4 and TLR5 have been implicated in recognizing and responding to CNTs, leading to the activation of the nuclear factor-kappa B (NF-κB) signaling pathway and subsequent cytokine production [24,25]. Studies have shown that the functionalization of the surface of CNTs can influence the magnitude of the innate immune response. Functionalized CNTs with hydrophilic coatings or targeting ligands tend to induce a lower inflammatory response compared to non-functionalized CNTs.

9.3.4.2 Adaptive Immune Response

CNTs can also have the potential to harmonize the adaptive immune responses leading to the activation or suppression of specific immune cells. Some studies have reported that CNT exposure can promote the production of antibodies and be characterized by the activation of T-lymphocytes and differentiation into effector T-cell subsets, such as T-helper 1 (Th1) and T-helper 17 (Th17) cells [26–28]. These cells are involved in pro-inflammatory responses and can contribute to tissue inflammation. However, the immunogenicity of CNTs appears to be influenced by factors such as their surface properties, impurities, and individual variability [29,30]. On the other hand, CNTs have also been shown to induce immunosuppressive effects by inhibiting the activity and function of immune cells, such as T-lymphocytes and natural killer (NK) cells [31,32].

9.4 INFLAMMATION

Prolonged or chronic exposure to CNTs can lead to persistent inflammation characterized by the recruitment of immune cells like neutrophils, macrophages, and dendritic cells at the site of exposure. The capacity of CNTs to cause oxidative stress, generate pro-inflammatory cytokines, trigger inflammasomes, and interfere with intracellular signaling pathways is thought to be responsible for their pro-inflammatory actions. Several studies have reported the release of cytokines such as interleukin-1β (IL-1β), interleukin-6 (IL-6), tumor necrosis factor-alpha (TNF-α), and chemokines in response to CNT exposure [33,34]. The distance and surface properties of CNTs play a crucial role in their inflammatory potential. It has been demonstrated that longer CNTs cause more inflammation than shorter ones. Surface alterations that change the inflammatory response include functionalization and coating. Different surface charges or functional groups can influence the interaction of CNTs with immune cells and modulate the inflammatory outcome [35].

9.5 FIBROSIS

Exposure to CNTs can cause fibrosis, which has been shown to generate fibrous tissues and is distinguished by the deposition of extracellular matrix proteins, especially collagen. Studies have shown that long and rigid CNTs can induce the activation of fibroblasts and myofibroblasts, leading to the secretion of profibrogenic cytokines such as transforming growth factor-beta (TGF-β1) [36]. The fibrotic response to CNTs is also associated with tissue remodeling and impaired lung function, resembling features of pulmonary fibrosis. The physicochemical features of CNTs, such as their length, diameter, surface area, surface charge, and surface functionalization, impact their fibrogenic potential. CNTs with longer lengths and higher aspect ratios have been reported to be more fibrogenic. Surface modifications can also impact the fibrotic response, as functionalization alters the interaction of CNTs with cells and tissues [37].

9.6 OXIDATIVE STRESS

The CNTs can interact with many parts of cells because of their large surface area and produce reactive oxygen species (ROS). Oxidative stress can be increased from excessive ROS production that upsets the equilibrium between oxidants and antioxidants. This oxidative stress can trigger inflammation, DNA damage, and cellular dysfunction. Studies have reported high levels of oxidative stress markers like malondialdehyde (MDA) and reactive nitrogen species (RNS) in response to CNT exposure, which is responsible for causing damage to cellular components and activation of redox-sensitive signaling pathways. This oxidative stress-mediated response contributes to the immunotoxicity of CNTs. Further research is needed to elucidate the underlying mechanisms and develop strategies for the safe use of CNTs [38,39].

9.7 IMMUNOMODULATION

CNTs can regulate the immune response by influencing the function of immune cells. These cylinder-shaped structures made of rolled graphene sheets are highly promising for multiple applications because of their remarkable mechanical strength, electrical conductivity, and thermal stability. A thorough study of the immune response linked to CNT exposure is necessary, nevertheless, as concerns about their possible immunological consequences have been raised. Phagocytosis, cytokine production, and antigen presentation have all been demonstrated to be affected by them. Additionally, CNT exposure can alter the balance between immune cell subsets, such as T-helper cells and regulatory T-cells (Tregs). These immunomodulatory effects of CNTs may have implications for autoimmune diseases and allergic responses [40].

9.8 GENOTOXICITY AND CARCINOGENICITY OF CNTs

CNTs, the novel drug delivery systems, pose risks to human health because safety assessment and evaluation of these systems are lengthy, time-consuming, and costly. The evaluation of CNT toxicity began in the early 2000s. Although CNTs have numerous applications in material science and delivery systems, they have raised alarms due to their distinct shape and durability. The similarity of CNTs with asbestos in shape, size, solubility, and high aspect ratio has caused a major concern. The length, diameter, biopersistence, and toughness of CNTs are considered crucial toxicological features. *In vivo* toxicity studies showed fibrosis, inflammation, malignancies, immunomodulation, and adverse cardiovascular and neurological consequences. Toxicity caused by CNTs can be classified as genetic toxicity and carcinogenicity. Heterogeneity in the structure of various CNTs has resulted in ambiguous toxicity outcomes [41–46].

9.8.1 Genetic Toxicity of CNTs

Genotoxicity means damage to the genetic material inside our DNA or mutations in genomic DNA, which includes breakage of strands, oxidative DNA damage, and the formation of bulky adducts [47,48]. Reactive oxygen species, reactive nitrogen species, aromatic compounds, and some fibers such as quartz, asbestos, and diesel engine exhaust act as genotoxic agents. The structural similarity between asbestos and CNTs has attracted attention to test the genotoxicity and carcinogenic potential of CNTs. Researchers have created a wide range of tests to evaluate the genotoxicity of substances and pharmaceuticals. In genotoxicity assays, different bacterial strains, eukaryotic cells, and whole animals are employed. Ames test (the bacterial mutagenicity test—an *in vitro* test), mutagenicity test for mammalian cells, comet assay test, chromosomal abnormality test, and micronucleus test (an *in vivo* test using cultured mammalian cells on animal models) are a few of the tests that can be used to determine the genotoxicity of CNTs. Various researchers have investigated the genotoxicity effects of CNTs and their possible mechanism for the same as given in Table 9.1.

9.9 CARCINOGENICITY OF CNTs

CNTs have similar size, shape, and biopersistence to asbestos, which leads to raising an alarm about its carcinogenic effects on living beings. Various scientists have studied CNTs for malignant mesothelioma, fibrosis, granuloma, and mesothelial proliferation [66]. A number of animal models were investigated in order to determine whether CNTs were harmful. These findings proved that CNTs cause angiogenesis, granulomas, lesions, fibrosis, oxidative stress, and inflammatory reactions in addition to tumor invasiveness. Furthermore, studies have documented toxicity related to growth, harm to the testicles, and toxicity to reproduction. CNTs have shown attributes similar to characteristic features of carcinogens. CNTs, similar to other nanoparticulate systems, generate reactive oxygen species that initiate a signaling cascade, which results in the conversion of normal cells to cancer cells. Additionally, CNTs also result in the release of pro-inflammatory cytokines from macrophages. However, the most popular method of CNTs toxicity is thought to be the breakdown of the balance between antioxidants and free radicals in healthy cells, which causes oxidative stress and a series of events that damage tissues or organs or cause toxicity in cells. Excessive production of ROS can stimulate pro-tumorigenic signals, augment tumor cell proliferation, and induce DNA damage [67–69]. In addition, CNTs activate an immunosuppressive atmosphere via immunosuppressive monocytic myeloid-derived suppressor cells and macrophages [70]. These macrophages release pro-inflammatory cytokines, i.e., IL-1β and TNFα that result in inflammation, mesothelial cell transformation, and ultimately the development of mesothelioma [71]. Various studies have reported the induction in macrophage number and cytokine capacity of IP-10, RANTES, IL-2, and IL-18 in MWCNT-treated rats [72]. An augmented release of cytokines by adjoining mesothelial cells has been observed in a research study when pro-inflammatory cytokines were exposed to MWCNTs [73]. Activation of NLRP3 inflammasome results in the release of IL-1β [74]. Additionally, long CNTs can also activate the NLRP3 inflammasome cascade via the production of reactive oxygen species (ROS) and activation of cathepsin B in macrophages [75]. Both SWCNTs and MWCNTs were found to be toxic as well as carcinogenic to living organisms (Table 9.2).

TABLE 9.1
Genotoxicity Studies of Single-Walled and Double-Walled CNTs

Type of Assay/ Test	Type of CNTs	Research/Investigation	Inference/Findings	References
Bacterial gene mutation test	SWCNT	Genotoxic and cytotoxic effects of nanoparticles were observed on lung fibroblast cell lines through comet assay and micronucleus test	DNA damage was observed in the comet assay test, however, no significant induction in micronucleus caused by SWCNT was observed. Following SWCNT exposure to two distinct strains of *Salmonella typhimurium*, no notable alterations were found	[49]
	SWCNT	*In vitro* Ames test, chromosomal aberration assay, and *in vivo* bone marrow micronucleus test were performed on *S. typhimurium* or *Escherichia coli* to assess the genotoxicity of SWCNT	SWCNT doesn't pose any genotoxic effect on human health	[50]
	MWCNT	The *Salmonella* microsome (Ames) test (OECD TG 471) was performed in *Salmonella typhimurium* to assess the genotoxic potential of Baytubes®	Under the conditions and concentrations utilized during the test, no mutagenic or clastogenic effects were seen	[51]
	MWCNT	To assess the mutagenicity of multi-walled carbon nanotubes, the Ames test was run on several strains of *Escherichia coli* and *Salmonella typhimurium*	No evidence of mutagenic activity was found	[52]
Comet assay on mammalian cells (*in vitro*)	SWCNT	Comet assay was performed on FE1-Muta™Mouse lung epithelial cells	SWCNT didn't show any significant mutagenicity after 576 hours of exposure to SWCNT	[53]
	SWCNT	Human mesothelial cells, both malignant and normal, were subjected to SWCNT, and the production of free oxygen radicals, cell viability, and notable DNA damage were assessed	It was discovered that SWCNTs altered mesothelial cells' typical biological responses by initiating oxidative stress	[54]

(*Continued*)

TABLE 9.1 (*Continued*)
Genotoxicity Studies of Single-Walled and Double-Walled CNTs

Type of Assay/Test	Type of CNTs	Research/Investigation	Inference/Findings	References
	SWCNT	Human bronchial epithelial cells were used in the comet assay and micronucleus assay in vitro	CNTs were found to be genotoxic to BEAS 2B cells	[55]
	MWCNT	Cellular response caused by MWCNT was observed in normal fibroblast cells found in the human dermis layer of skin	MWCNT-induced reduction in cell viability as a result of DNA damage and apoptosis was seen in a dose-dependent manner	[56]
	MWCNT	Epithelial cells of lung alveoli, i.e., A549, were treated with MWCNT and examined for the formation of 8-nitroguanine, a mutagenic agent	The results showed that inducible nitric oxide synthase (iNOS) was induced and that 8-nitroguanine was increased	[57]
Chromosomal assay (*in vitro*)	SWCNT	Primary and immortalized epithelial cells present in human airways were exposed to SWCNT to assess any mitotic disruption	Considerable harm to the mitotic spindle caused by SWCNT was seen, suggesting a higher likelihood of genetic alterations being passed on to daughter cells	[58]
Micronucleus assay (*in vitro*)	SWCNT and MWCNT	CNTs genotoxic qualities were evaluated using human lymphocytes that had been cultured	A concentration-dependent increase in micronuclei formation was observed	[59]
Phosphorylated H2AX assay and mutation assay (*in vitro*)	MWCNT	DNA damage caused by MWCNT was assessed using mouse embryonic stem cells	In embryonic stem cells, MWNTs have been shown to trigger both the tumor suppressor protein p53 and programmed cell death	[60]
	MWCNT	The genotoxic effect of MWCNTs on endothelial cells of the umbilical vein was assessed	MWCNTs have been shown to cause cytotoxicity as well as genotoxicity in umbilical vein cells via free radical damage	[61]

(*Continued*)

TABLE 9.1 (*Continued*)
Genotoxicity Studies of Single-Walled and Double-Walled CNTs

Type of Assay/Test	Type of CNTs	Research/Investigation	Inference/Findings	References
Genotoxicity studies (*in vivo*)	SWCNT	Changes in oxidative stress, mutagenesis, and inflammation in C57BL/6 mice after they inhaled as opposed to aspirated SWCNT	In the lungs of C57BL/6 mice, SWCNT inhalation was more harmful than aspiration in terms of inducing inflammation, oxidative stress, and alterations of the K-*ras* gene locus	[62]
	SWCNT	The genotoxicity following nanotube exposure in the lungs was assessed in mice lacking the apolipoprotein E gene ($ApoE^{-/-}$)	There was a significant increase in protein in BAL fluid, neutrophil proportion in BAL cells, and DNA damage in BAL cells	[63]
	SWCNT	Oxidative DNA damage was assessed post oral dose of SWCNT	It was discovered that SWCNT caused genotoxicity, as evidenced by elevated liver and lung levels of 8-oxodG	[64]
	MWCNT	The results showed that SWCNT was genotoxic, since 8-oxodG levels in the liver and lungs were found to be higher	It has been demonstrated that exposure to MWCNTs increases ($P<0.05$) the frequency of micronucleated cells, the number of chromosomal abnormalities, and the amount of DNA damage. Furthermore, when compared to non-functionalized MWCNTs, functionalized MWCNTs demonstrated a significantly higher potential for mutagenicity and genotoxicity	[65]

TABLE 9.2
Research Studies to Test Carcinogenicity of CNTs

Type of CNTs	Research/Investigation	Inference	References
MWCNT	Rod-shaped or fibrous CNTs were tested in p53(+/−) knockout mice to explore the carcinogenicity (for causing mesothelioma)	Intraperitoneal administration of MWCNT was found to be carcinogenic depending on its aspect ratio and persistence. Mechanisms were found to be similar as those of asbestos in causing mesothelioma	[76]
MWCNT	Carcinogenic potential of MWCNT in intact Fischer 344 DuCrlCrlj rats was assessed	MWCNT caused mesothelioma at a high frequency in intact rodents after a single intrascrotal injection	[77]
MWCNT	Carcinogenic potential of MWCNT was explored in rats based on previous research findings of lung inflammation, granulomas, fibrotic reactions, and mutations	MWCNT injection intraperitoneally did not result in any cancerous reaction. This could be because of how long the MWCNT was tested for and that the inflammation it generated wasn't constant	[78]
MWCNT	To assess the effects of diameter and rigidity of CNTs for their carcinogenic potential	While thick (diameter ~ 150 nm) or tangled (diameter ~ 2–20 nm) CNTs were shown to be less toxic, inflammogenic, and carcinogenic, highly crystalline thin MWCNTs with a diameter ~ 50 nm demonstrated cytotoxicity, inflammogenicity, and mesotheliomagenicity	[79]
MWCNT	Carcinogenic potential of CNTs in transgenic rasH2 mice	MWCNTs didn't result in the development of neoplasms in rasH2 mice	[80]
MWCNT	MWCNT's potential for carcinogenesis was examined after it was injected into the rat lung	When MWCNT was injected intratracheally into the lungs of rats, lung tumors and pleural malignant mesothelioma developed	[81]
MWCNT	CNTs' ability to cause cancer was evaluated in comparison to asbestos	Tumor suppressor gene disruption associated with mesothelioma was identified in Cdkn2a (Ink4a/Arf)	[82]
SWCNT	Carcinogenicity of CNTs was investigated after intratracheal instillation in F344 rats	There were cases of malignancy and bronchio-alveolar adenoma	[83]

9.10 BIODISTRIBUTION AND ACCUMULATION OF CNTS IN VARIOUS TISSUES

To evaluate the toxicity of CNTs, it is necessary to know about the biodistribution and clearance of CNTs throughout the body. This section will cover the biodistribution and biopersistency of CNTs by many modes of administration, including oral, GI, blood, peritoneum, and inhalation/instillation. Unintentional exposure to CNTs mainly occurs in industry through inhalational or dermal contact where manufacturing, harvesting, production, purification, and functionalization of CNTs take place [84].

9.10.1 Biodistribution after Inhalation

The pulmonary half-life of inhaled CNTs ranges from months to years and, therefore, remains in the alveoli and lung tissues for a longer duration after inhalation [85–87]. For this reason, the hazardous or toxic extrapulmonary effects of CNTs may sometimes emerge long after exposure. Studies have shown that within two weeks of inhalation, MWCNTs can reach the subpleural region of the lungs, where they have been detected in macrophages, mesenchymal cells, and within the collagen matrix [88]. Cortesi et al. found that 0.6% of CNTs have reached the subpleural region within a day post pharyngeal aspiration. Researchers observed frequent and sustained penetration of CNTs to the alveolar wall, alveolar macrophages, and visceral pleura. In addition, MWCNTs were found to be engulfed by macrophages as well as type 1 alveolar epithelial cells after one hour of intake. At the later stages of study, these nanotubes were observed within the alveolar interstitium, interstitial cells, or alveolar macrophages instead of the epithelial surface [89]. Mechanism of clearance of nanoparticles was suggested to be through lymph fluid, i.e., through narrow holes in the parietal pleura; the long fibers cannot migrate through the small stomata, leading to retention, local inflammation, and pleural pathology, for example, mesothelioma but longer fibrous CNTs are unable to pass through narrow stomatal holes resulting in retention, inflammation, and mesothelioma [90]. Pope et al. observed the presence of SWCNT in the blood of rodents after inhalation through Raman signal, presenting the possibility of secondary organ exposure. However, a study related to CNT excretion or accumulation was not performed [91]. The translocation of 14C-labeled taurine-functionalized MWCNT following rodent instillation was investigated by Deng et al. There was no 14C activity in the blood, liver, spleen, or kidneys, but the lungs had demonstrated the presence of the intravenous 14C-tau-MWCNT [90]. Around 30% of MWCNT was found to be cleared within one day of instillation; however, the remaining 70% was still found in the lungs after one year without any sign of reduction. The liver or brain showed an absence of nanotubes, whereas peri bronchial lymph nodes [92]. Williams et al. found that MWCNT was translocated to the lymph nodes of animals, observing similar outcomes. Nonetheless, they discovered a persistent decline in lung CNTs [93]. Mercer et al. experimented to analyze the presence of aerosolized nanotubes in various organs such as tissues, kidney, liver, heart, brain, chest wall, and diaphragm and found their presence in every analyzed organ in small quantities. Maximum burden was observed in extrapulmonary sites such as tracheobronchial lymph nodes [94]. A radio-imaging study performed by Czarny et al. showed a 90% reduction in pulmonary radioactivity within three months of the instillation of ^{14}C skeleton-labeled MWCNT. A small quantity of MWCNTs showed an increased amount in the liver, kidney, spleen, and bone marrow. The authors proposed that the mechanism of translocation post lung exposure occurred via pulmonary epithelium [95]. The bulk of deposited CNTs either stays in the lungs or is removed by the gastrointestinal system, according to a summary of current studies covered in the preceding paragraphs. Furthermore, a considerable fraction is noted to translocate to the neighboring pulmonary area, including the spleen and liver (less than 1%), lymph nodes (up to 7%), and subpleura/pleura (more than 1%).

9.10.2 Biodistribution after Oral Administration

One fundamental method of removing nanoparticles from the pulmonary region of the lungs is by upward mucociliary transit into the gastrointestinal tract for excretion. Such upward movement to pharynx results in either swallowing of nanoparticles into gastrointestinal tract or translocation to intestine. Heifang Wang et al. observed that low molecular weight (~600 kDa) hydroxylated ^{125}I-labeled SWCNTs were able to distribute/move freely to organs and compartments. SWCNTs were analyzed in 12 organs with maximum concentration observed in the stomach, kidney, lungs, and bone, whereas minimum concentration was observed in the brain, heart, and muscle after oral administration [96]. Gadolinium was observed throughout the gastrointestinal tract after oral administration of gadolinium embedded in SWCNT; however, it was not detected in blood, spleen, or liver. It was observed that around 80% of gadolinium was excreted within eight hours and nearly 100% within a day [90]. Deng et al. detected ^{14}C-tau-labeled small MWCNT in the gastrointestinal lumen (stomach, small intestine, and large intestine) ending up in feces after oral ingestion. They didn't observe any labeling in blood, indicating that ^{14}C-tau-labeled MWCNT was neither absorbed from the gastrointestinal tract nor entered the blood stream in measurable concentration range [90]. Czarny et al. determined the pathway of translocation of ^{14}C-enriched MWCNT and observed around 95% of the administered dose in the gastrointestinal tract and feces, however, no radioactivity was observed in the gastrointestinal tract after the fourth day from the day of exposure. In addition, they didn't detect any radioactivity in spleen and liver [95]. Contradictory results were observed in biodistribution of CNTs after oral administration in these research works, from zero biodistribution to complete detection in all analyzed organs. This could be because the nanotubes that were introduced to the GI tract had different sizes and shapes, or it could be because different detection or analysis techniques were employed. While the length of the nanotubes was roughly similar, the CNTs in various studies differed in thickness, shape, and flexibility. To summarize, it was observed that there was no tissue uptake of CNTs from the gastrointestinal tract except in one case. Similar observations have been found in previously studied nanomaterials such as ^{125}I–graphene and TiO_2 materials [90,97].

9.10.3 Biodistribution of CNTs after IV Exposure

The pharmacokinetics and biological effects of promising drug carriers and molecules should be well understood and explored before their use in the medical field for treatment or diagnostic purposes. After IV exposure or administration, the capillary network of the pulmonary system is the first one to get exposed. In this section, the biological distribution of unmodified, modified, single, and MWCNTs after IV injection will be discussed. Yang et al. observed the biodistribution of ^{13}C-enriched SWCNT for 28 days after injecting them into the tail vein of rodents. Excretion via urine or feces was insignificant, and the highest concentration was observed in the liver, followed by the spleen and lungs. Kupffer cells were the main deposition site of CNTs in the liver as observed by electron microscopy [98]. Another research from the same group has detected SWCNT in lung, liver, and spleen three months after IV administration. Maximum concentration was detected in the lungs even after 90 days of administration [90]. SWCNTs were detected in the liver of rabbits after IV injection. Injected SWCNT was found to be rapidly covered with blood proteins, and their blood concentration had a half-life of one hour [99]. After IV CNTs delivery, there are two excretory pathways: renal and hepatobiliary clearance. Small-sized CNTs get excreted through urine via perforations present in renal corpuscles [100]. Size plays an important role in whether these will get excreted through urine or get accumulated in body organs such as Kupffer cells of the liver. Chemically modified single-walled CNTs such as ^{64}Cu-radiolabeled-DOTA-PEG-SWCNT have been shown a half-life of 0.5–2 hours depending upon the PEG chain length. Maximum concentration was observed in liver followed by spleen and kidney, however, SWCNT with a larger PEG chain has shown comparatively lower accumulation in the liver as well as slower blood clearance [101]. The biodistribution of water-soluble 111In-labeled SWCNT was investigated by McDevitt et al. following intravenous delivery. They observed a rapid blood clearance after 1 and 20 hours, with the dose/gram blood of 3% and 0.4%, respectively. The kidney showed the greatest concentration of 111In, followed by the liver and then the spleen, and to a smaller level, the bone. An insignificant amount of radioactivity was observed in urine [102]. SWCNTs were mainly detected in the liver, kidney, and spleen, however, insignificant concentration was also detected in seven other organs [103]. Short-termed biodistribution study of hydroxylated ^{131}I-labeled SWCNT was performed in which radioactivity was detected in every organ except the brain within two minutes of IV administration. Major sites of accumulation of CNTs were kidneys followed by liver, bone, spleen, stomach, and lungs. Following intravenous administration of 488F-chitosan-functionalized SWCNTs to mice, rapid liver uptake and blood clearance were observed, with a half-life of 3–4 hours. Accumulated SWCNTs in the liver led to pathological changes, including blood coagulation, cell swelling, inflammation, and macrophage damage, with no significant reduction in SWCNT levels within 24 hours. In contrast, uptake in the kidney and spleen was minimal and did not result in any notable pathological changes [90]. Non-covalent PEG-functionalized SWCNT showed major accumulation took place in liver and spleen followed by bone, kidney, intestine, stomach, and lungs. In addition, the research group reported that SWCNT with larger PEG chains have shown reduced uptake by the liver and spleen as compared to SWCNT with the smaller PEG chains [104]. According to Yang et al., at one, four, and eight weeks after exposure, SWCNT functionalized with PEG1500N, acid-treated, and thermally annealed were found in the liver and spleen. PEG de-functionalization only happened in the liver between four and eight weeks [105]. From the above-mentioned research findings, it can be concluded that in case of SWCNT, blood clearance was determined comparatively fast except one case, however, half-life ranged from minutes to hours depending upon the type of material used and functionalization. Liver was observed to be major site of accumulation followed by kidneys and spleen based on the research findings in case of functionalized SWCNT whereas in unmodified SWCNT, lungs are additional site of accumulation. Related research has been given in Table 9.3.

TABLE 9.3
Biodistribution of CNTs through IV Administration

Type of CNTs	Functionalization and Labeling	Research Findings	References
SWCNT	Hydroxylated SWCNT labeled with ^{125}I	It was discovered that the biodistribution pattern following IV injection resembled that following oral administration. The concentration was highest in the stomach, kidneys, lungs, and bone, and it was incredibly negligible in the brain. Urine and feces contained about 80% of the dosage that was delivered	[106]
Water soluble SWCNT	DTPA SWCNT and labeled with indium (^{111}In)	$t_{1/2}$ of SWCNTs was found to be around 3.5 hours through kidneys. Liver or spleen accretion was not observed. Highest concentration of ^{111}In was observed in muscle tissue, followed by skin after half an hour of administration	[107]
SWCNT	Functionalized with lysine and DOTA labeled with [^{111}In]	Around 34% was observed in urine excretion. A very small amount got accumulated in kidneys and liver	[108]
MWCNT	^{14}C-tau-MWCNT	Approximately 80% of the dosage that was given was stored in the liver. TEM scans showed that MWCNT were localized in the Kupffer cells. The spleen and lungs were found to have concentrations between 1% and 5%	[109]
MWCNT	Nanotubes encased in Tween-80 and functionalized with taurine covalently, labeled with 125I (tau-MWCNT and tween-MWCNT)	The liver contained about 80% of the dosage that was given, with smaller quantities found in the spleen and lung. Additionally, some concentration was found in the kidney, big and small intestines, and stomach	[110]

(*Continued*)

TABLE 9.3 (*Continued*)
Biodistribution of CNTs through IV Administration

Type of CNTs	Functionalization and Labeling	Research Findings	References
MWCNT	^{14}C-MWCNT	Following intravenous injection, MWCNTs were dispersed throughout the liver, lungs, spleen, and kidneys. Over the course of the experiment, a drop in radioactivity was detected in all organs	[111]
MWCNT	Surface-modified with amines and labeled with radiotracer [^{111}In] DTPA	Rather than modifying, the diameter (thickness or thinness) of MWCNTs affects their biodistribution. The spleen and liver were the main organs where the thick MWCNT accumulated, with the lungs having the lowest concentration and the kidneys having the smallest amount of the other eight examined organs. Thin MWCNT collected in the lungs, with very little in the kidney and heart and less in the spleen and liver	[112]
MWT	Aminated and labeled with DTPA [^{111}In]	Five minutes after treatment, MWCNT accumulated in the liver and lungs. In skin, spleen, kidneys, muscle, and bone, there was a detection of around 1%–7% given dosage. Below 1% concentration was found in the stomach, heart, and brain	[113]
MWCNT	MWCNT ^{111}In-DTPA joined by amide bonds of pyrrolidine	Around 58% to 68% of the administered dose accumulated in the liver, whereas 2.6%–4% in spleen. Very low concentration was detected in lungs and kidneys	[114]
MWCNT	Technetium-labeled MWCNT (^{99m}Tc-oMWNT)	Within the first two hours following intravenous injection, a dosage of about 7% was found in the urine. Urine samples taken between 2 and 25 hours later showed 2.6%	[115]

Overall, various modifications in CNTs have shown the effect on biodistribution after IV injection. Large side chains sometimes cause urine excretion to be delayed or impossible, but changes have been demonstrated to increase the dispersibility of CNT in physiological conditions, leading to an increase in excretion of urine [116]. Quantitative assessment of biodistribution of CNTs still needs an accurate detection technique despite a decade of research. CNTs have been shown to get accumulated in the lung ranging from months to years. The majority of CNTs remain in the lung or are cleared through upward mucociliary clearance towards gastrointestinal tract. Additionally, CNTs that have collected in the lung not only go through circulation to local lymph nodes and pleura, but also, most likely, to distal organs like the liver and spleen.

9.10.4 Biodistribution of CNTs in Tumor Cells

Various medication delivery techniques have been developed recently to target malignant cells. Owing to the shorter residence time of some anticancer drugs, these don't reach the target organs. The distinctive physical attributes of CNTs can easily penetrate inside cancerous cells without being able to come out due to the ineffective lymphatic system of cancer cells [117]. Due to poor solubility and non-specific cell toxicity, paclitaxel (PXT), the most commonly used anticancer agent, doesn't reach the target organ. Liu et al. employed CNTs branched with PEG chains to target PTX to breast cancer cells. PTX uptake to tumor cells was enhanced through PEG conjugated SWCNTs with six times higher blood circulation time than taxol [118]. Wu et al. loaded new MWCNTs with hydroxycamptothecin (HCPT). It was discovered that the antitumor efficaciousness of currently available hydroxycamptothecin was inferior to that of hydroxycamptothecin-loaded CNTs. MWCNTs with HCPT showed high cancer specificity accumulation in tumor cells as demonstrated by *in vivo* single photon emission computed tomography imaging [119]. Due to the targeted delivery of siRNA via single-walled CNTs, Zhang et al. created siRNA connected to single-walled CNTs using CONH-(CH2)6-NH3+Cl-, which suppressed the expression of telomerase reverse transcriptase. Furthermore, *in vitro* and mouse models have demonstrated that these nanotubes suppress the growth and proliferation of cancer cells [120]. The targeted distribution of siRNA using immaculate SWCNTs has been shown in a recent study. When hypoxia-inducible factor 1 a (HIF-1a) was present, there was a significant suppression of the cellular HIF-1 activity of siRNA complexes. Furthermore, several cancer cell types have demonstrated a biological reaction to siRNA SWCNTs complexes. SWCNTs complexes injected intraperitoneally have been found to suppress tumor HIF-1a activity in mice with MiaPaCa-2/HRE tumors [121]. Various research findings have demonstrated the bright future of CNTs for cancer treatment owing to their targeted drug delivery to tumor cells, resulting in decreased side effects and toxicity to adjacent healthy cells [122]

9.11 EXCRETION AND ELIMINATION OF CNTs

CNTs have emerged as promising novel nano carrier in pharmaceutical and biomedical applications. However, besides the potential application, the pharmacokinetic behavior is an important issue that needs serious attention. The surface of CNTs has tremendous potential for functionalization and allows the impartation of extrinsic properties. Therefore, the surface chemistry of CNTs has an important role in determining *in vivo* behavior as functionalizing is the determining factor for pharmacokinetic behavior and biocompatibility of CNTs. The lipophilic nature of CNTs makes them more susceptible to RES uptake; therefore, these agents prefer hepatobiliary route for clearance unless the surface of CNTs is not modified. For instance, pegylation has been reported to reduce the RES uptake, resulting in reduced accumulation in the liver and spleen. On the other hand, pegylation leads to increased blood circulation of covalently attached pegylated single-wall CNTs. Furthermore, branching in pegylation leads to better coverage and increased hydrophilicity density around the SWCNTs causing increasing blood circulation time. Metabolism of CNTs *in vivo* is organ-dependent which means it requires individual evaluation for every accumulation organ in the body. The most possible metabolic organ for CNTs in the body is liver, which is indicated by hepatic de-functionalization of CNTs. The metabolism and biotransformation of CNTs mainly occur in the liver due to the presence of plenty of metabolism enzymes. Apart from the liver, the spleen is the major immunological organ where CNTs get biotransformed, but due to different enzymes in the liver and spleen, the CNTs get biotransformed differently. The metabolism of CNTs totally depends on whether it is covalently or non-covalently functionalized. The covalently functionalized CNTs are more stable than non-covalently functionalized CNTs [123–126]. As a consequence, CNTs show different stabilities in both organs. In a study, it has been observed evidently through Raman signals that fluorescent-pegylated SWCNTs showed fluorescence for more than eight weeks in the spleen; on the contrary, in the liver, the average time was observed to be four weeks after IV administration. This clearly indicates the de-functionalization of CNTs is more dominating in liver than spleen due to dominancy in metabolic enzymes. The major mechanisms involved for prominent de-functionalization in liver is the radical attack and acid hydrolysis. For the metabolism of CNTs, the major attention is given to de-functionalization of CNTs because the carbon skeleton of CNTs is very stable and difficult to digest (10.1021/bm900263z). In context to excretion of CNTs, again functionalization plays a key role to determine the clearance of CNTs. For instance, the chemical functionalization with antibody conjugation, glucosamine, hydroxylation, etc., affects the biodistribution in the blood compartment, liver, spleen, kidney, bones, etc., and consequently becomes the key determinant of rate of clearance. This can be further explained as the higher the degree of functionalization on carbon skeleton, the lesser will be the reticuloendothelial system retention and, as a result, the higher will be the rate of urinary excretion. Moreover, another key factor that influences the rate of CNTs excretion is the length of CNTs. It is suggested in various studies that the CNTs with short lengths (approximately <500 nm), along with appropriate chemical functionalization, have better urinary excretion as compared to less functionalized and large length CNTs [127–129].

REFERENCES

1. A. Aqel, K. M. M. Abou El-Nour, R. A. A. Ammar, and A. Al-Warthan, "Carbon nanotubes, science and technology part (I) structure, synthesis and characterisation," *Arab J Chem*, vol. 5, no. 1, pp. 1–23, 2012.
2. T. W. Odom, J.-L. Huang, P. Kim, and C. M. Lieber, "*Structure and Electronic Properties of Carbon Nanotubes*," 2000, ACS Publications, Washington, DC.
3. A. Eatemadi et al., "Carbon nanotubes: properties, synthesis, purification, and medical applications," *Nanoscale Res Lett*, vol. 9, pp. 1–13, 2014.
4. S. Mondal, "Carbon nanotube-reinforced polymer nanocomposite for biomedical applications," *Green Biocomposites for Biomedical Engineering*, vol. 2021, pp. 265–83, 2021.
5. R. Chen et al., "Subchronic toxicity and cardiovascular responses in spontaneously hypertensive rats after exposure to multiwalled carbon nanotubes by intratracheal instillation," *Chem Res Toxicol*, vol. 28, no. 3, pp. 440–50, Mar. 2015, doi: 10.1021/tx5004003.
6. X. Ma et al., "Multiwalled carbon nanotubes induced hypotension by regulating the central nervous system," *Adv Funct Mater*, vol. 28, no. 11, p. 1705479, Mar. 2018, doi: 10.1002/adfm.201705479.

7. W. Zheng, W. McKinney, M. Kashon, R. Salmen, V. Castranova, and H. Kan, "The influence of inhaled multi-walled carbon nanotubes on the autonomic nervous system," *Part Fibre Toxicol*, vol. 13, no. 1, p. 8, Dec. 2015, doi: 10.1186/s12989-016-0119-7.
8. P. D. Blanc et al., "The occupational burden of nonmalignant respiratory diseases. an official American thoracic society and european respiratory society statement," *Am J Respir Crit Care Med*, vol. 199, no. 11, pp. 1312–34, Jun. 2019, doi: 10.1164/rccm.201904-0717ST.
9. R. Chen et al., "Subchronic toxicity and cardiovascular responses in spontaneously hypertensive rats after exposure to multiwalled carbon nanotubes by intratracheal instillation," *Chem Res Toxicol*, vol. 28, no. 3, pp. 440–50, Mar. 2015, doi: 10.1021/tx5004003.
10. X. Ma et al., "Multiwalled carbon nanotubes induced hypotension by regulating the central nervous system," *Adv Funct Mater*, vol. 28, no. 11, p. 1705479, Mar. 2018, doi: 10.1002/adfm.201705479.
11. P. Møller et al., "Atherosclerosis and vasomotor dysfunction in arteries of animals after exposure to combustion-derived particulate matter or nanomaterials," *Crit Rev Toxicol*, vol. 46, no. 5, pp. 437–76, May 2016, doi: 10.3109/10408444.2016.1149451.
12. H. Y. Cheah et al., "Preclinical safety assessments of nano-sized constructs on cardiovascular system toxicity: a case for telemetry," *J Appl Toxicol*, vol. 37, no. 11, pp. 1268–85, Nov. 2017, doi: 10.1002/jat.3437.
13. J. M. Gutiérrez-Hernández et al., "Single-walled carbon nanotubes (SWCNTs) induce vasodilation in isolated rat aortic rings," *Toxicol in Vitro*, vol. 29, no. 4, pp. 657–62, Jun. 2015, doi: 10.1016/j.tiv.2015.02.001.
14. L. K. Vesterdal et al., "Pulmonary exposure to particles from diesel exhaust, urban dust or single-walled carbon nanotubes and oxidatively damaged DNA and vascular function in *apoE*$^{-/-}$ mice," *Nanotoxicology*, vol. 8, no. 1, pp. 61–71, Feb. 2014, doi: 10.3109/17435390.2012.750385.
15. M. D. Frame, A. M. Dewar, S. M. Chowdhury, and B. Sitharaman, "Vasoactive effects of stable aqueous suspensions of single walled carbon nanotubes in hamsters and mice," *Nanotoxicology*, vol. 8, no. 8, pp. 867–75, Dec. 2014, doi: 10.3109/17435390.2013.837209.
16. F. Dolati, Y. Yu, Y. Zhang, A. M. De Jesus, E. A. Sander, and I. T. Ozbolat, "*In vitro* evaluation of carbon-nanotube-reinforced bioprintable vascular conduits," *Nanotechnology*, vol. 25, no. 14, p. 145101, Apr. 2014, doi: 10.1088/0957–4484/25/14/145101.
17. N. B. Alsaleh, "Adverse cardiovascular responses of engineered nanomaterials: Current understanding of molecular mechanisms and future challenges," *Nanomedicine*, vol. 37, p. 102421, Oct. 2021, doi: 10.1016/j.nano.2021.102421.
18. Z. Li et al., "Cardiovascular effects of pulmonary exposure to single-wall carbon nanotubes," *Environ Health Perspect*, vol. 115, no. 3, pp. 377–82, Mar. 2007, doi: 10.1289/ehp.9688.
19. Y. Suzuki et al., "Single-and double-walled carbon nanotubes enhance atherosclerogenesis by promoting monocyte adhesion to endothelial cells and endothelial progenitor cell dysfunction," *Part Fibre Toxicol*, vol. 13, pp. 1–11, 2015.
20. Y. Cao et al., "Vascular effects of multiwalled carbon nanotubes in dyslipidemic ApoE$^{-/-}$ mice and cultured endothelial cells," *Toxicol Sci*, vol. 138, no. 1, pp. 104–16, Mar. 2014, doi: 10.1093/toxsci/kft328.
21. S. G. Han, D. Howatt, A. Daugherty, and G. Gairola, "Pulmonary and atherogenic effects of multi-walled carbon nanotubes (MWCNT) in apolipoprotein-E-deficient mice," *J Toxicol Environ Health A*, vol. 78, no. 4, pp. 244–53, 2015.
22. J. D. Beard et al., "Carbon nanotube and nanofiber exposure and sputum and blood biomarkers of early effect among U.S. workers," *Environ Int*, vol. 116, pp. 214–28, Jul. 2018, doi: 10.1016/j.envint.2018.04.004.
23. C. A. Poland et al., "Carbon nanotubes introduced into the abdominal cavity of mice show asbestos-like pathogenicity in a pilot study," *Nat Nanotechnol*, vol. 3, no. 7, pp. 423–8, Jul. 2008, doi: 10.1038/nnano.2008.111.

24. J. P. Ryman-Rasmussen et al., "Inhaled carbon nanotubes reach the subpleural tissue in mice," *Nat Nanotechnol*, vol. 4, no. 11, pp. 747–51, 2009.
25. M. Allegri et al., "Toxicity determinants of multi-walled carbon nanotubes: The relationship between functionalization and agglomeration," *Toxicol Rep*, vol. 3, pp. 230–43, 2016.
26. V. E. Kagan et al., "Carbon nanotubes degraded by neutrophil myeloperoxidase induce less pulmonary inflammation," *Nat Nanotechnol*, vol. 5, no. 5, pp. 354–9, 2010.
27. S. P. Mukherjee, M. Bottini, and B. Fadeel, "Graphene and the immune system: a romance of many dimensions," *Front Immunol*, vol. 8, p. 673, 2017.
28. H. Dai, "Carbon nanotubes: opportunities and challenges," *Surf Sci*, vol. 500, no. 1–3, pp. 218–41, 2002.
29. F. Zhou, S. Wu, S. Song, W. R. Chen, D. E. Resasco, and D. Xing, "Antitumor immunologically modified carbon nanotubes for photothermal therapy," *Biomaterials*, vol. 33, no. 11, pp. 3235–42, 2012.
30. B. Han et al., "Pulmonary inflammatory and fibrogenic response induced by graphitized multi-walled carbon nanotube involved in cGAS-STING signaling pathway," *J Hazard Mater*, vol. 417, p. 125984, 2021.
31. E.-J. Park, H. Kim, Y. Kim, J. Yi, K. Choi, and K. Park, "Inflammatory responses may be induced by a single intratracheal instillation of iron nanoparticles in mice," *Toxicology*, vol. 275, no. 1–3, pp. 65–71, Sep. 2010, doi: 10.1016/j.tox.2010.06.002.
32. T. Yamashita et al., "Carbon nanomaterials: efficacy and safety for nanomedicine," *Materials*, vol. 5, no. 2, pp. 350–63, Feb. 2012, doi: 10.3390/ma5020350.
33. J. Dong, "Signaling pathways implicated in carbon nanotube-induced lung inflammation," *Front Immunol*, vol. 11, p. 552613, 2020.
34. M. S. P. Boyles, L. C. Stoehr, P. Schlinkert, M. Himly, and A. Duschl, "The significance and insignificance of carbon nanotube-induced inflammation," *Fibers*, vol. 2, no. 1, pp. 45–74, 2014.
35. J. Zhang et al., "Surface chemistry induces mitochondria-mediated apoptosis of breast cancer cells via PTEN/PI3K/AKT signaling pathway," *Biochimica et Biophysica Acta (BBA)-Molecular Cell Res*, vol. 1865, no. 1, pp. 172–85, 2018.
36. A. A. Shvedova et al., "Unusual inflammatory and fibrogenic pulmonary responses to single-walled carbon nanotubes in mice," *Am J Physiol Lung Cell Mol Physiol*, vol. 289, no. 5, pp. L698–L708, 2005.
37. J. N. Mwangi et al., "Toxicity of carbon nanotubes to freshwater aquatic invertebrates," *Environ Toxicol Chem*, vol. 31, no. 8, pp. 1823–30, Aug. 2012, doi: 10.1002/etc.1888.
38. P. Møller et al., "Role of oxidative stress in carbon nanotube-generated health effects," *Arch Toxicol*, vol. 88, no. 11, pp. 1939–64, Nov. 2014, doi: 10.1007/s00204-014-1356-x.
39. E. Florek et al., "Oxidative stress in long-term exposure to multi-walled carbon nanotubes in male rats," *Antioxidants*, vol. 12, no. 2, p. 464, Feb. 2023, doi: 10.3390/antiox12020464.
40. W.-W. Cheng et al., "Single-wall carbon nanotubes induce oxidative stress in rat aortic endothelial cells," *Toxicol Mech Methods*, vol. 22, no. 4, pp. 268–76, May 2012, doi: 10.3109/15376516.2011.647112.
41. S. Toyokuni, "Genotoxicity and carcinogenicity risk of carbon nanotubes," *Adv Drug Deliv Rev*, vol. 65, no. 15, pp. 2098–110, Dec. 2013, doi: 10.1016/j.addr.2013.05.011.
42. K. Fraser et al., "Physicochemical characterization and genotoxicity of the broad class of carbon nanotubes and nanofibers used or produced in U.S. facilities," *Part Fibre Toxicol*, vol. 17, no. 1, p. 62, Dec. 2020, doi: 10.1186/s12989-020-00392-w.
43. Y. J. Wang, J. W. Gurd, and H. R. Mahler, "Topography of synaptosomal high affinity uptake systems.," *Life Sci*, vol. 17, no. 5, pp. 725–33, Sep. 1975, doi: 10.1016/0024-3205(75)90527-5.
44. A. B. Kane, R. H. Hurt, and H. Gao, "The asbestos-carbon nanotube analogy: an update," *Toxicol Appl Pharmacol*, vol. 361, pp. 68–80, Dec. 2018, doi: 10.1016/j.taap.2018.06.027.

45. K. Donaldson, F. Murphy, A. Schinwald, R. Duffin, and C. A. Poland, "Identifying the pulmonary hazard of high aspect ratio nanoparticles to enable their safety-by-design," *Nanomedicine*, vol. 6, no. 1, pp. 143–56, Jan. 2011, doi: 10.2217/nnm.10.139.
46. K. Donaldson, F. A. Murphy, R. Duffin, and C. A. Poland, "Asbestos, carbon nanotubes and the pleural mesothelium: a review and the hypothesis regarding the role of long fibre retention in the parietal pleura, inflammation and mesothelioma," *Part Fibre Toxicol*, vol. 7, no. 1, p. 5, 2010, doi: 10.1186/1743–8977-7-5.
47. B. van, M. Clift, C. Albrecht, and R. Schins, "Carbon nanotubes: an insight into the mechanisms of their potential genotoxicity," *Swiss Med Wkly*, vol. 142, p. 4546, Nov. 2012, doi: 10.4414/smw.2012.13698.
48. S. Toyokuni, "Genotoxicity and carcinogenicity risk of carbon nanotubes," *Adv Drug Deliv Rev*, vol. 65, no. 15, pp. 2098–110, Dec. 2013, doi: 10.1016/j.addr.2013.05.011.
49. E. R. Kisin et al., "Single-walled carbon nanotubes: geno- and cytotoxic effects in lung fibroblast V79 cells," *J Toxicol Environ Health* A, vol. 70, no. 24, pp. 2071–9, Nov. 2007, doi: 10.1080/15287390701601251.
50. M. Naya, N. Kobayashi, K. Mizuno, K. Matsumoto, M. Ema, and J. Nakanishi, "Evaluation of the genotoxic potential of single-wall carbon nanotubes by using a battery of in vitro and in vivo genotoxicity assays," *Regul Toxicol Pharmacol*, vol. 61, no. 2, pp. 192–8, Nov. 2011, doi: 10.1016/j.yrtph.2011.07.008.
51. U. Wirnitzer, B. Herbold, M. Voetz, and J. Ragot, "Studies on the in vitro genotoxicity of baytubes®, agglomerates of engineered multi-walled carbon-nanotubes (MWCNT)," *Toxicol Lett*, vol. 186, no. 3, pp. 160–5, May 2009, doi: 10.1016/j.toxlet.2008.11.024.
52. A. Di Sotto, M. Chiaretti, G. A. Carru, S. Bellucci, and G. Mazzanti, "Multi-walled carbon nanotubes: Lack of mutagenic activity in the bacterial reverse mutation assay," *Toxicol Lett*, vol. 184, no. 3, pp. 192–7, Feb. 2009, doi: 10.1016/j.toxlet.2008.11.007.
53. E. R. Kisin et al., "Single-walled carbon nanotubes: geno- and cytotoxic effects in lung fibroblast V79 cells," *J Toxicol Environ Health A*, vol. 70, no. 24, pp. 2071–9, Nov. 2007, doi: 10.1080/15287390701601251.
54. M. Pacurari et al., "Raw single-wall carbon nanotubes induce oxidative stress and activate MAPKs, AP-1, NF-κB, and Akt in normal and malignant human mesothelial cells," *Environ Health Perspect*, vol. 116, no. 9, pp. 1211–7, Sep. 2008, doi: 10.1289/ehp.10924.
55. H. K. Lindberg et al., "Genotoxicity of nanomaterials: DNA damage and micronuclei induced by carbon nanotubes and graphite nanofibres in human bronchial epithelial cells in vitro," *Toxicol Lett*, vol. 186, no. 3, pp. 166–73, May 2009, doi: 10.1016/j.toxlet.2008.11.019.
56. A. Patlolla, B. Knighten, and P. Tchounwou, "Multi-walled carbon nanotubes induce cytotoxicity, genotoxicity and apoptosis in normal human dermal fibroblast cells" *Ethn Dis*, vol. 20, no. 1, p. S1–173, 2010.
57. F. Guo, N. Ma, Y. Horibe, S. Kawanishi, M. Murata, and Y. Hiraku, "Nitrative DNA damage induced by multi-walled carbon nanotube via endocytosis in human lung epithelial cells," *Toxicol Appl Pharmacol*, vol. 260, no. 2, pp. 183–92, Apr. 2012, doi: 10.1016/j.taap.2012.02.010.
58. L. M. Sargent et al., "Single-walled carbon nanotube-induced mitotic disruption," *Mutat Res Genet Toxicol Environ Mutagen*, vol. 745, no. 1–2, pp. 28–37, Jun. 2012, doi: 10.1016/j.mrgentox.2011.11.017.
59. J. Cveticanin, G. Joksic, A. Leskovac, S. Petrovic, A. V. Sobot, and O. Neskovic, "Using carbon nanotubes to induce micronuclei and double strand breaks of the DNA in human cells," *Nanotechnology*, vol. 21, no. 1, p. 015102, Jan. 2010, doi: 10.1088/0957–4484/21/1/015102.
60. L. Zhu, D. W. Chang, L. Dai, and Y. Hong, "DNA damage induced by multiwalled carbon nanotubes in mouse embryonic stem cells," *Nano Lett*, vol. 7, no. 12, pp. 3592–7, Dec. 2007, doi: 10.1021/nl071303v.

61. Y.-Y. Guo, J. Zhang, Y.-F. Zheng, J. Yang, and X.-Q. Zhu, "Cytotoxic and genotoxic effects of multi-wall carbon nanotubes on human umbilical vein endothelial cells in vitro," *Mutat Res Genet Toxicol Environ Mutagen*, vol. 721, no. 2, pp. 184–91, Apr. 2011, doi: 10.1016/j.mrgentox.2011.01.014.
62. A. A. Shvedova et al., "Inhalation vs. aspiration of single-walled carbon nanotubes in C57BL/6 mice: inflammation, fibrosis, oxidative stress, and mutagenesis," *Am J Physiol Lung Cell Mol Physiol*, vol. 295, no. 4, pp. L552–65, Oct. 2008, doi: 10.1152/ajplung.90287.2008.
63. N. R. Jacobsen et al., "Lung inflammation and genotoxicity following pulmonary exposure to nanoparticles in ApoE$^{-/-}$ mice," *Part Fibre Toxicol*, vol. 6, no. 1, p. 2, Dec. 2009, doi: 10.1186/1743–8977-6-2.
64. Z. Li et al., "Cardiovascular effects of pulmonary exposure to single-wall carbon nanotubes," *Environ Health Perspect*, vol. 115, no. 3, pp. 377–82, Mar. 2007, doi: 10.1289/ehp.9688.
65. A. K. Patlolla, S. M. Hussain, J. J. Schlager, S. Patlolla, and P. B. Tchounwou, "Comparative study of the clastogenicity of functionalized and nonfunctionalized multiwalled carbon nanotubes in bone marrow cells of Swiss-Webster mice," *Environ Toxicol*, vol. 25, no. 6, pp. 608–21, Dec. 2010, doi: 10.1002/tox.20621.
66. C. Zhang, L. Wu, M. de Perrot, and X. Zhao, "Carbon nanotubes: a summary of beneficial and dangerous aspects of an increasingly popular group of nanomaterials," *Front Oncol*, vol. 11, p. 693814, Jul. 2021, doi: 10.3389/fonc.2021.693814.
67. M. Pacurari et al., "Raw single-wall carbon nanotubes induce oxidative stress and activate MAPKs, AP-1, NF-kappaB, and Akt in normal and malignant human mesothelial cells.," *Environ Health Perspect*, vol. 116, no. 9, pp. 1211–7, Sep. 2008, doi: 10.1289/ehp.10924.
68. J. N. Moloney and T. G. Cotter, "ROS signalling in the biology of cancer," *Semin Cell Dev Biol*, vol. 80, pp. 50–64, Aug. 2018, doi: 10.1016/j.semcdb.2017.05.023.
69. A. Nel, T. Xia, L. Mädler, and N. Li, "Toxic potential of materials at the nanolevel," *Science*, vol. 311, no. 5761, pp. 622–7, Feb. 2006, doi: 10.1126/science.1114397.
70. F. Huaux et al., "Mesothelioma response to carbon nanotubes is associated with an early and selective accumulation of immunosuppressive monocytic cells," *Part Fibre Toxicol*, vol. 13, no. 1, p. 46, Dec. 2015, doi: 10.1186/s12989-016-0158-0.
71. Y. J. Arnoldussen et al., "Inflammation in the pleural cavity following injection of multi-walled carbon nanotubes is dependent on their characteristics and the presence of IL-1 genes," *Nanotoxicology*, vol. 12, no. 6, pp. 522–38, Aug. 2018, doi: 10.1080/17435390.2018.1465139.
72. J. Xu et al., "Size- and shape-dependent pleural translocation, deposition, fibrogenesis, and mesothelial proliferation by multiwalled carbon nanotubes," *Cancer Sci*, vol. 105, no. 7, pp. 763–9, Jul. 2014, doi: 10.1111/cas.12437.
73. F. A. Murphy, A. Schinwald, C. A. Poland, and K. Donaldson, "The mechanism of pleural inflammation by long carbon nanotubes: interaction of long fibres with macrophages stimulates them to amplify pro-inflammatory responses in mesothelial cells," *Part Fibre Toxicol*, vol. 9, no. 1, p. 8, Dec. 2012, doi: 10.1186/1743-8977-9-8.
74. S. L. Cassel et al., "The Nalp3 inflammasome is essential for the development of silicosis," *Proc Natl Acad Sci*, vol. 105, no. 26, pp. 9035–40, Jul. 2008, doi: 10.1073/pnas.0803933105.
75. J. Palomäki et al., "Long, needle-like carbon nanotubes and asbestos activate the NLRP3 inflammasome through a similar mechanism," *ACS Nano*, vol. 5, no. 9, pp. 6861–70, Sep. 2011, doi: 10.1021/nn200595c.
76. A. Takagi et al., "Induction of mesothelioma in p53+/– mouse by intraperitoneal application of multi-wall carbon nanotube," *J Toxicol Sci*, vol. 33, no. 1, pp. 105–16, Feb. 2008, doi: 10.2131/jts.33.105.
77. Y. Sakamoto et al., "Induction of mesothelioma by a single intrascrotal administration of multi-wall carbon nanotube in intact male Fischer 344 rats," *J Toxicol Sci*, vol. 34, no. 1, pp. 65–76, 2009, doi: 10.2131/jts.34.65.

78. J. Muller, M. Delos, N. Panin, V. Rabolli, F. Huaux, and D. Lison, "Absence of carcinogenic response to multiwall carbon nanotubes in a 2-year bioassay in the peritoneal cavity of the rat," *Toxicol Sci*, vol. 110, no. 2, pp. 442–8, Aug. 2009, doi: 10.1093/toxsci/kfp100.
79. H. Nagai et al., "Diameter and rigidity of multiwalled carbon nanotubes are critical factors in mesothelial injury and carcinogenesis," *Proc Natl Acad Sci U S A*, vol. 108, no. 49, pp. E1330–8, Dec. 2011, doi: 10.1073/pnas.1110013108.
80. J. E. Rohde, "Beyond survival: promoting healthy growth," *Indian J Pediatr*, vol. 55, no. 1 Suppl, pp. S3–8, 1988, doi: 10.1007/BF02810383.
81. H. Baier and H. D. Taubert, "Investigations on the effects of epimestrol and stilboestrol upon the release of FSH and LH," *Horm Metab Res*, vol. 1, no. 6, pp. 309–13, Nov. 1969, doi: 10.1055/s-0028–1095134.
82. T. Chernova et al., "Long-fiber carbon nanotubes replicate asbestos-induced mesothelioma with disruption of the tumor suppressor gene Cdkn2a (Ink4a/Arf)," *Curr Biol*, vol. 27, no. 21, pp. 3302–314, Nov. 2017, doi: 10.1016/j.cub.2017.09.007.
83. K. Honda, M. Naya, H. Takehara, H. Kataura, K. Fujita, and M. Ema, "A 104-week pulmonary toxicity assessment of long and short single-wall carbon nanotubes after a single intratracheal instillation in rats," *Inhal Toxicol*, vol. 29, no. 11, pp. 471–82, Sep. 2017, doi: 10.1080/08958378.2017.1394930.
84. N. R. Jacobsen et al., "Biodistribution of carbon nanotubes in animal models," *Basic Clin Pharmacol Toxicol*, vol. 121, no. S3, pp. 30–43, Sep. 2017, doi: 10.1111/bcpt.12705.
85. J. Pauluhn, "Multi-walled carbon nanotubes (Baytubes®): approach for derivation of occupational exposure limit," *Regul Toxicol Pharmacol*, vol. 57, no. 1, pp. 78–89, Jun. 2010, doi: 10.1016/j.yrtph.2009.12.012.
86. H. E. Williams and T. R. Wandzilak, "Oxalate synthesis, transport and the hyperoxaluric syndromes," *J Urol*, vol. 141, no. 3 Pt 2, pp. 742–9, Mar. 1989, doi: 10.1016/s0022–5347(17)40999-2.
87. I. Rostenberg, M. Jiménez, M. Daltabuit, and S. Armendares, "Somatometry of the face in preschool children. 1. In the 5 and 6-year-age groups," *Arch Invest Med (Mex)*, vol. 4, no. 3, pp. 167–74, 1973.
88. O. F. Dent, B. Richardson, S. Wilson, K. J. Goulston, and C. W. Murdoch, "Postwar mortality among Australian World War II prisoners of the Japanese," *Med J Aust*, vol. 150, no. 7, pp. 378–82, Apr. 1989, doi: 10.5694/j.1326–5377.1989.tb136529.x.
89. N. Cortesi, A. Manenti, G. Gibertini, and A. Rossi, "Emergency laparoscopy in multiple trauma patients: experience with 106 cases," *Acta Chir Belg*, vol. 87, no. 4, pp. 239–41, 1987.
90. X. Deng et al., "Translocation and fate of multi-walled carbon nanotubes in vivo," *Carbon N Y*, vol. 45, no. 7, pp. 1419–24, Jun. 2007, doi: 10.1016/j.carbon.2007.03.035.
91. R. A. Pope, Y. Okumura, A. S. Baker, and J. S. Laughlin, "Transverse axial tomography for treatment planning," *Phys Med Biol*, vol. 14, no. 2, p. 340, Apr. 1969.
92. I. Rostenberg, M. Jiménez, M. Daltabuit, and S. Armendares, "Somatometry of the face in preschool children. 1. In the 5 and 6-year-age groups," *Arch Invest Med (Mex)*, vol. 4, no. 3, pp. 167–74, 1973.
93. H. E. Williams and T. R. Wandzilak, "Oxalate synthesis, transport and the hyperoxaluric syndromes.," *J Urol*, vol. 141, no. 3 Pt 2, pp. 742–9, Mar. 1989, doi: 10.1016/s0022-5347(17)40999-2.
94. L. L. Shimkevich, B. M. Kostiuchenok, V. G. Rudneva, and M. I. Chernov, "Effect of hemosorption on blood coagulation system factors in patients with severe suppurative infection," *Vestn Khir Im I I Grek*, vol. 136, no. 4, pp. 111–4, Apr. 1986.
95. B. Czarny et al., "Carbon nanotube translocation to distant organs after pulmonary exposure: insights from *in Situ* ^{14}C-radiolabeling and tissue radioimaging," *ACS Nano*, vol. 8, no. 6, pp. 5715–24, Jun. 2014, doi: 10.1021/nn500475u.
96. H. Wang et al., "Biodistribution of carbon single-wall carbon nanotubes in mice," *J Nanosci Nanotechnol*, vol. 4, no. 8, pp. 1019–24, Nov. 2004, doi: 10.1166/jnn.2004.146.

97. N. R. Jacobsen et al., "Biodistribution of carbon nanotubes in animal models," *Basic Clin Pharmacol Toxicol*, vol. 121, no. S3, pp. 30–43, Sep. 2017, doi: 10.1111/bcpt.12705.
98. S. Yang et al., "Biodistribution of pristine single-walled carbon nanotubes in vivo," *Am J Phys Chem C*, vol. 111, no. 48, pp. 17761–4, Dec. 2007, doi: 10.1021/jp070712c.
99. M. L. Orlova, "Alzheimer's disease and vascular dementia: comparative pathologic vascular structure of the cerebral cortex and white matter of the brain," *Zh Nevropatol Psikhiatr Im S S Korsakova*, vol. 91, no. 10, pp. 28–30, 1991.
100. B. Haraldsson and J. Sörensson, "Why do we not all have proteinuria? An update of our current understanding of the glomerular barrier," *Physiology*, vol. 19, no. 1, pp. 7–10, Feb. 2004, doi: 10.1152/nips.01461.2003.
101. Z. Liu et al., "In vivo biodistribution and highly efficient tumour targeting of carbon nanotubes in mice," *Nat Nanotechnol*, vol. 2, no. 1, pp. 47–52, Jan. 2007, doi: 10.1038/nnano.2006.170.
102. M. R. McDevitt et al., "Tumor targeting with antibody-functionalized, radiolabeled carbon nanotubes," *J Nucl Med*, vol. 48, no. 7, pp. 1180–9, Jul. 2007, doi: 10.2967/jnumed.106.039131.
103. D. F. Branson, G. J. Picha, and J. Desprez, "Expanded polytetrafluoroethylene as a microvascular graft: a study of four fibril lengths," *Plast Reconstr Surg*, vol. 76, no. 5, pp. 754–63, Nov. 1985, doi: 10.1097/00006534–198511000-00017.
104. Z. Liu, C. Davis, W. Cai, L. He, X. Chen, and H. Dai, "Circulation and long-term fate of functionalized, biocompatible single-walled carbon nanotubes in mice probed by Raman spectroscopy," *Proc Natl Acad Sci*, vol. 105, no. 5, pp. 1410–5, Feb. 2008, doi: 10.1073/pnas.0707654105.
105. S.-T. Yang, H. Wang, M. J. Meziani, Y. Liu, X. Wang, and Y.-P. Sun, "Biodefunctionalization of functionalized single-walled carbon nanotubes in mice," *Biomacromolecules*, vol. 10, no. 7, pp. 2009–12, Jul. 2009, doi: 10.1021/bm900263z.
106. H. Wang et al., "Biodistribution of carbon single-wall carbon nanotubes in mice," *J Nanosci Nanotechnol*, vol. 4, no. 8, pp. 1019–24, Nov. 2004, doi: 10.1166/jnn.2004.146.
107. M. Sohn, S. H. Yu, and Y. H. Kim, "Network analysis of Korean health insurance policy-making process," *Yonsei Med J*, vol. 33, no. 2, pp. 121–36, Jun. 1992, doi: 10.3349/ymj.1992.33.2.121.
108. V. Janata and J. Wagner, "The difference between the mechanism of the effect of hydroxylamine and of phenylhydroxylamine on the metabolism of erythrocytes," *Folia Haematol Int Mag Klin Morphol Blutforsch*, vol. 83, no. 4, pp. 526–31, 1965.
109. X. Deng et al., "Translocation and fate of multi-walled carbon nanotubes in vivo," *Carbon N Y*, vol. 45, no. 7, pp. 1419–24, Jun. 2007, doi: 10.1016/j.carbon.2007.03.035.
110. X. Deng, S. Yang, H. Nie, H. Wang, and Y. Liu, "A generally adoptable radiotracing method for tracking carbon nanotubes in animals," *Nanotechnology*, vol. 19, no. 7, p. 075101, Feb. 2008, doi: 10.1088/0957–4484/19/7/075101.
111. D. Georgin et al., "Preparation of (14)C-labeled multiwalled carbon nanotubes for biodistribution investigations," *J Am Chem Soc*, vol. 131, no. 41, pp. 14658–9, Oct. 2009, doi: 10.1021/ja906319z.
112. J. T.-W. Wang et al., "The relationship between the diameter of chemically-functionalized multi-walled carbon nanotubes and their organ biodistribution profiles in vivo," *Biomaterials*, vol. 35, no. 35, pp. 9517–28, Nov. 2014, doi: 10.1016/j.biomaterials.2014.07.054.
113. H. Kafa et al., "The interaction of carbon nanotubes with an in vitro blood-brain barrier model and mouse brain in vivo," *Biomaterials*, vol. 53, pp. 437–52, 2015, doi: 10.1016/j.biomaterials.2015.02.083.
114. K. T. Al-Jamal et al., "Degree of chemical functionalization of carbon nanotubes determines tissue distribution and excretion profile," *Angew Chem Int Ed Engl*, vol. 51, no. 26, pp. 6389–93, Jun. 2012, doi: 10.1002/anie.201201991.

115. M. Rautalahti, E. O. Terho, I. Vohlonen, and K. Husman, "Atopic sensitization of dairy farmers to work-related and common allergens," *Eur J Respir Dis Suppl*, vol. 152, pp. 155–64, 1987.
116. K. T. Al-Jamal et al., "Degree of chemical functionalization of carbon nanotubes determines tissue distribution and excretion profile," *Angewandte Chemie – International Edition*, vol. 51, no. 26, pp. 6389–93, Jun. 2012, doi: 10.1002/anie.201201991.
117. G. G. Zakarian, "Results of treatment of trichocephaliasis with diphesyl," *Med Parazitol (Mosk)*, vol. 38, no. 5, p. 614, 1969.
118. Z. Liu et al., "Drug delivery with carbon nanotubes for in vivo cancer treatment," *Cancer Res*, vol. 68, no. 16, pp. 6652–60, Aug. 2008, doi: 10.1158/0008-5472.CAN-08-1468.
119. W. Wu et al., "Covalently combining carbon nanotubes with anticancer agent: preparation and antitumor activity," *ACS Nano*, vol. 3, no. 9, pp. 2740–50, Sep. 2009, doi: 10.1021/nn9005686.
120. Z. Zhang et al., "Delivery of telomerase reverse transcriptase small interfering RNA in complex with positively charged single-walled carbon nanotubes suppresses tumor growth," *Clin Cancer Res*, vol. 12, no. 16, pp. 4933–9, Aug. 2006, doi: 10.1158/1078-0432.CCR-05-2831.
121. J. Recht, H. Evrard, and C. Guillaume, "Treatment of fractures of the carpal scaphoid with Herbert's bone screw. Review of 21 clinical cases," *Acta Orthop Belg*, vol. 55, no. 2, pp. 183–90, 1989.
122. W. Zhang, Z. Zhang, and Y. Zhang, "The application of carbon nanotubes in target drug delivery systems for cancer therapies," *Nanoscale Res Lett*, vol. 6, no. 1, p. 555, Oct. 2011, doi: 10.1186/1556-276X-6-555.
123. M. Longmire, P. L. Choyke, and H. Kobayashi, "Clearance properties of nano-sized particles and molecules as imaging agents: Considerations and caveats," *Nanomedicine*, vol. 3, no. 5, pp. 703–17, Oct. 2008. doi: 10.2217/17435889.3.5.703.
124. A. Ruggiero et al., "Paradoxical glomerular filtration of carbon nanotubes," *Proc Natl Acad Sci U S A*, vol. 107, no. 27, pp. 12369–74, Jul. 2010, doi: 10.1073/pnas.0913667107.
125. S. T. Yang, J. Luo, Q. Zhou, and H. Wang, "Pharmacokinetics, metabolism and toxicity of carbon nanotubes for bio-medical purposes," *Theranostics*, vol. 2, no. 3, p. 271, Mar. 2012. doi: 10.7150/thno.3618.
126. Z. Liu, C. Davis, W. Cai, L. He, X. Chen, and H. Dai, "Circulation and long-term fate of functionalized, biocompatible single-walled carbon nanotubes in mice probed by Raman spectroscopy," *Proc Natl Acad Sci*, vol. 105, no. 5, pp. 1410–5, Feb. 2008. Available: www.pnas.org/cgi/content/full/
127. H. Ali-Boucetta and K. Kostarelos, "Pharmacology of carbon nanotubes: toxicokinetics, excretion and tissue accumulation," *Adv Drug Deliv Rev*, vol. 65, no. 15, pp. 2111–9, Dec. 2013. doi: 10.1016/j.addr.2013.10.004.
128. M. R. McDevitt et al., "PET imaging of soluble yttrium-86-labeled carbon nanotubes in mice," *PLoS One*, vol. 2, no. 9, p. 907, Sep. 2007, doi: 10.1371/journal.pone.0000907.
129. J. Guo, X. Zhang, Q. Li, and W. Li, "Biodistribution of functionalized multiwall carbon nanotubes in mice," *Nucl Med Biol*, vol. 34, no. 5, pp. 579–83, Jul. 2007, doi: 10.1016/j.nucmedbio.2007.03.003.

10 Regulatory Consideration of Carbon Nanotubes

Biswajit Rout, Devaraj, Sandeep Kr Maharana, Naveen Singh, Dimpal Suthar, and Aakanchha Jain

ABBREVIATIONS

CNTs carbon nanotubes
ECHA European Chemicals Agency
HSE Health and Safety Executive
OSHA Occupational Safety and Health Administration
REACH Registration, Evaluation, Authorization, and Restriction of Chemicals

10.1 INTRODUCTION

Carbon nanotubes (CNTs) are one-dimensional cylindrical nanomaterials composed of single-layered carbon sheets arranged in a hexagonal lattice, known as graphene. CNTs possess distinctive hollow tubular shapes, capped with a hemispherical fullerene structure. This configuration arises from the curling of multiple graphene layers through π–π interaction. The CNTs are a new era of nanomaterial-based drug delivery systems with many unique and advantageous physicochemical properties like high length-to-diameter ratio, good mechanical strength, economic feasibility, and versatility [1–3].

CNTs can be classified into three main types: single-walled carbon nanotubes (SWCNTs), multi-walled carbon nanotubes (MWCNTs), and double-walled carbon nanotubes (DWCNTs). SWCNTs, composed of a single layer of graphene sheet seamlessly rolled into a cylindrical shape with a diameter fluctuating between 0.5 and 2 nm and a length of 2–5 μm, frequently aggregate into bundles. Within these bundles, the SWCNTs arrange themselves in hexagonal patterns, resulting in an open, branching, crystalline structure [4]. These distinct structures result in special SWCNT characteristics, such as high mechanical strength that surpasses that of iron, a low density lesser than aluminum, and impressive thermal stability. Because of these special qualities, SWCNTs are now considered a promising material for applications in physics, chemistry, and biology [5]. In contrast, MWCNTs are composed of several layers of graphene sheets wrapped around the central axis with a diameter as little as 0.7 nm and lengths of some micrometers [6]. Though both SWCNTs and MWCNTs are made from graphene sheets and have many similarities, they differ from each other in several ways. For example, MWCNTs have more graphene layers, are easier to make, are purer, and are less likely to exhibit synthesis flaws. In addition, these MWCNTs exhibit higher electrical conductivity than SWCNTs, better thermal and

DOI: 10.1201/9781003386537-10

chemical stability, which makes them more appropriate for use in high-temperature or harsh chemical environments, and superior mechanical strength, which allows them to offer greater resistance to mechanical stress [7,8]. DWCNTs are a specific type of MWCNT and can be defined as a synthetic blend of both SWCNTs and MWCNTs. These materials exhibit enhanced chemical stability while maintaining the same form and characteristics as SWCNTs. This characteristic is particularly crucial when functionality is needed to give the nanotubes new characteristics [9,10].

CNTs are generally made by using several methods like chemical vapor deposition (CVD), laser ablation, arc discharge, electrolysis, hydrothermal synthesis, and flame synthesis. The CVD is a standard method that can produce both SWCNTs and MWCNTs. In this process, a carbon-containing gas is decomposed at elevated temperatures (600–1,200°C) amid a metal catalyst. The liberated carbon atoms then assemble on the catalyst particles, forming CNTs [11]. The laser ablation technique includes the use of laser radiation to synthesize nanotubes. In this technique, a pulsed or continuous laser beam is directed onto a graphite target. The high-intensity laser vaporizes the target material, generating a trail of carbon atoms and ions that subsequently condense and coalesce to form CNTs. Despite offering several advantages, such as higher yield, fewer metallic impurities, and precise diameter control, laser ablation is limited by its high cost and the production of undesired branched CNTs. The arc discharge technique involves creating an electric arc in an environment of inert gas between two carbon electrodes. This causes the vaporization of carbon materials that subsequently condense to create CNTs with special structural characteristics [12]. The electrolysis method uses a graphite cathode and anode to electrolyze a molten halide salt, which produces CNTs. During electrolysis, the graphite cathode undergoes erudition and releases carbon residue into the molten salt solution, which ultimately causes CNT formation [13]. The hydrothermal method produces CNTs by crystallizing compounds in aqueous solutions at high temperatures and vapor pressures [14]. The flame method is a simple, flame-based method to produce CNTs, but the chances of soot formation limit its application [15–17].

Overly, CNTs exhibit a range of remarkable physicochemical properties, making them highly versatile and valuable in numerous applications. CNTs possess high tensile strength characterized by their elastic modulus, allowing them to stretch and bend easily. The high tensile strength of CNTs makes them a perfect material for the reinforcement of polymers, metals, and steel [18–21]. CNTs also excel in thermal conductivity, as confirmed by the analysis of the thermal bridge method (TBM). They remain stable up to 750°C under normal conditions and up to 2,800°C in vacuum environments. This high thermal stability gives a possible application of CNTs as thermal interface materials (TIMs) in temperature sensors and thermal insulators [22]. Moreover, CNTs have excellent electric conductivity, making them suitable for electronic and electrochemical applications. Their nanoscale size results in a large surface-to-volume ratio, a trait that enhances their performance in fields such as drug delivery and catalysis. The CNTs also exhibit photoluminescence and demonstrate high immobilization efficiency, which is advantageous in biosensing and imaging technologies. Additionally, CNTs are easy to functionalize. The functionalization of CNTs with hydrophilic moieties can enhance their water solubility, making them more suitable for biomedical applications. Their elevated surface area, porous structure, aptitude to penetrate physiological barriers, and ability to get functionalized with antibodies or specific ligands make them potential nanocarrier systems for targeted drug delivery [23–25]. Also, CNTs are implemented in biosensing, tissue engineering, and imaging owing to their biocompatibility and functionalization potential. Additionally, CNTs have numerous environmental applications. For instance, in wastewater treatment, CNTs act as efficient adsorbents and catalysts for removing pollutants and contaminants from water, enhancing purification processes [26,27]. Similarly, CNTs can be used to remediate air, soil, and groundwater contamination (Figure 10.1) [28].

Despite their promising potential, the application of CNTs is still limited because of their toxic effect and environmental concerns. CNTs can cause potential toxicity through oxidative stress, DNA damage, and other mechanisms. The CNTs can accumulate in various large organs of the body. It can persist in the lungs for a considerable time, leading to persistent inflammation and injury. The long persistent exposure of CNTs leads to tumor cell growth. CNTs can pose toxicity risks to both living organisms and the environment. For instance, MWCNTs have been shown to induce genotoxic effects and disrupt the transcription of genes involved in apoptosis. CNTs can also influence the bioavailability and biodegradation of other chemicals in the environment. They can bind to natural organic matter, potentially competing with metal ions, which may increase the bioavailability and toxicity of metals. The environmental impact and human exposure to CNTs can vary depending on their properties and the product's life cycle stage. For example, pristine CNTs are

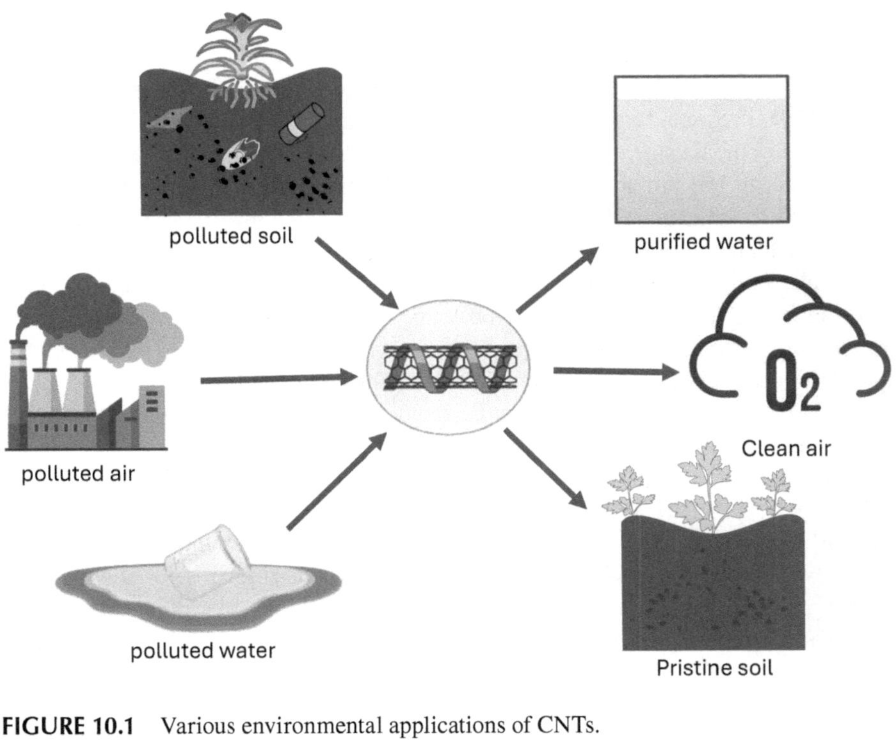

FIGURE 10.1 Various environmental applications of CNTs.

lipophilic, meaning they can be absorbed by microbial communities and plant roots, leading to accumulation in plant tissues [29–31].

Due to this potential toxicity of CNTs, regulatory considerations are essential. Regulatory frameworks require precise classification to differentiate CNTs based on their toxicity profile. The Environmental Protection Agency (EPA) has made mandatory requirements for testing CNTs under the Toxic Substances Control Act (TSCA) in the United States. In Europe, CNTs must comply with the Registration, Evaluation, Authorization, and Restriction of Chemicals (REACH) regulation, which requires manufacturers to register their substances by providing safety and environmental impact data to get marketing approval. The International Agency for Research on Cancer (IARC) has classified certain types of CNTs as probably oncogenic, which necessitates stringent health risk evaluation before their widespread use [32,33].

10.2 BIOMEDICAL APPLICATION, TOXICITIES, AND REGULATORY CONCERN OF CNTs

CNTs have emerged as a significant innovation in biomedical applications by virtue of their unique properties and versatility. These cylindrical nanostructures possess remarkable mechanical, electrical, and thermal characteristics, making them ideal for various medical uses. Their unique characteristics have drawn significant attention for their applications in cancer treatment, vaccine administration, biosensors, tissue engineering, and others [34–36].

In recent years, several research works have shown that CNTs can act as a promising carrier system for controlled and targeted drug delivery [37,38]. It provides a perfect platform for the targeted delivery of drug molecules to diseased tissues, owing to their miniature size, large surface, outstanding strength, and ease of modification [39]. They can target diseased cells both actively and passively. For instance, due to their compact size, CNTs can exploit the enhanced permeability and retention (EPR) effect of the tumor microenvironment, allowing them to target cancer cells passively [40]. Similarly, CNTs can be readily functionalized with targeting moieties to improve their precision in targeting diseased cells [41,42]. At the same time, they can be altered to enhance their compatibility within the body and reduce toxicity, allowing for the attachment of drugs, proteins, or genes [43]. Besides targeted delivery, CNTs can be employed to control the release of drug molecules over extended periods. This sustained release is crucial for maintaining therapeutic levels of drugs in the bloodstream while minimizing the frequency of drug administration [44].

Furthermore, in addition to conventional drugs, CNTs have intriguing uses in immunotherapy since they can be utilized to encapsulate and deliver genetic elements and antigens [45,46]. They not only provide an effective platform for encapsulating proteins and genetic components but also reduce their degradation and enhance their targeted uptake [47,48]. CNTs beyond their role as carrier systems for chemicals or genetic materials, can serve as highly effective biosensors. As one-dimensional nanomaterials, CNTs offer high surface-to-volume ratios and excellent conductivity making them ideal for developing sensitive and selective biosensors [49]. These sensors are crucial for detecting biomolecules in diverse applications, from medical diagnostics to environmental monitoring [41,50]. CNT-based biosensors are renowned for their speedy response times, high sensitivity, and aptitude to detect numerous biomolecules under various conditions [51]. Similar to biosensors, CNTs can serve as scaffolds in tissue engineering by providing a compatible microenvironment that closely resembles the physical structure and chemical composition of the biological extracellular matrix [52]. Finally, CNTs can be used as a platform for MRI and radiotracer contrast agents, enabling the visualization and tracking of tissue formation processes [53].

Nanoparticles are now well-recognized for their enhanced chemical and biological activities, primarily attributed to their small size. While these properties make nanoparticles advantageous in various applications, they also pose significant risks, as their increased activity correlates with a higher potential for toxicity to both humans and the environment. Similarly, as a type of nanomaterial, CNTs also carry a substantial risk of toxicity [54]. The toxicity of CNTs largely depends on their physicochemical properties, types, functionalization, presence of impurities, size, shape, and route of administration [55]. Numerous studies have elucidated molecular mechanisms associated with CNT toxicity, with oxidative stress emerging as the primary pathway. This oxidative stress is induced by the formation of reactive oxygen species (ROS) and other free radicals, which interrupt the cell cycle, damage mitochondria, and ultimately lead to cell death. While many aspects of CNT toxicity remain unclear, these properties significantly influence their interactions with biological systems. Prolonged exposure to CNTs has been linked to acute and chronic inflammation, granuloma formation, interstitial fibrosis, immunotoxicity, genotoxicity, and, in some cases, cancer [55,56]. According to Liu et al., their chemical reactivity encourages the development of free radicals, which can oxidize DNA, proteins, and lipids inside cells [57]. A major reason for CNT toxicity is due to their ability to accumulate in cells and the inability of cells to eliminate them. This often leads to CNT-induced hepatic, renal, pulmonary, neuronal, cardiovascular, embryonic, and spermatocyte cell toxicity (Figure 10.2) [54,58].

Given the toxic effects of CNTs, it is crucial to establish appropriate regulatory guidelines tailored to each biomedical application of these materials. Regulatory guidelines for CNT applications, particularly in targeted and controlled drug delivery, are continually evolving to address their unique properties and potential risks. Key regulatory priorities include biocompatibility, thorough safety evaluations, and toxicity pathway analysis.

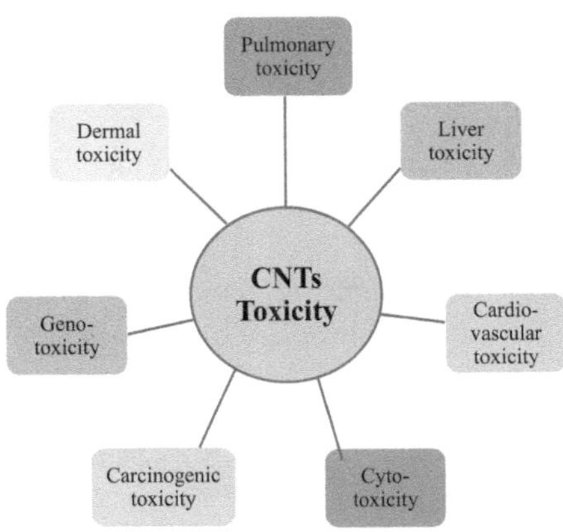

FIGURE 10.2 Toxic effects of carbon nanotubes on human health.

International bodies such as the FDA, EMA, and OECD are crafting frameworks to address safety and efficacy, including dosage standards and assessments of long-term exposure effects. Clinical trials and stringent quality control standards are essential to advance CNT-based therapeutics safely. These guidelines aim to protect public health by ensuring that nanotechnology products undergo rigorous evaluations, reflecting the regulatory body's commitment to evidence-based decisions for managing the risks associated with nanomaterials [59,60]. Despite the regulatory progress, commercialization of nanomaterials, including CNTs, remains limited, largely due to a lack of unified global regulatory standards of safety evaluation and rising public concerns over potential toxicity [61]. Additionally, regulatory and ethical considerations in vaccine development are critical to ensuring vaccines are safe, effective, and accessible. The vaccine approval process is governed by stringent regulatory requirements that aim to uphold safety and efficacy standards [62,63]. The regulatory landscape for CNT-based genes and vaccine delivery is complex. Agencies such as the USFDA and EMA impose stringent requirements for pre-clinical evaluations and toxicological studies. The variability in CNTs synthesis methods, surface modifications, and batch consistency complicates standardization and quality control processes [64]. Similarly, extensive and ongoing research into the nanotoxicity of CNTs is essential before regulatory bodies can approve their clinical use as functional biomaterials and in biomedical devices [65]. Currently, the absence of specific regulatory guidelines for nano biosensors presents a major obstacle to their clinical application. Establishing well-defined regulatory pathways will be crucial in bridging the gap between laboratory research and practical healthcare applications, enabling the safe and effective use of these advanced technologies in clinical settings [66,67].

10.3 REGULATORY FRAMEWORK FOR CNTs

10.3.1 In Europe

In 2007, one of the regulatory bodies in Europe, named Registration, Evaluation, Authorization, and Restriction of Chemicals (REACH), came into existence, which covers all aspects related to chemical nanomaterial systems. It primarily comprises the following elements [68].

- Firstly, "*Registration*" denotes the data collection related to chemical use and toxicity.
- Secondly, "*Evaluation*" implies the government's examination of the requirement for additional testing to be carried out along with the regulation of chemicals.
- Thirdly, "*Authorization of chemicals*" denotes the requirements for firms to grant permission for using chemicals of higher concern.
- Lastly, "*Restrictions of chemicals*" implies a complete restriction on certain chemicals that cannot be used safely.

The European Chemicals Agency (ECHA) is primarily accountable for the proper REACH execution. The primary objective of REACH is to safeguard both health and the environment against harmful substances. It also ensures the safety of the chemical substances and helps in promoting innovation as well as competitiveness. Since 2007, REACH has evolved to reflect the advancements related to nanomaterials along with their chemical properties. Thus, REACH has a more focused database where all the chemicals produced or imported in the EU in 1 ton/year or more quantity per manufacturer or importer have been registered. REACH also suggested that if the quantity exceeds 10 tons, a chemical safety report would be needed. Nanomaterials are regarded as chemical substances whose size is <100nm. In 2011, the REACH Commission, which adopted the recommendation on the definition of a nanomaterial, revised the same in 2022 while retaining its principal features. Carbon was initially considered to be a material of less concern, according to REACH. However, in 2008, the above consideration was removed due to the existence of various nanoscale forms of carbon [68]. It has been reported, according to ECHA, in Classification, Labelling, and Packaging (CLP) notifications that CNTs might lead to eye and skin irritation along with other respiratory problems [69]. Furthermore, in the case of MWCNTs, REACH registrations have stated that this substance, through prolonged or repeated exposure, may cause damage to organs [70].

10.3.2 In the United States of America

In the United States, the Environmental Protection Agency or EPA, under the "Toxic Substances Control Act (TSCA)", plays a vital role in the regulation of CNTs. A four-pronged regulatory approach is pursued by the EPA. Firstly, it includes the pre-manufacture notifications under the "TSCA Section 5". The EPA, in compliance with "TSCA Section 5", evaluates various notices (like Pre-Manufacture Notice (PMN), Microbial Commercial Activity Notice (MCAN), or Significant New Use Notice (SNUN)) and thus determines subsequent actions that may include regulation or restriction of the chemical substance. The EPA updated its working approach in December 2019 under "TSCA Section 5" to prepare new chemical determinations. Therefore, "TSCA Section 5" Significant New Use Rules (SNURs) might be implemented to mandate prior notification to the EPA for the application of new chemical substances or mixtures that could potentially pose high risks [71]. Similarly, under the "TSCA Section 5(a)(2)", the EPA includes the SNURs. The EPA has the authority to evaluate whether the use of a specific chemical substance is significant or not based on the provisions outlined in the relevant section. It considers the following factors before coming to a certain determination [71].

- the anticipated volume of a chemical substance that will be manufactured or processed.
- the degree to which a particular use of a chemical substance alters the type or amount of exposure of humans or the environment.
- the degree to which a specific use of a chemical substance increases human or environmental exposure, both in terms of intensity and duration.
- the methods by which a certain chemical substance is manufactured, processed, distributed in commerce, and finally discarded.

Thirdly, "TSCA Section 8(a)" includes the informational gathering rules. This section generally authorizes the EPA to promulgate new rules that require various entities, other than the small manufacturers who manufacture the chemical substance, to keep certain records and submit all the generated reports as the EPA Administrator may reasonably require. However, this section enables the EPA to require small manufacturers to report under certain circumstances, such as when the chemical is the subject of a "TSCA Section 6" proposed or final rule. Under "TSCA Section 8(a)", EPA may require the reporting and recordkeeping of the following [72].

- the trade name, molecular structure, chemical identity of each substance/mixture,
- the category/proposed categories of using each substance/mixture,
- the total manufactured quantity of each substance/mixture, the quantity manufactured/processed for each of the categories of use, and the estimation that is closely related to the quantity to be manufactured for each of its categories/proposed categories of use,
- the byproducts description that results when a substance is manufactured, processed, used, or disposed of,
- all the existing information on each substance or mixture, whether it might cause environmental and health effects,
- estimated worker exposure, duration, and total potential workplace exposure to the substance/mixture,
- potential disposal methods for each substance/mixture and any changes in these methods.

Similarly, another TSCA section, "TSCA Section 4", comprises the test rules. Under the TSCA, the EPA has the authority to issue certain regulations that are designed to require manufacturers or processors to test chemical substances and mixtures under the "TSCA Section 4: Test Requirements" [73]. "TSCA Section 4" allows the EPA to enquire of the chemical manufacturers (like importers and processors) for developing the information related to chemicals and submit the same to the EPA. EPA, under the "TSCA Section 4", might generate an order that requires information about a certain chemical substance if the EPA makes a certain statutorily required finding. Furthermore, under "Section 4(a)", the EPA might also issue a certain order that requires the development of information for supporting certain federal activities. "TSCA Section 4(a)(1)" authorizes the EPA to issue an order where there exists insufficient information, therefore the testing is necessary to get that information, and that chemical substance might cause or produce an unreasonable risk, or is produced in substantial quantities, thereby causing significant exposures both to the environment and humans.

Similarly, under "TSCA Section 4(a)(2)", an order may also be issued by the EPA that requires the development of information related to [74]:

- to review a submitted notice under the "TSCA Section 5",
- to perform a risk-based evaluation under the "TSCA Section 6(b)",
- for implementing a requirement that is imposed in an order, rule, or consent agreement under the "TSCA Section 5(e) or (f)" or "TSCA Section 6(a)",

- on being approved from the relevant federal authority, the chemical substance must thereby undergo regulatory toxicity and exposure testing as mandated by that concerned authority,
- to assess whether a chemical substance or mixture manufactured, processed, or distributed solely for export to the United States may pose a potential risk to health and environment as per "TSCA Section 12(a)(2)",
- for prioritizing a chemical substance under the "TSCA Section 6(b)" which is subject to certain limitations.

The EPA effectively regulates the CNTs under the TSCA, similar to its regulation of asbestos. Thus, the EPA employs the four policy-making models (i.e., rational, incremental, interest group, and process streams). Therefore, it examines the working process of the policy in developing the statutory laws and rules for the CNTs. The EPA has the authority to regulate CNTs during the steps involved in their manufacture, use, and disposal [75]. "15 U.S.C. Section 2601–2629" is a section of the US Code that outlines the TSCA. This act, passed in 1976, empowers the EPA to regulate various chemical substances and mixtures, thereby addressing potential human health and environmental risks that are raised by chemicals such as CNTs. The TSCA requires the manufacturers of chemicals to submit information (i.e., data toxicity, environmental fate, and other relevant properties) to the EPA about the chemicals they produce or use. The EPA thus uses this information for assessing the risks that are posed by chemicals and it might take certain appropriate actions to protect public health and the environment [75]. The EPA has thus identified significant regulatory concerns related to the toxicity and environmental safety of CNTs. Since 2005, there have been various chemical notices under TSCA for CNTs for which it has undertaken suitable actions to limit their use, environmental release, and the requirement of the use of personal protective equipment [76]. The EPA has issued notifications to manufacturers regarding the potential requirement for CNTs to comply with "TSCA Section 5". It has considered CNTs from graphite or other allotropes of carbon. In 2007 ("72 FR 38081", "FRL-8139-9") and 2008 ("73 FR 4861" "FRL-8344–5"), the EPA issued a legal notice referenced as "TSCA Inventory Status of Nanoscale Substances—General Approach". Here, it is explained that the CNTs may not possess similar molecular identity as various non-nanoscale carbon allotropes. EPA recommends that persons associated with CNT manufacturing and commercialization should investigate beforehand whether their manufactured CNTs are on the "TSCA Inventory" and whether they comply with the requirements set in "TSCA Section 5" [77]. Health and Safety Executive (HSE) in 2009, classified the CNTs as substances of significant concern. Despite the limited toxicological data available for specific CNT types, there should be a precautionary approach that must be adopted for all CNT-based products in order to minimize health and safety risks. Stringent safety protocols and controls must be implemented in cases where CNT exposure cannot be fully eliminated. For CNTs, the British Standards Institution (BSI) has proposed some draft specifications. HSE, however, highlighted some of the challenges that are associated for accurate quantification of the extremely thin and lightweight CNTs. The BSI in the specifications suggested "0.01 fibers/ml" for CNTs as the exposure limit; however, implementation of the above limit will require time-consuming investigations [78].

Worker safety and health are primarily important, as they are the ones who are directly exposed to nanomaterials in the workplace. Consequently, there is a critical need for guidelines to safeguard worker safety. The Occupational Safety and Health Administration (OSHA) oversees these safety regulations. Currently, there are no regulations that look after the selection of protective clothing or other apparel, thereby preventing CNT exposure. However, under the "OSHA 1910.138(a)", OSHA states that employers should provide hand protection to the employees when they are being exposed to the hazards. Only limited data is available to predict the result of CNT exposure to the skin, along with its other potential health hazards. In one study, dermal exposure to SWCNT in workers was carried out to determine the potential airborne exposure of these CNTs during manufacturing and handling by placing cotton gloves over rubber gloves [79]. It has also been found that injecting CNT intraperitonially in mice results in the inflammation from the long MWCNTs (i.e., >5 µm in length), whereas the above was not observed in the case of short MWCNTs (i.e., <1 µm in length) or tangled CNT. The pulmonary toxicity of CNTs in rats and mice has also been reported during the instillation or aspiration studies. Thus, OSHA plays a major role in setting the appropriate standards for the permissible exposure limit (PEL) for the occupational exposures to minimize this CNTs pulmonary toxicity [80]. Respiratory protection is only provided to the workers in case the work practices and engineering controls are unable to reduce the CNTs exposure below the REL. OSHA's respiratory protection standard, i.e., "29 CFR 1910.134", outlines the requirements for both mandatory and voluntary respirator use. While voluntary respirator use is permitted, it must adhere to the specific provisions outlined in sections "CFR 1910.134(c)(2)(i)" and "CFR 1910.134(c)(2)(ii)" of the standard [79]. According to "CFR 1910.134(c)(2)(i)", the employers may provide the respirators upon prior request of employees or allow them to use their respirators, provided these respirators do not pose biohazards. If voluntary respirator use is permitted, employers must provide the users with the required information outlined in "Appendix D" of the standard [81]. "Appendix D" explains that the respirators are effective in protecting the employees from any type of hazards, especially when they are properly selected and worn. The respirator use is permitted to provide additional levels of protection to the workers. However, the respirator might lead to hazards if it is not kept clean. Mostly, the workers might wear respirators to avoid any hazards, even if the hazardous substance does not exceed the OSHA standard limits. If the employer provides the employees the respirators for voluntary use, they must follow certain precautions to prevent any hazard. Similarly, according to "CFR 1910.134(c)(2)(ii)", employers must implement the essential elements of a written respiratory protection program. Thus, this ensures that employees voluntarily using the respirators are medically fit and that the respirators are properly cleaned, stored, and maintained to prevent health risks [81]. A strict occupational exposure limit (i.e., <7 µg of CNTs/m^3 of air) has been proposed in the United States by the National Institute for Occupational Safety and Health (NIOSH), which is much lower in comparison to EU-OSHA's exposure limit for airborne graphite dust (i.e., 5 mg/m^3). Thus, the above shows that CNTs are smaller and lighter as compared to graphite dust particles. NIOSH also proposed a method, "Method 5040" in 1999 to measure CNT exposure for ultra-fine carbon dust that involves drawing air through a quartz filter. Captured carbon particles are thereafter analyzed using a thermal-optical analyzer. The proposed exposure limit of 7 μg/m^3 represents the minimum detectable concentration [78].

10.3.3 In Japan, Brazil, Australia, and China

In Australia, previously, there was only the voluntary submission of information on nano-substances, but later, the SWCNTs and MWCNTs were considered as new chemicals according to the new regulation published. Furthermore, according to the Industrial Chemicals Notification and Assessment Act 1989, pre-import authorization is needed [82]. In Japan, in the past, there have been significant discussions and considerations that focus on the establishment of occupational exposure limits (OELs) for CNTs. All of these discussions have been driven by health concerns associated with CNT exposure, particularly respiratory toxicity [82]. Various regulatory agencies of Japan, like the Ministry of Economy, Trade, and Industry (METI), the Ministry of Health, Labour, and Welfare (MHLW), and the Ministry of the Environment (MOE), regulate the nanomaterials like CNTs. METI possesses the authority to regulate the production, sale, and import of CNTs and primarily looks after the product's safety as well as quality, which also includes nanomaterials. MHLW plays a role in setting standards for the safe use of nanomaterials and in monitoring potential health risks [83]. Similarly, China has a relatively well-developed regulatory framework for CNTs, which includes a combination of national standards, technical specifications, and industry guidelines. In China, there are various regulatory bodies (as the State Administration for Market Regulation (SAMR), the National Health Commission (NHC), and the Ministry of Ecology and Environment (MEE)), which have well-defined roles in the regulation of CNTs. The role of SAMR is to look after the safety and efficacy of products containing nanomaterials like CNT along with the production and sale of CNT [84]. Furthermore, Brazil has a relatively well-developed regulatory framework for CNTs. Regulatory agencies (like the National Institute of Metrology, Quality, and Technology (INMETRO), the Brazilian Institute for the Environment and Renewable Natural Resources (IBAMA), and ANVISA) are actively looking after the various regulations related to CNTs. The IBAMA is primarily responsible for environmental protection, and they regulate the release of pollutants into the environment. Therefore, they also ensure the production and use of CNTs comply with the environmental regulations. ANVISA, another regulatory agency, is also responsible for public health and safety, including the regulation of hazardous substances, thereby playing a major role in setting the standards for the safe use of nanomaterials like CNTs, and in monitoring potential health risks. Furthermore, INMETRO looks after the safety and quality of nanomaterials like CNTs. They also actively regulate the production, sale, and import of CNTs [85].

10.3.4 ISO Guidelines

The International Organization for Standardization (ISO) is considered a worldwide federation of national standards organizations. It is primarily responsible for developing various international standards through its technical committees [86]. The ISO has set its technical specifications and standards related to CNTs.

All these ISO guidelines present a general framework for the CNTs' manufacture, characterization, and safe handling. "ISO/TR 10929" identifies the characteristic features of MWCNTs and the impurity content, which characterize the MWCNT bulk samples. The above guideline also focuses on the method of determining the impurity content in MWCNT bulk samples available to the industry for determining the above parameters [87]. Furthermore, "ISO 10797" explains the characterization of SWCNTs using transmission electron microscopy (TEM). It also explains the TEM measurement for thin samples. The guideline thus gives us a brief idea regarding the different sample preparation (i.e., liquid suspension sample, composite sample, dry powder, and dry film sample) for TEM analysis and their measurement procedure (i.e., small and large field view) [88]. Mesoscopic shape factors quantify the average dimensions and morphology of individual MWCNTs. Hence, "ISO/TS 11888" provides the technical specifications for characterizing the mesoscopic shape factors of these MWCNTs, and it also comprises the techniques employed, like SEM, TEM, and light scattering analysis. The specification also explains the various methods of MWCNT preparation [89]. Nanoparticle suspensions, which are stabilized by suitable fluids and additives, prevent agglomeration and loss to the environment during handling. In industry, this pre-treatment is common in manufacturing, particularly for preparing CNT suspensions. Considering the widespread availability of MWCNT suspensions, it is necessary to develop standardized specifications for their effective communication, commercialization, and consistent product performance. One of the guidelines, "ISO/TS 19808:2020", specifies the characteristic features that need to be measured of MWCNT suspensions, which includes the essential and additional features of the CNT suspension and their corresponding measurement methods [86].

10.4 CONCLUSION

Although CNTs have the potential to positively impact multiple fields, the several toxicities associated with them are still restricting their extensive use. The use of carbon nanomaterials is growing daily, increasing the likelihood of exposure to toxicity. In the current world, governments, environmentalists, and the scientific community are all quite concerned about CNT toxicity. A robust and efficient regulatory framework is urgently needed to combat this issue. Unfortunately, CNT's regulatory landscape is complex and still evolving, though it is getting better. Current regulations offer some guidance, but they often don't include any limits tailored to CNTs. This necessitates a careful balancing act between encouraging innovation and safety. As research and uses expand, a comprehensive regulatory framework will be required to manage possible hazards and maximize the benefits of this innovative technology.

DECLARATION OF COMPETING INTERESTS

The authors declare that they have no known competing financial interests or personal relationships that could have appeared to influence the work reported in this paper.

ACKNOWLEDGMENTS

All the authors are grateful to the Director of the National Institute of Pharmaceutical Education and Research, Ahmedabad (NIPER-A) for providing all the necessary facilities.

Funding: None

REFERENCES

1. A.N.K.V. Sravani, N. Chandrasekaran, J. Thomas, A. Mukherjee, Formulation and characterization of cisplatin-loaded hydroxyl functionalized single-walled carbon nanotubes for targeting gastric cancer stem cells, *Heliyon* 9 (2023) e18798. https://doi.org/10.1016/j.heliyon.2023.e18798.
2. P. Saudagar, V.K. Dubey, Carbon nanotube based betulin formulation shows better efficacy against Leishmania parasite, *Parasitol. Int.* 63 (2014) 772–776.
3. M.D. Angione, R. Pilolli, S. Cotrone, M. Magliulo, A. Mallardi, G. Palazzo, L. Sabbatini, D. Fine, A. Dodabalapur, N. Cioffi, Carbon based materials for electronic bio-sensing, *Mater. Today* 14 (2011) 424–433.
4. S. Iijima, Carbon nanotubes: Past, present, and future, *Phys. B Condens. Matter* 323 (2002) 1–5. https://doi.org/10.1016/S0921-4526(02)00869-4.
5. A.C. Dillon, K.M. Jones, T.A. Bekkedahl, C.H. Kiang, D.S. Bethune, M.J. Heben, Storage of hydrogen in single-walled carbon nanotubes, *Nature* 386 (1997) 377–379. https://doi.org/10.1038/386377a0.
6. P. Zhang, J. Su, J. Guo, S. Hu, Influence of carbon nanotube on properties of concrete: A review, *Constr. Build. Mater.* 369 (2023) 130388. https://doi.org/10.1016/j.conbuildmat.2023.130388.
7. N. Saifuddin, A.Z. Raziah, A.R. Junizah, Carbon nanotubes: A review on structure and their interaction with proteins, *J. Chem.* 2013 (2013) 676815.
8. C. Zhao, J. Kang, Y. Li, Y. Wang, X. Tang, Z. Jiang, Carbon-based stimuli-responsive nanomaterials: Classification and application, *Cyborg. Bionic. Syst.* 4 (2023) 22.
9. H. Rahman, B. Krishnamoorthy, N. Tamilselvan, K. Siram, S. Karthik, R. Hariprasad, Chapter 7- Nanomaterials in drug delivery: Existing scenario and potential scope, in: A. M. Grumezescu (Ed.), *Nanobiomaterials in Drug Delivery*, William Andrew Publishing, Norwich, NY, 2016: pp. 197–228. https://doi.org/10.1016/B978-0-323-42866-8.00007-1.
10. M.S. Hasnain, A.K. Nayak, M.S. Hasnain, A.K. Nayak, Classification of carbon nanotubes, *Carbon Nanotub. Target. Drug Deliv.* 106 (2019) 11–15.
11. N. Gupta, S.M. Gupta, S.K. Sharma, Carbon nanotubes: Synthesis, properties and engineering applications, *Carbon Lett.* 29 (2019) 419–447. https://doi.org/10.1007/s42823-019-00068-2.
12. V.K. Pandey, A. Tripathi, A. Taufeeq, A.H. Dar, A. V Samrot, S. Rustagi, S. Malik, T. Bhattacharya, B. Kovacs, A.M. Shaikh, Significance and applications of carbon dots in anti cancerous nanodrug conjugate development: A review, *Appl. Surf. Sci. Adv.* 19 (2024) 100550. https://doi.org/10.1016/j.apsadv.2023.100550.
13. A. Gacem, S. Modi, V.K. Yadav, S. Islam, A. Patel, V. Dawane, M. Jameel, G.K. Inwati, S. Piplode, V.S. Solanki, Recent advances in methods for synthesis of carbon nanotubes and carbon nanocomposite and their emerging applications: A descriptive review, *J. Nanomater.* 2022 (2022) 7238602.
14. S. Murali, S. Perumal, Synthesis of nanotubes under carbon environment at low temperature using hydrothermal method, *Optik (Stuttg).* 162 (2018) 81–85. https://doi.org/10.1016/j.ijleo.2018.01.106.

15. A. Szabó, C. Perri, A. Csató, G. Giordano, D. Vuono, J.B. Nagy, Synthesis methods of carbon nanotubes and related materials, *Materials (Basel).* 3 (2010) 3092–3140.
16. B. Safaei, H.C. How, G. Scribano, A computational study on synthesis of carbon nanotubes in a sooty inverse diffusion flame, *Int. J. Environ. Sci. Technol.* 20 (2023) 1–10. https://doi.org/10.1007/s13762-022-04143-6.
17. J. Tomada, K. Amsharov, Single-walled carbon nanotube's end-cap engineering—molecular seeds for controlled synthesis of chirality-pure SWCNT, *Mol. Sci. & Chem. Eng.* 2018. https://doi.org/10.1016/B978-0-12-409547-2.14210-6.
18. A. Bhatt, A. Jain, E. Gurnany, R. Jain, A. Modi, A. Jain, 17- Carbon nanotubes: A promising carrier for drug delivery and targeting, in: A. M. Holban, D. T. Grumezescu (Eds.), *Nanoarchitectonics for smart Delivery and drug targeting*, William Andrew Publishing, Norwich, NY, 2016: pp. 465–501. https://doi.org/10.1016/B978-0-323-47347-7.00017-3.
19. N. Anzar, R. Hasan, M. Tyagi, N. Yadav, J. Narang, Carbon nanotube - A review on synthesis, properties and plethora of applications in the field of biomedical science, *Sensors Int.* 1 (2020) 100003. https://doi.org/10.1016/j.sintl.2020.100003.
20. S.P. Sharma, S.C. Lakkad, Effect of CNTs growth on carbon fibers on the tensile strength of CNTs grown carbon fiber-reinforced polymer matrix composites, *Compos. Part A Appl. Sci. Manuf.* 42 (2011) 8–15. https://doi.org/10.1016/j.compositesa.2010.09.008.
21. A. V Radhamani, H.C. Lau, S. Ramakrishna, CNT-reinforced metal and steel nanocomposites: A comprehensive assessment of progress and future directions, *Compos. Part A Appl. Sci. Manuf.* 114 (2018) 170–187. https://doi.org/10.1016/j.compositesa.2018.08.010.
22. B. Kumanek, D. Janas, Thermal conductivity of carbon nanotube networks: A review, *J. Mater. Sci.* 54 (2019) 7397–7427. https://doi.org/10.1007/s10853-019-03368-0.
23. D. Rudakiya, Y. Patel, U. Chhaya, A. Gupte, Carbon nanotubes in agriculture: Production, potential, and prospects, in: D. Panpatte, Y. Jhala (Eds.), *Nanotechnology for Agriculture*, Springer, Singapore, 2019: pp. 121–130. https://doi.org/10.1007/978-981-32-9370-0_8.
24. K. Mustafa, J. Kanwal, S. Musaddiq, Waste plastic-based nanomaterials and their applications, in: A.S.H. Makhlouf, G.A.M. Ali (Eds.), *Waste Recycling Technologies for Nanomaterials Manufacturing. Topics in Mining, Metallurgy and Materials Engineering*, Springer, Cham, 2021: pp. 781–803. https://doi.org/10.1007/978-3-030-68031-2_27.
25. A. Eatemadi, H. Daraee, H. Karimkhanloo, M. Kouhi, N. Zarghami, A. Akbarzadeh, M. Abasi, Y. Hanifehpour, S.W. Joo, Carbon nanotubes: Properties, synthesis, purification, and medical applications, *Nanoscale Res. Lett.* 9 (2014) 1–13.
26. N. Jain, E. Gupta, N.J. Kanu, Plethora of carbon nanotubes applications in various fields – a state-of-the-art-review, *Smart Sci.* 10 (2022) 1–24. https://doi.org/10.1080/23080477.2021.1940752.
27. Z. Shariatinia, Chapter 7- Applications of carbon nanotubes, in: S. Thomas, C. Sarathchandran, S.A. Ilangovan, J.C. Moreno-Piraján (Eds.), *Micro and Nano Technologies*, Elsevier, Amsterdam, The Netherlands, 2021: pp. 321–364. https://doi.org/10.1016/B978-0-12-821996-6.00016-6.
28. S.S. Ogunsola, M.E. Oladipo, P.O. Oladoye, M. Kadhom, Carbon nanotubes for sustainable environmental remediation: A critical and comprehensive review, *Nano-Structures & Nano-Objects* 37 (2024) 101099. https://doi.org/10.1016/j.nanoso.2024.101099.
29. A. Helland, P. Wick, A. Koehler, K. Schmid, C. Som, Reviewing the environmental and human health knowledge base of carbon nanotubes, *Environ. Health Perspect.* 115 (2007) 1125–1131. https://doi.org/10.1289/ehp.9652.
30. P. Jackson, N.R. Jacobsen, A. Baun, R. Birkedal, D. Kühnel, K.A. Jensen, U. Vogel, H. Wallin, Bioaccumulation and ecotoxicity of carbon nanotubes, *Chem. Cent. J.* 7 (2013) 154. https://doi.org/10.1186/1752-153X-7-154.

31. Z. Peng, X. Liu, W. Zhang, Z. Zeng, Z. Liu, C. Zhang, Y. Liu, B. Shao, Q. Liang, W. Tang, X. Yuan, Advances in the application, toxicity and degradation of carbon nanomaterials in environment: A review, *Environ. Int.* 134 (2020) 105298. https://doi.org/10.1016/j.envint.2019.105298.
32. Y. Morimoto, N. Kobayashi, Evaluations of the carcinogenicity of carbon nanotubes, fluoro-edinite, and silicon carbide by the international agency for research on cancer (IARC), *Nihon Eiseigaku Zasshi.* 71 (2016) 252–259. https://doi.org/10.1265/jjh.71.252.
33. A. Wardak, M.E. Gorman, N. Swami, D. Rejeski, Environmental regulation of nanotechnology and the TSCA, *IEEE Technol. Soc. Mag.* 26 (2007) 48–56.
34. T. Saliev, The advances in biomedical applications of carbon nanotubes, *C* 5 (2019) 29. https://doi.org/10.3390/c5020029.
35. V.R. Raphey, T.K. Henna, K.P. Nivitha, P. Mufeedha, C. Sabu, K. Pramod, Advanced biomedical applications of carbon nanotube, *Mater. Sci. Eng. C.* 100 (2019) 616–630. https://doi.org/10.1016/j.msec.2019.03.043.
36. T. Agnihotri, S.K. Prajapati, S.S. Gomte, A. Jain, Chemically modified carbon nanotubes in cancer therapy, *Chem. Modif. Carbon Nanotub. Commer. Appl.* 13 (2023) 299–330.
37. M. Rahamathulla, R.R. Bhosale, R.A.M. Osmani, K.C. Mahima, A.P. Johnson, U. Hani, M. Ghazwani, M.Y. Begum, S. Alshehri, M.M. Ghoneim, F. Shakeel, H. V. Gangadharappa, Carbon nanotubes: Current perspectives on diverse applications in targeted drug delivery and therapies, *Materials (Basel).* 14 (2021) 14216707. https://doi.org/10.3390/ma14216707.
38. A. Cifuentes-Rius, N.R.B. Boase, I. Font, N. Coronas, V. Ramos-Perez, K.J. Thurecht, S. Borrós, In vivo fate of carbon nanotubes with different physicochemical properties for gene delivery applications, *ACS Appl. Mater. Interfaces* 9 (2017) 11461–11471. https://doi.org/10.1021/acsami.7b00677.
39. W. Zhang, Z. Zhang, Y. Zhang, The application of carbon nanotubes in target drug delivery systems for cancer therapies, *Nanoscale Res. Lett.* 6 (2011) 1–22. https://doi.org/10.1186/1556-276X-6-555.
40. C. Fabbro, H. Ali-Boucetta, T. Da Ros, A. Bianco, K. Kostarelos, M. Prato, Targeting carbon nanotubes against cancer, *Chem. Commun.* 48 (2012) 3911–3926. https://doi.org/10.1039/c2cc17995d.
41. S.K. Prajapati, T. Agnihotri, S.S. Gomte, A. Jain, Surface modification strategies for the carbon nanotubes, *Chem. Modif. Carbon Nanotub. Commer. Appl.* 2023 (2023) 51–74.
42. S. Salave, D. Rana, J. Vitore, A. Jain, Functionalized carbon nanotubes for cell tracking, *Funct. Carbon Nanotub. Biomed. Appl.* 2023 (2023) 319–338.
43. M.N. Al-Qattan, P.K. Deb, R.K. Tekade, Molecular dynamics simulation strategies for designing carbon-nanotube-based targeted drug delivery, *Drug Discov. Today* 23 (2018) 235–250. https://doi.org/10.1016/j.drudis.2017.10.002.
44. S. Sau, Y. Law, G. Liou, Y. Nagai, J. Giménez-dejoz, A. Tateishi, Y. Kodama, T. Fujigaya, K. Numata, Polymer-coated carbon nanotube hybrids with functional peptides for gene delivery into plant mitochondria, *Nat. Commun.* 13 (2022) 2417.
45. E.N. de C. Lima, R.S. Diaz, J.F. Justo, J.R.C. Piqueira, Advances and perspectives in the use of carbon nanotubes in vaccine development, *Int. J. Nanomedicine* 16 (2021) 5411–5435. https://doi.org/10.2147/IJN.S314308.
46. H.A.F.M. Hassan, S.S. Diebold, L.A. Smyth, A.A. Walters, G. Lombardi, K.T. Al-Jamal, Application of carbon nanotubes in cancer vaccines: Achievements, challenges and chances, *J. Control. Release* 297 (2019) 79–90. https://doi.org/10.1016/j.jconrel.2019.01.017.
47. T. Agnihotri, T. Shinde, M. Gitte, P.K. Paradia, R.K. Tekade, A. Jain, Functionalized carbon nanotubes for gene therapy, *Funct. Carbon Nanotub. Biomed. Appl.* 2023 (2023) 139–170.

48. M. Mohajeri, B. Behnam, A. Sahebkar, Biomedical applications of carbon nanomaterials: Drug and gene delivery potentials, *J. Cell. Physiol.* 234 (2018) 298–319. https://doi.org/10.1002/jcp.26899.
49. S. Ranjbari, M. Bolourinezhad, P. Kesharwani, M. Rezayi, A. Sahebkar, Applications of carbon nanotube biosensors: Sensing the future, *J. Drug Deliv. Sci. Technol.* 97 (2024) 105747. https://doi.org/10.1016/j.jddst.2024.105747.
50. B. Han, Y. Chen, H. Wang, J. Yan, G. Liu, Z. Huang, C. Zhou, A biosensor for simultaneous detection of epinephrine and ascorbic acid based on Fe(III)–polyhistidine-functionalized multi-wall carbon nanotube composites, *Int. J. Mol. Sci.* 25 (2024) 25147883. https://doi.org/10.3390/ijms25147883.
51. H.M. Dewey, A. Lamb, J. Budhathoki-Uprety, Recent advances on applications of single-walled carbon nanotubes as cutting-edge optical nanosensors for biosensing technologies, *Nanoscale* 16 (2024) 16344–16375. https://doi.org/10.1039/d4nr01892c.
52. S.H. Ku, M. Lee, C.B. Park, Carbon-based nanomaterials for tissue engineering, *Adv. Healthc. Mater.* 2 (2013) 244–260. https://doi.org/10.1002/adhm.201200307.
53. H.A. Katsuhisa Kitano, S.Hamaguchi, (12) Patent Application Publication (10) Pub. No.: US 2010/0292345 A1 Patent Application Publication, 1 (2010) 1–5.
54. S.K. Prajapati, A. Malaiya, P. Kesharwani, D. Soni, A. Jain, Biomedical applications and toxicities of carbon nanotubes, *Drug Chem. Toxicol.* 45 (2022) 435–450. https://doi.org/10.1080/01480545.2019.1709492.
55. M.A. Saleemi, M.H Fouladi, P.V.C. Yong, K. Chinna, N.K. Palanisamy, E.H. Wong, Toxicity of carbon nanotubes: Molecular mechanisms, signaling cascades, and remedies in biomedical applications, *Chem. Res. Toxicol.* 34 (2021) 24–46. https://doi.org/10.1021/acs.chemrestox.0c00172.
56. D. Mohanta, S. Patnaik, S. Sood, N. Das, Carbon nanotubes: Evaluation of toxicity at biointerfaces, *J. Pharm. Anal.* 9 (2019) 293–300. https://doi.org/10.1016/j.jpha.2019.04.003.
57. Z. Liu, S. Tabakman, K. Welsher, H. Dai, Carbon nanotubes in biology and medicine: In vitro and in vivo detection, imaging and drug delivery, *Nano Res.* 2 (2009) 85–120. https://doi.org/10.1007/s12274-009-9009-8.
58. M. Valko, D. Leibfritz, J. Moncol, M.T.D. Cronin, M. Mazur, J. Telser, Free radicals and antioxidants in normal physiological functions and human disease, *Int. J. Biochem. Cell Biol.* 39 (2007) 44–84. https://doi.org/10.1016/j.biocel.2006.07.001.
59. B. Gupta, P.K. Sharma, R. Malviya, Carbon nanotubes for targeted therapy: Safety, efficacy, feasibility and regulatory aspects, *Curr. Pharm. Des.* 30 (2024) 81–99.
60. H.C. Çetintaş, H. Tonbul, A. Şahin, Y. Çapan, Regulatory guidelines of the US food and drug administration and the european medicines agency for actively targeted nanomedicines, in: Y. Çapan, A. Sahin, H. Tonbul (Eds.), *Drug Delivery with Targeted Nanoparticles*, Jenny Stanford Publishing, New York, 2021: pp. 725–741.
61. F and D Administration, Considering whether an FDA-regulated product involves the application of nanotechnology, *U.S. Food Drug Adm.* (2014) 1–14.
62. M. Brisse, S.M. Vrba, N. Kirk, Y. Liang, H. Ly, Emerging concepts and technologies in vaccine development, *Front. Immunol.* 11 (2020) 1–22. https://doi.org/10.3389/fimmu.2020.583077.
63. R.B. Kennedy, I.G. Ovsyannikova, P. Palese, G.A. Poland, Current challenges in vaccinology, *Front. Immunol.* 11 (2020) 1181.
64. E.B. Souto, C. Blanco-Llamero, K. Krambeck, N.S. Kiran, C. Yashaswini, H. Postwala, P. Severino, R. Priefer, B.G. Prajapati, R. Maheshwari, Regulatory insights into nanomedicine and gene vaccine innovation: Safety assessment, challenges, and regulatory perspectives, *Acta Biomater.* 180 (2024) 1–17. https://doi.org/10.1016/j.actbio.2024.04.010.

65. C. Fisher, A.E. Rider, Z. Jun Han, S. Kumar, I. Levchenko, K. Ostrikov, Applications and nanotoxicity of carbon nanotubes and graphene in biomedicine, *J. Nanomater.* 2012 (2012) 315185. https://doi.org/10.1155/2012/315185.
66. F. Karimi, H. Karimi-Maleh, J. Rouhi, N. Zare, C. Karaman, M. Baghayeri, L. Fu, S. Rostamnia, E.N. Dragoi, A. Ayati, Revolutionizing cancer monitoring with carbon-based electrochemical biosensors, *Environ. Res.* 239 (2023) 117368.
67. S.E. Martínez Tolibia, A. Galdámez-Martínez, R.A. Salinas, A. Dutt, Anticipating challenges in optical nanobiosensors for global detection of respiratory viruses and emerging threats, *ECS Sensors Plus* 2 (2023) ad08d5. https://doi.org/10.1149/2754-2726/ad08d5.
68. R. Fries, S. Grebler, M. Simko, Carbon nanotubes – Part II: Risks and regulations, *NanoTrust Doss.* 24 (2012) 2–7.
69. ECHA, (n.d.). https://echa.europa.eu/substance-information/-/substanceinfo/100.109.645 (accessed October 28, 2024).
70. ECHA, (n.d.). https://echa.europa.eu/substance-information/-/substanceinfo/100.217.898 (accessed October 28, 2024).
71. EPA, (n.d.). https://www.epa.gov/reviewing-new-chemicals-under-toxic-substances-control-act-tsca/actions-under-tsca-section-5 (accessed November 4, 2024).
72. TSCA Section 8(a)(1) Reporting and Recordkeeping Requirements for Asbestos, (n.d.). https://www.epa.gov/assessing-and-managing-chemicals-under-tsca/tsca-section-8a1-reporting-and-recordkeeping#:~:text=TSCA section 8(a) generally, EPA Administrator may reasonably require (accessed November 4, 2024).
73. Industry Testing Requirements under TSCA Section 4, (n.d.). https://www.epa.gov/assessing-and-managing-chemicals-under-tsca/industry-testing-requirements-under-tsca-section-4 (accessed November 4, 2024).
74. TSCA Section 4 Test Orders, (n.d.). https://www.epa.gov/assessing-and-managing-chemicals-under-tsca/tsca-section-4-test-orders (accessed November 4, 2024).
75. R.C. Slate, Environmental Protection Agency Regulation of Asbestos and Carbon Nanotubes Under the Toxic Substances Control Act: Investigating the Role of Politics, Science, and Policy in Administrative Rulemaking and Implementation, (2014).
76. Control of Nanoscale Materials under the Toxic Substances Control Act, (n.d.). https://www.epa.gov/reviewing-new-chemicals-under-toxic-substances-control-act-tsca/control-nanoscale-materials-under (accessed November 4, 2024).
77. Toxic Substances Control Act Inventory Status of Carbon Nanotubes, (n.d.). https://www.federalregister.gov/documents/2008/10/31/E8-26026/toxic-substances-control-act-inventory-status-of-carbon-nanotubes#:~:text=EPA strongly recommends that persons, consistent with its other priorities. (accessed November 4, 2024).
78. M.S. Hasnain, A.K. Nayak, Regulatory considerations of carbon nanotubes, *Appl. Sci. Technol.* 15 (2019) 103–106. https://doi.org/10.1007/978-981-15-0910-0_15.
79. A.D. Maynard, P.A. Baron, M. Foley, A.A. Shvedova, E.R. Kisin, V. Castranova, Exposure to carbon nanotube material: Aerosol release during the handling of unrefined single-walled carbon nanotube material, *J. Toxicol. Environ. Heal. Part A* 67 (2004) 87–107.
80. C.W. Lam, J.T. James, R. McCluskey, S. Arepalli, R.L. Hunter, A review of carbon nanotube toxicity and assessment of potential occupational and environmental health risks, *Crit. Rev. Toxicol.* 36 (2006) 189–217. https://doi.org/10.1080/10408440600570233.
81. Occupational Safety and Health Administration, (n.d.). https://www.osha.gov/laws-regs/regulations/standardnumber/1910/1910.134.
82. N. Kobayashi, Risk Assessment of Manufactured Nanomaterials-Carbon Nanotubes (CNTs)-Interim Report issued on October 16, 2009 Exective Summary, http://good-nanoguide.org/tiki-download_wiki_attachment.php?attId=31 (2009).

83. K. Rasmussen, B. Sokull-Klüttgen, I.J. Yu, J. Kanno, A. Hirose, M.R. Gwinn, Regulation and legislation, *Advers. Eff. Eng. Nanomater.* 81 (2017) 159–188.
84. R. Kumari, K. Suman, S. Karmakar, V. Mishra, S.G. Lakra, G.K. Saurav, B.K. Mahto, Regulation and safety measures for nanotechnology-based agri-products, *Front. Genome Ed.* 5 (2023) 1200987.
85. A. Jorio, I.O. Maciel, P.T. Araujo, P.B.C. Pesce, M.A. Pimenta, The fundamental aspects of carbon nanotube metrology, *Phys. Status Solidi* 244 (2007) 4011–4015.
86. SO/TS 19808:2020(en) Nanotechnologies — Carbon nanotube suspensions — Specification of characteristics and measurement methods, (n.d.). https://www.iso.org/obp/ui/en/#iso:std:iso:ts:19808:ed-1:v1:en (accessed November 4, 2024).
87. ISO/TR 10929:2012(en) Nanotechnologies — Characterization of multiwall carbon nanotube (MWCNT) samples, (n.d.). https://www.iso.org/obp/ui/en/#iso:std:iso:tr:10929:ed-1:v1:en (accessed November 4, 2024).
88. ISO/TS 10797:2012(en) Nanotechnologies — Characterization of single-wall carbon nanotubes using transmission electron microscopy, (n.d.). https://www.iso.org/obp/ui/en/#iso:std:iso:ts:10797:ed-1:v1:en (accessed November 4, 2024).
89. ISO/TS 11888:2017(en) Nanotechnologies — Characterization of multiwall carbon nanotubes — Mesoscopic shape factors, (n.d.). https://www.iso.org/obp/ui/en/#iso:std:iso:ts:11888:ed-2:v1:en (accessed November 4, 2024).

Index

FSC
www.fsc.org
MIX
Papier | Fördert
gute Waldnutzung
FSC® C083411

Kolibri 360 GmbH
Ferdinand-Jühlke-Straße 7
99095 Erfurt, Deutschland
produktsicherheit@kolibri360.de